Metabolomics in Neurodegenerative Disease

Metabolomics in Neurodegenerative Disease

Special Issue Editor

Brian Green

MDPI • Basel • Beijing • Wuhan • Barcelona • Belgrade

Special Issue Editor
Brian Green
University Road
UK

Editorial Office
MDPI
St. Alban-Anlage 66
4052 Basel, Switzerland

This is a reprint of articles from the Special Issue published online in the open access journal *Metabolites* (ISSN 2218-1989) from 2018 to 2019 (available at: https://www.mdpi.com/journal/metabolites/special_issues/Metabolomics_Neurodegenerative_Disease).

For citation purposes, cite each article independently as indicated on the article page online and as indicated below:

LastName, A.A.; LastName, B.B.; LastName, C.C. Article Title. *Journal Name* **Year**, *Article Number*, Page Range.

ISBN 978-3-03928-040-7 (Hbk)
ISBN 978-3-03928-041-4 (PDF)

Cover image courtesy of Brian Green.

Contents

About the Special Issue Editor vii

Brian D. Green
The Application of Metabolomic Techniques in Research Investigating Neurodegenerative Diseases
Reprinted from: *Metabolites* **2019**, *9*, 283, doi:10.3390/metabo9120283 1

Massimo S. Fiandaca, Thomas J. Gross, Thomas M. Johnson, Michele T. Hu, Samuel Evetts, Richard Wade-Martins, Kian Merchant-Borna, Jeffrey Bazarian, Amrita K. Cheema, Mark Mapstone and Howard J. Federoff
Potential Metabolomic Linkage in Blood between Parkinson's Disease and Traumatic Brain Injury
Reprinted from: *Metabolites* **2018**, *8*, 50, doi:10.3390/metabo8030050 3

Raúl González-Domínguez, Ana Sayago and Ángeles Fernández-Recamales
High-Throughput Direct Mass Spectrometry-Based Metabolomics to Characterize Metabolite Fingerprints Associated with Alzheimer's Disease Pathogenesis
Reprinted from: *Metabolites* **2018**, *8*, 52, doi:10.3390/metabo8030052 23

Muhammad L. Nasaruddin, Xiaobei Pan, Bernadette McGuinness, Peter Passmore, Patrick G. Kehoe, Christian Hölscher, Stewart F. Graham and Brian D. Green
Evidence That Parietal Lobe Fatty Acids May Be More Profoundly Affected in Moderate Alzheimer's Disease (AD) Pathology Than in Severe AD Pathology
Reprinted from: *Metabolites* **2018**, *8*, 69, doi:10.3390/metabo8040069 32

Kim A. Caldwell, Jennifer L. Thies and Guy A. Caldwell
No Country for Old Worms: A Systematic Review of the Application of *C. elegans* to Investigate a Bacterial Source of Environmental Neurotoxicity in Parkinson's Disease
Reprinted from: *Metabolites* **2018**, *8*, 70, doi:10.3390/metabo8040070 48

Stewart F. Graham, Nolwen L. Rey, Zafer Ugur, Ali Yilmaz, Eric Sherman, Michael Maddens, Ray O. Bahado-Singh, Katelyn Becker, Emily Schulz, Lindsay K. Meyerdirk, Jennifer A. Steiner, Jiyan Ma and Patrik Brundin
Metabolomic Profiling of Bile Acids in an Experimental Model of Prodromal Parkinson's Disease
Reprinted from: *Metabolites* **2018**, *8*, 71, doi:10.3390/metabo8040071 67

Connor N. Brown, Brian D. Green, Richard B. Thompson, Anneke I. den Hollander and Imre Lengyel
Metabolomics and Age-Related Macular Degeneration
Reprinted from: *Metabolites* **2019**, *9*, 4, doi:10.3390/metabo9010004 77

Zeynep Alpay Savasan, Ali Yilmaz, Zafer Ugur, Buket Aydas, Ray O. Bahado-Singh and Stewart F. Graham
Metabolomic Profiling of Cerebral Palsy Brain Tissue Reveals Novel Central Biomarkers and Biochemical Pathways Associated with the Disease: A Pilot Study
Reprinted from: *Metabolites* **2019**, *9*, 27, doi:10.3390/metabo9020027 114

Anuri Shah, Pei Han, Mung-Yee Wong, Raymond Chuen-Chung Chang and Cristina Legido-Quigley
Palmitate and Stearate are Increased in the Plasma in a 6-OHDA Model of Parkinson's Disease
Reprinted from: *Metabolites* **2019**, *9*, 31, doi:10.3390/metabo9020031 130

Kevin Chen, Dodge Baluya, Mehmet Tosun, Feng Li and Mirjana Maletic-Savatic
Imaging Mass Spectrometry: A New Tool to Assess Molecular Underpinnings of Neurodegeneration
Reprinted from: *Metabolites* **2019**, *9*, 135, doi:10.3390/metabo9070135 . . . . **144**

Jesper F. Havelund, Kevin H. Nygaard, Troels H. Nielsen, Carl-Henrik Nordström, Frantz R. Poulsen, Nils. J. Færgeman, Axel Forsse and Jan Bert Gramsbergen
In Vivo Microdialysis of Endogenous and ^{13}C-labeled TCA Metabolites in Rat Brain: Reversible and Persistent Effects of Mitochondrial Inhibition and Transient Cerebral Ischemia
Reprinted from: *Metabolites* **2019**, *9*, 204, doi:10.3390/metabo9100204 . . . . **162**

About the Special Issue Editor

Brian Green graduated with a degree in biochemistry from the Faculty of Medicine, Health and Life Sciences at Queen's University Belfast (QUB) and completed a PhD in biomedical sciences at Ulster University (UU). He is currently a senior lecturer in molecular nutrition at Queen's University Belfast, within the School of Biological Sciences (SBS) and the Institute for Global Food Security (IGFS). He has published more than 200 papers (h-index of 35) in international peer-reviewed journals and serves as academic lead for the Core Technology Unit (CTU) in mass spectrometry. His metabolomics research focuses on the investigation of human diet and disease, particularly diet–metabolite interactions and perturbation in specific biochemical pathways, as well as the discovery of novel metabolite biomarkers.

metabolites

MDPI

Editorial

The Application of Metabolomic Techniques in Research Investigating Neurodegenerative Diseases

Brian D. Green

School of Biological Sciences, Institute for Global Food Security, Queen's University Belfast, Biological Sciences Building, Chlorine Gardens, Stranmillis, Northern Ireland BT9 5DL, UK; b.green@qub.ac.uk

Received: 1 November 2019; Accepted: 11 November 2019; Published: 20 November 2019

We live in a world posing many new and different challenges for human health, and one such challenge is the rapidly expanding number of cases of human neurodegenerative disease. Many of the most common neurodegenerative diseases are dementias affecting cognitive and behavioural functions, and it is very concerning that treatment options remain extremely limited. The unmet medical needs for many conditions are extremely high because, unlike many of the other common non-communicable diseases (NCDs), such as cardiovascular diseases, cancers and diabetes, few disease-modifying therapies exist. The causes are multifactorial and the potential disease drivers are numerous. Aside from the rising age profile of the global population and the known genetic risk factors, there are many potential modifiable risk factors, ranging from hypertension, obesity, hearing loss, smoking, depression, physical inactivity, social isolation, diabetes and years of education [1]. The progressive and terminal nature of these conditions places a considerable personal burden on the individual affected. Additionally, there is a growing economic and public health burden, forcing governments and health services to make difficult choices concerning the allocation of medical resources. Tens of millions of people are indiscriminately affected by various dementias, which are rising at an alarming rate [2].

It has been emphasised that the quantity of basic science in dementia research lags behind many other diseases [3]. So in order to make progress here, our fundamental understanding of how biochemical processes are affected by these chronic, complex and seemingly stealthy diseases needs to improve. There is a need for new disease classification strategies and early diagnostic tools. Metabolomics still represents a relatively new field of analytical science, which can be extremely useful in the early diagnosis of disease. The relatively unique feature of metabolites is that they sit at the intersection between the genetic background of an organism and its environment. Since many neurodegenerative diseases are not genetically inherited (instead having a range of known genetic risk factors and also a large number of unknown environmental triggers), metabolomics offers great promise for the discovery of new, biologically and clinically relevant biomarkers for neurodegenerative disorders. It is already bringing forward new knowledge in terms of the mechanisms of neurodegenerative diseases. For instance, work of our own indicates that, viewed longitudinally, the metabolic impact of Alzheimer's pathology is transient, perhaps with distinct phases [4], and is undoubtedly affected by severity [5]. The last 10 years of metabolomics research has brought forward a considerable amount of new biochemical knowledge about diseases such as Alzheimer's disease (AD), however, many other diseases are underrepresented and new collaborations and initiatives are needed for metabolomics to better penetrate these research areas.

Overall, this Special Issue of Metabolites presents a collection of cutting-edge studies and review articles demonstrating the application of metabolomics for the investigation of neurodegenerative diseases. The issue covers a broad range of disease areas, including AD, Parkinson's disease (PD), Cerebral palsy (CP) and age-related macular degeneration (AMD), but also includes conditions such as traumatic brain injury (TBI) and transient ischemic attacks (TIA). Within the research articles, metabolomic methods include ^{1}H NMR, direct injection liquid chromatography-tandem mass spectrometry (DI/LC-MS/MS), gas chromatography-mass spectrometry (GC-MS) and LC-MS following perfusion with ^{13}C-labelled compounds. There are also reviews of the different method types that can be utilised in neurodegenerative disease research, including imaging mass spectrometry (IMS) and direct mass spectrometry-based approaches. Finally, the articles feature the analysis and review of data from clinical samples, various rodent models and also more fundamental models such as *C. elegans*.

I hope that you enjoy reading this special issue.

Funding: The author is currently in receipt of funding from Alzheimer's Research UK (ARUK-NC2019-NI), Medical Research Council (MRC) (CIC-CD1718-CIC25), US-Ireland Health and Social Care NI (HSC R&D ST/5460/2018) and InvestNI (RD101427 11-01-17-008).

Acknowledgments: I extend my thanks to all contributing authors for this Special issue.

Conflicts of Interest: The author is currently appointed as the Coordinator of the Alzheimer's Research UK Network Centre for Northern Ireland.

References

1. Livingston, G.; Sommerlad, A.; Orgeta, V.; Costafreda, S.G.; Huntley, J.; Ames, D.; Ballard, C.; Banerjee, S.; Burns, A.; Cohen-Mansfield, J.; et al. Dementia prevention, intervention, and care. *Lancet* **2017**, *390*, 2673–2734. [CrossRef]
2. Alzheimer's Association. 2016 Alzheimer's disease facts and figures. *Alzheimers Dement.* **2016**, *12*, 459–509. [CrossRef] [PubMed]
3. Alzheimer's Disease International. World Alzheimer Report 2018. Available online: https://www.alz.co.uk/research/WorldAlzheimerReport2018.pdf (accessed on 31 October 2019).
4. Pan, X.; Nasaruddin, M.B.; Elliott, C.T.; McGuinness, B.; Passmore, A.P.; Kehoe, P.G.; Hölscher, C.; McClean, P.L.; Graham, S.F.; Green, B.D. Alzheimer's disease-like pathology has transient effects on the brain and blood metabolome. *Neurobiol. Aging* **2016**, *38*, 151–163. [CrossRef] [PubMed]
5. Nasaruddin, M.L.; Pan, X.; McGuinness, B.; Passmore, P.; Kehoe, P.G.; Hölscher, C.; Graham, S.F.; Green, B.D. Evidence That Parietal Lobe Fatty Acids May Be More Profoundly Affected in Moderate Alzheimer's Disease (AD) Pathology Than in Severe AD Pathology. *Metabolites* **2018**, *8*, 69. [CrossRef] [PubMed]

metabolites

MDPI

Article

Potential Metabolomic Linkage in Blood between Parkinson's Disease and Traumatic Brain Injury

Massimo S. Fiandaca [1,2,3,*], Thomas J. Gross [1,3], Thomas M. Johnson [4], Michele T. Hu [5,6], Samuel Evetts [5], Richard Wade-Martins [7], Kian Merchant-Borna [8], Jeffrey Bazarian [8], Amrita K. Cheema [9,10], Mark Mapstone [1] and Howard J. Federoff [1,*]

1 Translational Laboratory and Biorepository, Department of Neurology, University of California Irvine School of Medicine, Irvine, CA 92697-3910, USA; tjgross@uci.edu (T.J.G.); mark.mapstone@uci.edu (M.M.)
2 Department of Neurological Surgery, University of California Irvine School of Medicine, Irvine, CA 92697-3910, USA
3 Department of Anatomy & Neurobiology, University of California Irvine School of Medicine, Irvine, CA 92697-3910, USA
4 Intrepid Spirit Concussion Recovery Center, Naval Medical Center Camp Lejeune, Jacksonville, NC 28540, USA; thomas.m.johnson74.mil@mail.mil
5 Nuffield Department of Clinical Neurosciences, University of Oxford, 01865 Oxford, UK; michele.hu@ndcn.ox.ac.uk (M.T.H.); samuel.evetts@ndcn.ox.ac.uk (S.E.)
6 Department of Neurology, John Radcliffe Hospital, Oxford University Hospitals Trust, Oxford 01865, UK
7 Department of Physiology, Anatomy and Genetics, Oxford Parkinson's Disease Centre, University of Oxford, Oxford 01865, UK; richard.wade-martins@dpag.ox.ac.uk
8 Department of Emergency Medicine, University of Rochester School of Medicine and Dentistry, Rochester, NY 14604, USA; Kian_Merchant-Borna@URMC.Rochester.edu (K.M.-B.); jeff_bazarian@urmc.rochester.edu (J.B.)
9 Department of Oncology, Lombardi Comprehensive Cancer Center, Georgetown University Medical Center, Washington, DC 20001, USA; amrita.cheema@georgetown.edu
10 Department of Biochemistry and Molecular & Cellular Biology, Georgetown University Medical Center, Washington, DC 20001, USA
* Correspondence: mfiandac@uci.edu (M.S.F.); federoff@uci.edu (H.J.F.); Tel.: +1-949-824-5579 (M.S.F.)

Received: 24 July 2018; Accepted: 4 September 2018; Published: 7 September 2018

Abstract: The etiologic basis for sporadic forms of neurodegenerative diseases has been elusive but likely represents the product of genetic predisposition and various environmental factors. Specific gene-environment interactions have become more salient owing, in part, to the elucidation of epigenetic mechanisms and their impact on health and disease. The linkage between traumatic brain injury (TBI) and Parkinson's disease (PD) is one such association that currently lacks a mechanistic basis. Herein, we present preliminary blood-based metabolomic evidence in support of potential association between TBI and PD. Using untargeted and targeted high-performance liquid chromatography-mass spectrometry we identified metabolomic biomarker profiles in a cohort of symptomatic mild TBI (mTBI) subjects (n = 75) 3–12 months following injury (subacute) and TBI controls (n = 20), and a PD cohort with known PD (n = 20) or PD dementia (PDD) (n = 20) and PD controls (n = 20). Surprisingly, blood glutamic acid levels in both the subacute mTBI (increased) and PD/PDD (decreased) groups were notably altered from control levels. The observed changes in blood glutamic acid levels in mTBI and PD/PDD are discussed in relation to other metabolite profiling studies. Should our preliminary results be replicated in comparable metabolomic investigations of TBI and PD cohorts, they may contribute to an "excitotoxic" linkage between TBI and PD/PDD.

Keywords: Parkinson's disease; Parkinson's disease dementia; subacute mild traumatic brain injury; glutamic acid; excitotoxicity; metabolomics

1. Introduction

Compelling epidemiological observations associate moderate and severe traumatic brain injury (TBI) and Parkinson's disease (PD) [1]. Whether mild TBI (mTBI) is a significant risk factor for the development of PD (and other neurodegenerative disorders) has been more difficult to prove, due to fewer controlled investigations [2–4], conflicting results [5], and a lack of agreement on diagnostic criteria [6]. We anticipate that molecular phenotyping may ultimately resolve the latter discrepancies in the definition of mTBI. Recent studies [7,8], however, have more strongly endorsed an association between PD and TBI (including mTBI) sustained both early or later in life. Absent a consensus regarding a potential post-traumatic etiology for PD (or dementing conditions), the future definition of such relationships likely requires comprehensive longitudinal investigations and novel biomarkers [9]. Despite the limitations in current knowledge, there is emerging agreement that chronic neuroinflammatory conditions are associated with clinical parkinsonism and/or dementia, if not true PD or Alzheimer's disease (AD), and significant pathobiologic overlap exists (i.e., neuroinflammation, oxidative stress response, mitochondrial dysfunction, cognitive decline, and clinical depression) between neurodegenerative disorders (e.g., AD and D) and TBI [10,11]. The mechanisms underlying a precipitating event such as TBI to those downstream dysregulated networks associated with neurodegenerative diseases remains unknown.

For this article, as well as our previous report on acute mild brain trauma biomarkers [12], we based our diagnosis of mTBI (including the term concussion) on diagnostic criteria provided by our medical co-authors and medical doctors involved in the assessment of study participants. We have reported a set of human plasma metabolites associated with acute mTBI (within 6 h of injury) that accurately classify concussed individuals from non-concussed controls [12]. In this extension of our mTBI biomarker efforts we sought to define metabolomic similarities and differences between plasma specimens from a subacute cohort that includes subjects 3 to 12 months following mTBI, the previously reported acute mTBI biomarker panel, and in a cross-sectional design, whether plasma metabolites with TBI provide novel insights related to potential future risk of PD.

2. Results

2.1. Study Population Differences

A comparison of the demographics for the study cohorts is provided in Table 1. Our TBI cohort consisted of 75 cases and 20 controls. Described values are provided as the mean and standard deviation (S.D.). Frequency distribution of ages for the cases and controls in the TBI cohort did not follow a normal distribution, while ages in the PD cohort did. The TBI cases had a mean age of 24.9 ± 5.2 years, with 71 males and 4 females represented, and all of whom sustained a TBI during a three to twelve month interval prior to phlebotomy. The TBI controls (n = 20) had a mean age of 18.7 ± 0.8 years, included 8 males and 12 females, and did not have a history of a witnessed concussion or mTBI during the previous year prior to blood draw. Statistically significant age and sex differences existed between cases and controls in the TBI cohort. All TBI case and control participants attained the minimum of a high school graduate level of education. The number of injuries sustained by the TBI cases ranged from 1 to 9, with a mean of 2.0 ± 1.5. The severity of the last medically documented injury was a mTBI or concussion in 71 cases and moderate TBI in the other 4 cases. Individuals with TBIs prior to the last one reported injuries 12 months to 11 years prior, with a mean of 3.8 ± 3.7 years. Subjects in the PD cohort (n = 60) consisted of the PD (n = 20) and PD dementia (PDD) (n = 20) cases (combined n = 40), and the PD controls (n = 20). The PD cohort was approximately 40 years older than the TBI cohort. Mean ages (±S.D.) for the PD cohort, as well as the PD/PDD, PD, PDD, and PD control groups were 66.8 ± 11.0, 67.2 ± 11.4, 62.9 ± 10.4, 71.6 ± 10.9, and 65.9 ± 10.3 years, respectively. The mean age of the PD/PDD cases and the PD controls were not significantly different. Commensurate with previous studies, a male to female preponderance was noted across the PD cohort (overall 33 males and 27 females, with 22 males and 18 females making up the PD/PDD cases, and 11 males and 9 females being PD controls). There were no statistically significant sex differences between cases and controls in the PD cohort. At the time of blood collections on average, PD and PDD subjects were 2.9 ± 1.2 and 3.4 ± 1.2 years, respectively, from their original PD diagnosis. The TBI cohort subjects provided plasma for metabolomic analysis, while the PD cohort subjects provided serum.

Table 1. Demographic differences of study cohorts.

Population Characteristic	Subacute TBI Cases	TBI Controls	PD Cases (PD/PDD)	PD Controls
Number of subjects (n)	75	20	40	20
Age in years (mean ± S.D.)	24.9 ± 5.2 *	18.7 ± 0.8 *	67.2 ± 11.4 NS	65.9 ± 10.3 NS
Sex (n; M/F)	71/4 **	8/12 **	22/18 NS	11/9 NS

S.D. = standard deviation. * Statistically significant via Mann-Whitney U test ($p < 0.025$, Bonferroni corrected). ** Statistically significant via chi-square ($p < 0.025$, Bonferroni corrected). NS indicates no significant difference.

2.2. Subacute mTBI Plasma Metabolomic Biomarkers–MetaboAnalyst 4.0 Method

Of the top 15 preliminarily annotated metabolites derived using each of the unbiased feature selection algorithms within the Explorer module of MetaboAnalyst 4.0, the top nine are presented in Table 2, along with their qualitative differences between controls and cases. The metabolites are designated by their preliminarily annotated names followed by an appropriate structural symbol (as required) and finally a letter designation of whether identified in (N)egative or (P)ositive electrospray ionization (ESI) mode. Three of the top 9 metabolites (denoted by asterisk) were common to each of the four possible unbiased feature selection methods available. Of the nine, six specific metabolites combined in a classification model provided highly accurate receiver operating characteristic area under the curve (ROC AUC) results for distinguishing control subjects from those with subacute mTBI (Table 3). This 6-member panel provided classification AUCs of ≥0.9 for each of the analytic methods evaluated. Similar classification ROC AUC results were obtained using least absolute shrinkage and selection operator (LASSO) feature selection and a disparate group of 9 of the top 10 metabolites (data not shown), that also excluded the top-ranked Monoacylglycerol (MG) C16:0_N, but did include Creatinine_N and Glutamic Acid_N. Inclusion of MG C16:0_N alone, or in combination with other metabolites, provided ROC AUC values approaching 1.0, but did not allow model convergence required to provide ROC AUC and sensitivity and specificity results associated with the LR + 10FCV algorithm within MetaboAnalyst 4.0.

Table 2. Top 9 common metabolites derived using unbiased feature selection methods.

Preliminary Annotation	RVU in TBI Controls	RVU in Subacute mTBI Cases
* Monoacylglycerol (MG) C16:0_N	Low	High
Taurine_N	Low	High
Sphingosine 1 Phosphate_P (S1P_P)	Low	High
*** Glutamic Acid_N**	Low	High
Glucosylceramide (GlcCer) d18:1/26:0_N	High	Low
*** Creatinine_N**	High	Low
GlcCer d18:0/26:0_N	High	Low
Phosphatidylcholine (PC) ae C41:1_N	High	Low
PC ae C44:5_N	Low	High

Common metabolites were derived from the top 15 of each feature selection methodology, including linear support vector machine (LinSVM), partial least squares discriminant analysis (PLS-DA), and random forest (RandFor) unbiased algorithms. Comparisons of relative metabolite RVU abundances in TBI controls and cases are presented for each metabolite. * Denotes a top-15 metabolite via the LinSVM, PLS-DA, RandFor, and LASSO feature selection methods. RVU = relative value unit. LASSO = least absolute shrinkage and selection operator. The six metabolites in **bold** combined to provide a convergent logistic regression model. The ae designations for the two PCs indicate that acyl- and alkyl- side chains were represented. Final metabolite identifications will require additional tandem mass spectrometry (MS/MS) analyses. Metabolites confirmed via MS/MS are considered fully validated, to a high degree of confidence.

Table 3. Classification results for the convergent 6-metabolite subacute mTBI panel.

Classification Algorithm for Model	ROC AUC	95% CI	Sensitivity/Specificity
LinSVM	0.968	0.945–0.992	-
PLS-DA	0.977	0.945–0.992	-
RandFor	0.965	0.882–1.00	-
LR	0.939	0.734–0.984	-
LR + 10FCV Discovery	0.993	0.984–1.00	0.981/0.939
LR + 10FCV Internal Validation	0.893	0.789–0.996	0.947/0.850

mTBI = mild traumatic brain injury. ROC AUC = receiver operating characteristic area under the curve. CI = confidence interval. LinSVM = linear support vector machine. PLS-DA = partial least squares discriminant analysis. RandFor = random forests. LR = logistic regression. LR + 10FCV = logistic regression with 10-fold cross validation.

2.3. Subacute Plasma mTBI Metabolomic Biomarkers–mixOmics, sPLS-DA Method

The subacute mTBI cases and controls could readily be distinguished using graphical sparse partial least squares discriminant analysis (sPLS-DA) plots (Figure 1) within *mixOmics*, showing a complete group separation on the two component axes. Ten repetitions of 10-fold cross validation provided a final sPLS-DA 2 component model that provided error-free classification via 20 metabolites (Figure 2) that included the most significant Monoacylglycerol C16:0_N, which was excluded from all the convergent MetaboAnalyst 4.0-derived results.

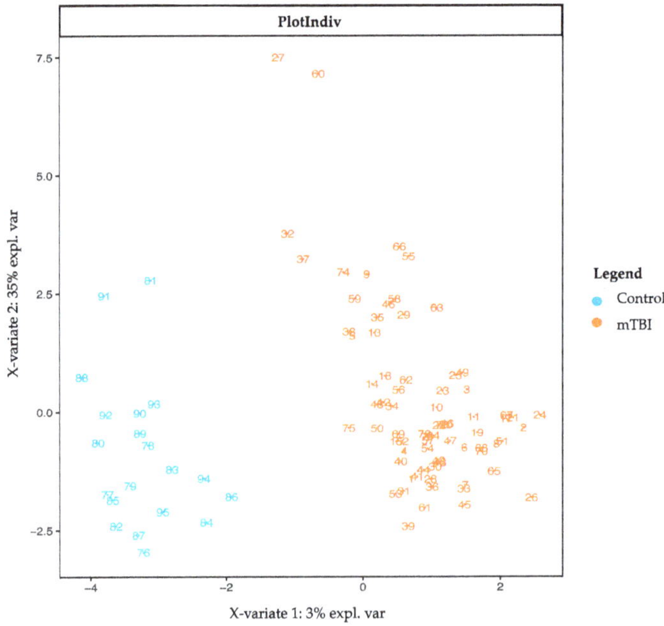

Figure 1. Sparse partial least squares discriminant analysis (sPLS-DA) plot. Note separation of subacute mTBI compared to TBI control data, as determined by metabolites making up the first two analytic components. The separation of the case and control groups is complete, without overlap. sPLS-DA = sparse partial least squares-discriminant analysis. Control = TBI control. mTBI = mild traumatic brain injury.

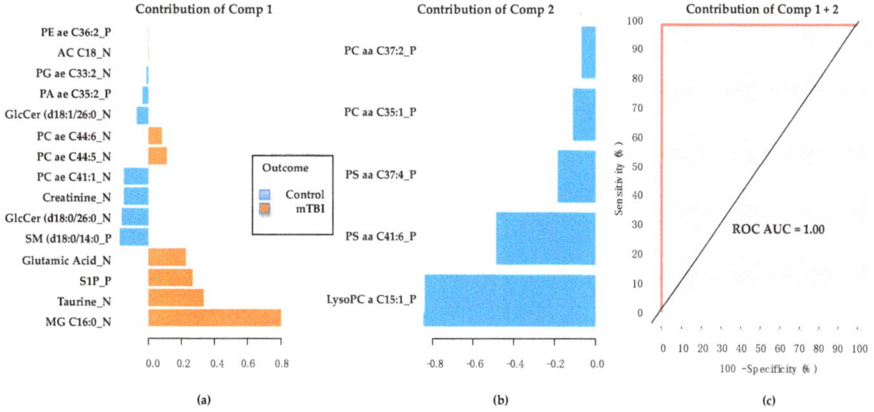

Figure 2. Metabolites associated with first two discriminant components. (**a**) The first component provides 15 metabolites, and the bottom 4 listed providing the greatest contributions (all higher in TBI cases) to classification accuracy. (**b**) The second principal component provides 5 metabolites (all lower in TBI controls). (**c**) Receiver operating characteristic area under the curve (ROC AUC) provides result of 1.00 using 20 metabolites from the two components in the classifier model. Comp = sPLS-DA model component. PE = phosphatidylethanolamine. AC = acylcarnitine. PG = Phosphatidylglycerol. PA = Phosphatidic acid. GlcCer = glucosylceramide. PC = phosphatidylcholine. SM = sphingomyelin. S1P = sphingosine-1-phosphate. MG = Monoacylglycerol. PS = phosphatidylserine. LysoPC = lysophosphatidylserine. Final metabolite identifications will require additional tandem mass spectrometry (MS/MS) analyses. Metabolites confirmed via MS/MS are considered fully validated, to a high degree of confidence.

2.4. Subacute mTBI Plasma Metabolomic Biomarkers–Targeted Analysis via mixOmics

Targeted metabolite (Biocrates AbsoluteIDQ®p180 kit, Biocrates Life Sciences AG, Innsbruck, Austria) values were developed into an optimal classification model using 10 repetitions of 10-fold cross validation through sPLS-DA in *mixOmics*. The final model featured 15 metabolites and metabolite ratios (Figure 3a) that provided perfect classification of the groups (Figure 3b). Of interest, both Taurine and Glutamic Acid were top contributors to the panel, thereby indirectly supporting their putative identities and importance derived from the untargeted analyses previously presented, with both elevated in the subacute mTBI cases, as opposed to controls.

In summary, we discovered and internally validated several plasma metabolomic biomarker panels using both untargeted and targeted metabolomic approaches and using two different analytic platforms, MetaboAnalyst 4.0 and *mixOmics*. The final biomarker panels derived by the untargeted methods featured several of the same top metabolites as the targeted analysis, and suggested potential relevance for both Glutamic Acid and Taurine in subacute TBI. Of interest, the top 4 metabolites resulting from unbiased feature selection via MetaboAnalyst 4.0 and *mixOmics* were identical (see Table 2 and Figure 2a). Additional investigations are required to confirm the identification of the preliminarily annotated plasma biomarkers proposed in this study using untargeted methods. While tandem MS (MS/MS) is typically required, a preliminary confirmation of both Taurine and Glutamic Acid can be proposed given the confirmed identities provided by the targeted metabolomic results. It remains important, however, that the preliminarily annotated plasma metabolomic panels for subacute mTBI be externally replicated utilizing similar groups of cases and controls.

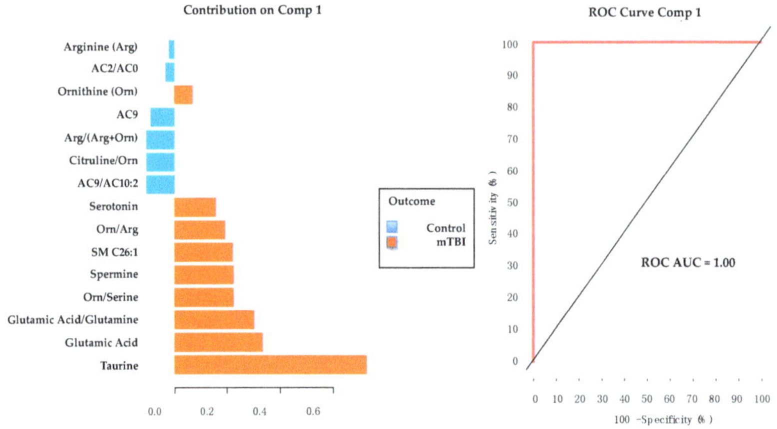

(a) (b)

Figure 3. Targeted metabolomic panel and classification performance. Using the sPLS-DA methods in *mixOmics*, this 15-member metabolite panel was derived (**a**) featuring primarily amino acids, biogenic amines and specific metabolite ratios. This particular targeted metabolite panel classified subacute mTBI subjects from TBI controls with a ROC AUC = 1.0. (**b**) Note the two metabolites with the highest contribution are Taurine and Glutamic Acid. Comp 1 = feature selection component 1. mTBI = mild traumatic brain injury. ROC = receiver operating characteristic. AUC = area under the curve. AC = acylcarnitine. SM = sphingomyelin.

2.5. PD/PDD Serum Metabolomic Biomarkers–Utilizing the mixOmics-Derived sPLS-DA Top 20 Metabolites from Subacute mTBI Analysis

Metabolite matching using *MSFmetabolomics*, between the 20 sPLS-DA-derived plasma subacute mTBI metabolite biomarkers and the serum-derived PD/PDD/Control metabolomic data, indicated that only nine of the 20 metabolites were also present in preliminarily annotated metabolites from the PD/PDD/Control specimens (Figure 4a). Despite such a limitation in numbers of matched metabolites between the two datasets, the performance of the 9-metabolite panel in a *mixOmics* PLS-DA classifier model provided respectable ROC AUC (0.8488) results (Figure 4b). Importantly, Glutamic Acid was again a prominent contributor to the model's performance, although this time it was notably increased in control subjects in comparison to the PD/PDD group. Taurine was not present as a member of this panel. These findings suggest a relative loss of serum Glutamic Acid concentration in those with PD/PDD compared to age-matched controls, while the absence of Taurine from the panel likely represents an insignificant difference in levels between PD/PDD and control subjects.

Utilizing the subacute mTBI metabolite panel members in a group of much older PD/PDD/ Control subjects provided very good classification accuracy for discriminating PD/PPD from matched controls, and despite using only 9 of the original 20 metabolites in the model. Although encouraging, these findings are limited by the relatively small group sizes in the PD/PDD/Control cohort, with only 20 individuals represented in each diagnostic category. Larger numbers of subjects may provide alternative impressions, as well as analyzing the PD cohort's plasma specimens rather than serum. Impressively, however, Glutamic Acid remained the most significant metabolite differentiating cases from controls in the PD cohort analysis, with the opposite relative abundance (higher in controls rather than cases) to that found in the subacute mTBI subjects.

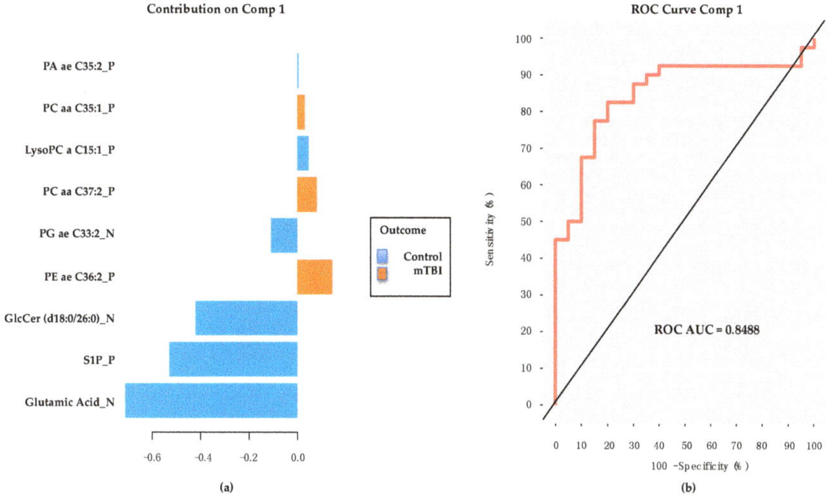

Figure 4. Contribution plot and performance of 9 common subacute TBI biomarkers classifying the PD/PPD subjects from PD controls. (**a**) Note prominence of the Glutamic Acid contribution, but with relative abundance values reduced in PD/PDD and compared to controls. (**b**) Respectable performance (ROC AUC = 0.8488) of 9 member panel in classifying PD/PDD subjects from controls. Comp = PLS-DA model component. TBI = traumatic brain injury. PD = Parkinson's disease. PDD = PD dementia. ROC AUC = receiver operating characteristic area under the curve. PA = Phosphatidic acid. PC = phosphatidylcholine. LysoPC lysophosphatidylcholine. PG = Phosphatidylglycerol. PE = phosphatidylethanolamine. GlcCer = glucosylceramide. S1P = sphingosine-1-phosphate. Final metabolite identifications will require additional tandem mass spectrometry (MS/MS) analyses. Metabolites confirmed via MS/MS are considered fully validated, to a high degree of confidence.

2.6. PD/PDD Serum Metabolomic Biomarkers–New Discovery Using mixOmics sPLS-DA

Utilizing the *mixOmics* platform and sPLS-DA, unbiased feature selection was used to discover an optimal classification model when comparing the PD/PDD group to PD controls. Using 10 repetitions of 10-fold cross-validation a model utilizing a single component composed of 10 metabolites was developed (Figure 5a). The model's classification contribution was significantly weighted toward Glutamic Acid, which was again higher in the serum of control subjects than in those with PD/PDD. As in the previous section, performance of this 10 member panel provided an ROC AUC of 0.85 (Figure 5b).

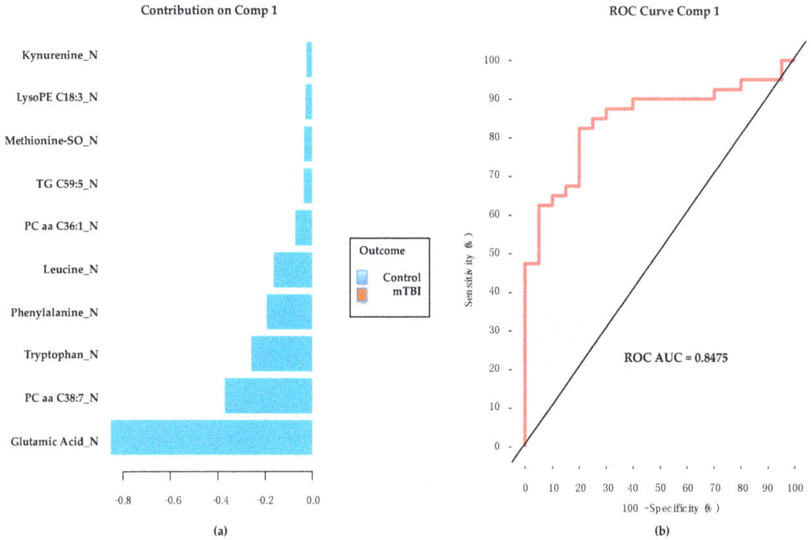

Figure 5. Contribution plot and classification performance of 10 metabolites derived via sPLS-DA from PD/PPD/Control subjects. (**a**) Glutamic Acid continues to provide the major contribution to the classification performance of this preliminarily annotated 10-metabolite panel. (**b**) A similar ROC AUC is obtained in new discovery with these data as had been obtained using the subacute TBI biomarker panel's 9 preliminarily annotated common metabolites (see Figure 4). Of interest, the only common metabolite between these results and those from the TBI panel is Glutamic Acid. Comp = sPLS-DA model component. TBI = traumatic brain injury. PD = Parkinson's disease. PDD = PD dementia. ROC AUC = receiver operating characteristic area under the curve. LysoPE = lysophosphatidylethanolamine. SO = sulfoxide. TG = triglyceride. P = phosphatidylcholine. Final metabolite identifications will require additional tandem mass spectrometry (MS/MS) analyses. Metabolites confirmed via MS/MS are considered fully validated, to a high degree of confidence.

2.7. Evaluation of Glutamic Acid's Performance as Sole Metabolite in mixOmics PLS-DA Classifier Models for Subacute mTBI and PD Cohorts

Relative abundance values for Glutamic Acid were higher in the TBI cases as opposed to TBI controls (Figure 6a), while controls provided higher abundance values than cases in the PD cohort (Figure 6b) We tested the classification ability of Glutamic Acid as a sole classifier for both of our cohorts, the subacute mTBI and PD. Using the *mixOmics* PLS-DA algorithm, and Glutamic Acid alone, comparable classification ROC AUC results were attained in both cohorts (Figure 6c,d), despite the opposite relative abundance measures noted between cases and controls.

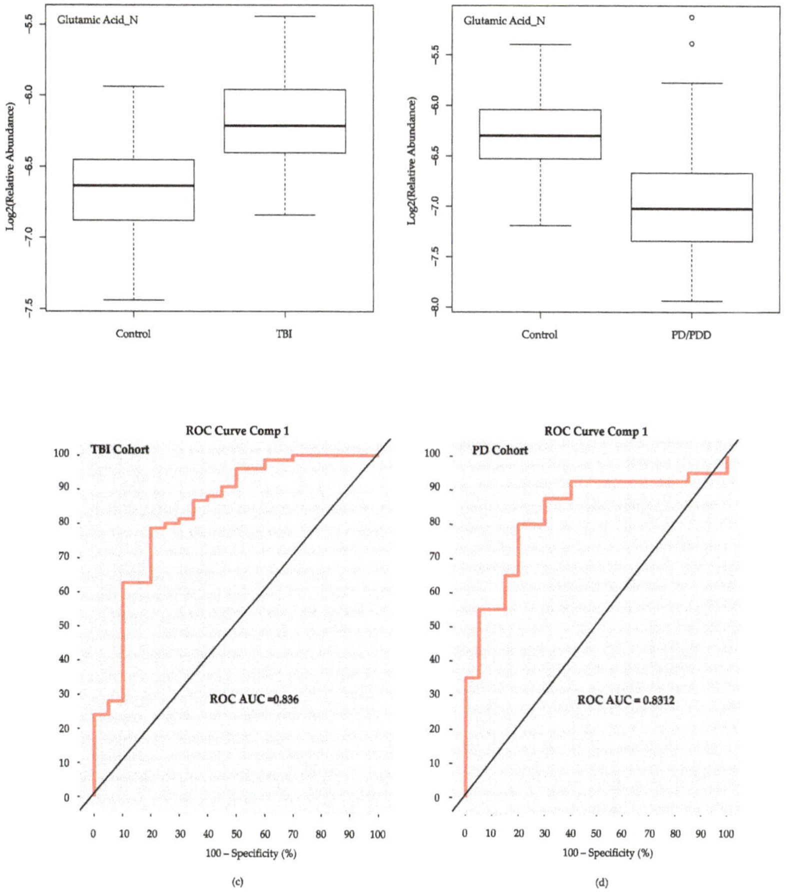

Figure 6. Classification of cohort groups using Glutamic Acid as the sole metabolite. Log2 (relative abundance) values for Glutamic Acid in the two study cohorts are depicted via boxplots in panels (**a**,**b**). For the subacute TBI cohort (**a**) Glutamic Acid is elevated in TBI cases compared to controls. In the PD cohort (**b**) Glutamic Acid was elevated in controls compared to the PD/PDD cases. ROC AUC results are nearly identical using the *mixOmics* PLS-DA model with only Glutamic Acid as classifier of the TBI (subacute mTBI) cases from TBI controls (**c**), as well as the PD (PD/PDD) cases from PD controls (**d**). Comp = PLS-DA model component. TBI = traumatic brain injury. mTBI = mild TBI. PD = Parkinson's disease. PDD = PD dementia. ROC AUC = receiver operating characteristic area under the curve.

3. Discussion

In addition to the prominence of Taurine and Glutamic Acid in blood specimens from our subacute mTBI subjects, elevations of the Glutamic Acid/Glutamine ratio in the targeted metabolomic results suggests potential alterations in the cycling of these two species during the subacute recovery from TBI. Such an altered ratio has been previously noted in both children and adults following acute TBI [13,14]. Interestingly, Taurine is known to function as an osmoregulator [15], neuromodulator, calcium regulator [16], antioxidant [17], and neuroprotectant from excitotoxic cell death [18]. It does not require much extrapolation, therefore, to see how these two metabolites may be integral responses to continuing, subacute processes in response to a TBI.

Glutamic acid, or glutamate is the most abundant excitatory neurotransmitter in brain tissue [19,20]. Physiologically, glutamate helps mediate cellular function through binding to glutamate receptor (GluR) proteins localized to the external face of the plasma membrane [21] and thereby activating a variety of ion channels or intracellular networks via G-proteins and other membrane and cytoplasmic mediators. Levels of glutamate in the brain's extracellular fluid (ECF) are typically maintained within a tight range, and thus, concentrations that are too low or too high may produce negative consequences [20]. Regulation of glutamate levels in the brain ECF, therefore, is important in preventing cellular toxicity. Synthesis of brain glutamate involves uptake of peripherally derived branched chain amino acids (BCAA) from blood, with their uptake and intracellular processing to form glutamine taking place within astrocytes, and release of glutamine from astrocytes and uptake by neurons finally resulting in production of glutamate from glutamine, and glutamate's eventual release as a neurotransmitter [22]. Since there are no glutamate degrading enzymes within the ECF, regulation of glutamate levels is controlled via cellular release and cellular uptake. The primary mechanism that controls the brain's ECF glutamate levels under normal physiological conditions is via uptake/transport mechanisms associated with local neurons, astroglia, and the endothelial cell components of the blood-brain barrier (BBB) [20,23]. Specific abluminal transporters on brain vascular endothelium vessels and within the choroid plexus aid in regulating ECF and cerebrospinal fluid (CSF) glutamate levels by transport of excess glutamate into endothelial cells of the neurovasculature and thereby released into circulating blood [24,25]. In contrast, direct uptake of glutamate from blood is insignificant [25–27]. In known cases of toxic glutamate levels in brain, such following ischemic stroke or TBI, such unidirectional flux may potentially be modulated to help restore homeostasis [28] and improve outcomes. In TBI, glutamate is known to increase acutely within the ECF as a result of the associated cellular injury and BBB dysfunction/disruption, with abnormalities in the physiologic uptake/transport mechanisms [29]. While there is robust evidence for changes in ECF glutamate following acute TBI [13,14,30,31], similar analyses during the subacute stage of TBI recovery are limited. Our current analyses suggest that glutamate might remain increased in plasma for at least 3–12 months following mTBI, at least in those symptomatic individuals within our subacute mTBI group. We speculate that elevated levels of Taurine might be an attempt at physiological buffering of what might otherwise be considered a relatively "excitotoxic" environment [32] if elevated Glutamic Acid levels in plasma reflect similar conditions in the brain ECF of the subacute mTBI subjects. Finally, we propose that glutamate elevations in brain ECF may be a direct expression of the degree of parenchymal injury sustained, while elevations in Taurine may indicate an intrinsic attempt to mitigate progressive secondary injury effects.

Concussion (or mTBI) produces early ECF increases in free fatty acids [33], via activation of phospholipases [34]. Elevations in the free saturated fatty acids (SFA) palmitate and stearate rapidly increase in brain following experimental TBI, achieve concentrations 2–3 times those of the polyunsaturated fatty acid arachidonic acid, and remain elevated beyond 6 h following injury [35]. Release of membrane SFAs is the result of phospholipase A_1 (PLA_1) activity on the plasma membrane [36,37]. We have recently been able to link levels of palmitate to dysregulated expression of a bioenergetic regulator in PD [38], which may ultimately prove significant in other neurological conditions, including TBI. Both PD [39] and AD [40] are characterized by reductions in brain levels of the peroxisome proliferator-activated receptor gamma coactivator 1-alpha (PGC-1α), a major regulator of mitochondrial number and function, also known to control cellular lipid metabolism, glucose metabolism, electron transport, and certain anti-inflammatory effects. We have shown that elevated levels of the free palmitate reduce PGC-1α gene expression through the epigenetic non-canonical promoter hypermethylation, both in vitro and in vivo [38]. The relation of cellular PGC-1α levels to TBI does not appear to be as straightforward as in PD, but may result from the previously noted elevations in palmitate following brain trauma and secondary effects that maintain abnormally elevated free palmitate levels. Whether elevated levels of palmitate result in additional complex reactions [41] impacting brain parenchyma remains to be determined. Importantly, however, a link between abnormal lipid levels in brain (and likely blood) together with downregulation of PGC-1α gene expression have been made in PD [42] and AD [43], and appear related to metabolism and epigenetic controls on gene expression. Whether such links can be made in TBI remains to be seen.

As opposed to TBI, the susceptibility of brain to glutamate toxicity has been primarily demonstrated following hypoxic/ischemic insults [44,45], commonly associated with excessive increases in measured ECF glutamate levels. Although similar elevations in ECF glutamate levels have been associated with severe TBI, significant elevations in mTBI (i.e., concussions) or neurodegenerative disorders (e.g., PD or AD) have been considered less likely [46]. Much more likely, however, is that in PD and other neurodegenerative disorders, and in more diminished brain insults (e.g., mTBI), less dramatic ECF glutamate levels may somehow become toxic and perpetuate a cellular injury cascade. Varying susceptibilities to toxicity from physiological glutamate levels may involve mitochondrial energy metabolism, and the energy-dependent maintenance of neuronal membrane polarization. Since energy-dependent ion channels and pumps are primarily responsible for sustaining the resting membrane potential of neurons, a depletion of adenosine triphosphate (ATP), associated with mitochondrial dysfunction (as in PD, accompanied by reduced PGC-1α expression), will result in a reduced membrane potential or actual depolarization [47]. Alternative hypotheses for glutamate cytotoxicity in neurodegenerative disorders have been proposed [48,49] and provide a much more solid experimental foundation [50]. These postulates highlight the synergistic interaction [51] between bioenergetic defects and glutamate toxicity at physiologic levels.

The Glutamic Acid and Taurine elevations seen in our subacute mTBI cases were absent in the PD/PDD cases assessed. Subjects with PD (and PDD) likely suffer from varying degrees of brain mitochondrial dysfunction, featuring aberrant lipid and glucose metabolism, and altered energy production as a result of epigenetic downregulation of PGC-1α, among other mechanisms [38]. While such susceptibilities in PD may be limited to modulation of subcortical motor pathways, resulting from deficits related to dopaminergic nigrostriatal degeneration, as the pathobiology progresses to include PDD, the susceptible brain regions may expand to involve cortical gray matter neuronal populations critical to higher order cognition. Metabolite profiling in PD remains challenging, with the most common findings related to alterations associated with mitochondrial dysfunction [52]. In this analysis between age-matched controls and PD/PDD cases, we found relative elevations in serum Glutamic Acid levels in the control subjects compared to cases, as previously noted using nuclear magnetic resonance (NMR) analyses of in vitro PD models [53] and CSF in human PD [54], along with PD plasma using similar MS methods to our own [55]. Such results, however, contradict prior blood-based results using less sensitive methodologies [56]. Our Glutamic Acid findings in PD/PDD serum are supported by our use of UPLC-MS, and may reflect a pathological reduction of brain ECF glutamate levels, a relative exhaustion of glutamate production in PD/PDD, or an attempted compensatory reduction of serum Glutamic Acid levels (resulting from reduced excitatory neurotransmitter tone within the brain) reducing the susceptibility of "excitotoxicity".

We acknowledge specific limitations associated with our biomarker investigation. A common specimen collection and processing protocol for both cohorts would have been ideal, but were not possible for this study. Given our past experience collecting, processing, and analyzing metabolomic specimens, we only accepted and analyzed specimens that we felt met strict collection and processing standards. We acknowledge, however, that differences in whether the specimens were collected fasting or not, and processed within 4 or 24 h from collection, to produce either plasma or serum, may have adversely impacted our ability to adequately interpret the results. Measuring the oxidation of lipids, especially phosphatidylcholines (PCs) in biospecimens, as a determinant of specimen integrity or enhanced disease-related phospholipase activity, have been reported in humans [57–59]. In relation to AD biospecimens [57,58], the ratio of Lysophosphatidylcholines (LysoPCs)/PCs has been proposed as useful in differentiating between control subjects and those with prodromal or manifest disease, possibly reflecting pathologic membrane oxidation. Such ratios can more accurately be determined using quantitative targeted MS results, with such ratios increasingly provided in analytic outputs.

Although the metabolomic field, to our knowledge, has yet to adopt routine use of such ratios as determinants of specimen integrity, we are in support of an eventual consensus measure that would allow discrimination of specimen integrity [60]. For this investigation, comparison of targeted metabolomic output results, and calculation of LysoPC/PC ratios was not possible between our TBI and PD cohort specimens, since targeted analyses were only available from the former. Despite an adequate number of available subacute mTBI specimens, we admit including a less than optimally matched set of TBI control group specimens, both in number and comparable characteristics. As presented in Table 1 and described in Results 2.1, there were significant age and sex differences noted within the TBI cases compared to TBI control groups. These dissimilarities likely resulted from the inclusion of subjects from two separate, independent investigations in our study, with one featuring military personnel and the other made up of college athletes. Ideally a military non-TBI cohort from the same military institution as the TBI cases would have likely provided a better-matched control group for the subacute mTBI cases. Additional TBI controls would have strengthened the analysis as well, especially with the inclusion of a number of non-TBI, trauma controls (e.g., orthopedic injuries), to attempt differentiation of TBI-specific biomarkers from those related to a more generalized post-traumatic state [61]. It remains possible that the age and sex differences between the mTBI cases and controls may have somehow contributed to the observed metabolomic differences. Although the PD cohort's groups were much better balanced and matched on all parameters, we believe that larger numbers in each of the subgroups could provide added weight to the results. The addition of subjects from the preclinical PD spectrum, including those diagnosed with rapid eye movement (REM) sleep behavior disorder (RBD) [62–64] without PD, considered a preclinical non-motor stage of PD, might have allowed blood-based Glutamic Acid assessment during this transition to the clinically evident motor stages of PD. Our goal for future investigations assessing biomarkers in both TBI and PD will include evaluating larger, more comparable cohorts of subjects (including matching ages and sex in cases and controls) and specimens (using the same collection protocols and blood matrix for analysis). Increased detail should be paid to lifelong histories of TBI experiences, the ethnicity of participants, environmental exposures (e.g., rural versus urban living), and mitigating any geographical bias between groups. While an initial homogeneity of cohorts might be helpful in defining significant classifiers of specific conditions, once such classifiers are determined and replicated under the same settings, stress testing of such panels in more disparate subject groups would be a requisite next step toward biomarker development and more widespread utility. Finally, we believe it is important to avoid analyses of disparate blood matrices whenever possible. For this study we did not have the option to evaluate only plasma or serum within both cohorts, as the TBI cohort only provided plasma while the PD cohort had collected serum. Such comparisons, we believe, are not as ideal, as we have raised previously [60]. Evidence of differences between serum and plasma metabolites within the same subjects has been documented [65], and such differences are especially notable for certain glycerophospholipids [66]. Despite these limitations, we believe the information developed through this study provides relevant preliminary guidance related to potential pathobiologic links between subacute mTBI and PD, with the prominence of Glutamic Acid in blood, or lack thereof, in both cohorts. Replicative investigations are necessary to assess the significance, if any, of blood Glutamic Acid as a biomarker for subacute mTBI, and possibly for staging PD. Such investigations will detail whether there are consequential ties between mTBI and PD.

4. Materials and Methods

4.1. Study Populations

The institutional review board (IRB) at Naval Medical Center Portsmouth, VA, approved study protocol, informed consent documents, and participation for all consenting subacute mTBI participants from Naval Medical Center Camp Lejeune in compliance with all applicable Federal regulations governing the protection of human subjects. The Research Subjects Review Board at the University of

Rochester and Rochester Institute of Technology provided approval for human subject participation for the TBI control participants, all of whom provided written informed consent prior to entering the study [67] and providing specimens. Control, PD, and PDD subjects giving informed consent for study participation and collection of blood specimens were part of an Oxford Parkinson's Disease Centre (OPDC) Discovery Cohort study, Oxford, UK approved by the Oxfordshire A Research Ethics Committee (10/H0505/71, Version 5, 23/07/14), with transfer of specimens to Georgetown University and University of California Irvine (UCI) approved by the OPDC Data Access Committee. In addition, all protocols, consents, and relevant documents for each individual study and combined storage and analyses of de-identified specimens and study protocols were approved by the IRBs at Georgetown University and the University of California, Irvine, and by the Department of Defense Human Research Protection Office.

The subacute mTBI cohort was made up of 75 active duty sailors and marines cared for at the Intrepid Spirit Concussion Recovery Center, Naval Medical Center Camp Lejeune, Jacksonville, NC. All study participants had sustained a TBI within the 3–12 month interval (subacute period of recovery) prior to blood collection, and all were being followed for persistent neuropsychological symptoms following their TBIs. Control TBI subjects (n = 20) were asymptomatic, non-concussed collegiate athletes participating in an acute sports-related mTBI study [67] in Rochester, NY. Included athletes for this study provided blood specimens prior to their participation in their respective sports season and had no history of a recent TBI (within the previous 12 months). PD subjects were recruited from the longitudinally assessed, population-ascertained Oxford Discovery Cohort [68]. The clinical diagnosis of PD and PDD was made according to UK PD Society Brain Bank diagnostic criteria [69], and Movement Disorders Society level 1 criteria for PDD [70] during 18 month longitudinal clinical evaluation by a trained neurologist. The PD cohort was made up of PD controls (n = 20), PD (n = 20), PDD (n = 20) subjects followed and diagnosed via the OPDC. Demographic details for cases and controls in both cohorts are provided within the Supplementary Materials.

For this investigation, the blood collection protocols differed between cohorts. Our approaches to blood collection and specimen processing methods for human investigations have been previously detailed [60,71–74] and were used for our subacute mTBI specimens. Collection and processing that differed in this study included the lack of fasting for our TBI control group [12], and lack of fasting, medication withholding, and collection of serum rather than plasma in the PD cohort. Input blood specimens for the TBI cohort were collected in ethylenediaminetetraacetic acid (EDTA) tubes, thoroughly mixed, and kept on ice until separated by centrifugation into components (e.g., plasma, leukocytes, erythrocytes), typically within 4 h of collection, except the subacute mTBI group. The latter group's EDTA tubes were shipped on ice for processing and separation within 24 h of collection, at Georgetown University. Separated plasma was aliquoted into cryovials and placed into −80 °C freezer until analyzed. The PD cohort blood specimens featured collection into BD Vacutainer SSTII tubes. Each tube was mixed via inversion and left at room temperature for 10 min to allow clot formation. Clot tubes then underwent centrifugation, with serum collected into cryovials kept on dry ice until placed into −80 °C freezer for later analyses.

4.2. Metabolomic Analyses and Data

For metabolomic analyses utilizing ultraperformance liquid chromatography-mass spectrometry (UPLC-MS), all collected specimens for this study were shipped frozen as individual ≥100 μL aliquots of plasma or serum to the Metabolomics Shared Resource at Georgetown University. All specimens were processed and analyzed using untargeted and targeted methods previously detailed for human studies of preclinical AD [71,73], optimal cognitive aging [74], and acute mTBI [12]. In brief, after sequential extraction [75], untargeted metabolomic profiling of all the plasma specimens was carried out utilizing ultra-performance liquid chromatography-electrospray ionization-quadrupole time of flight-mass spectrometry (UPLC-ESI-QTOF-MS)-based data acquisition and state of the art instrumentation (Acquity H-class UPLC system and Xevo G2 QTOF, Waters Corporation, Milford, MA,

USA), with strict adherence to quality control (QC) protocols. Pooled QC samples were run every ten injections. This methodology is conducive to the extraction of a broad range of metabolites, including lipids. Metabolomic relative abundance data output was provided in two ESI modes (negative, NEG, _N; or positive, POS, _P) for each analyzed sample. The Xevo G2 QTOF MS instrument was set up to scan the 50-1200 *m/z* mass range for each ESI mode, for each plasma specimen in the data set. Each ESI mode typically provides up to 3500 unique *m/z* features. The UPLC-MS raw data files were initially pre-processed using the XCMS software [76,77] (Scripps Institute, La Jolla, CA, USA). The untargeted Excel output files produced were populated with mode-specific *m/z* values corresponding to preliminarily annotated metabolites (www.msfmetabolomics.com, [78]) and their relative abundance values within the sample. The untargeted metabolomic approach used in this investigation is considered semi-quantitative [79], and requires an additional step to confirm analyte identification and quantification, typically via tandem mass spectrometry (MS/MS) [80]. The untargeted metabolomic (see Supplementary Material) data in this study has only been preliminarily annotated and has not undergone confirmation of identities via MS/MS. The untargeted metabolomics data were normalized to the intensity of internal standards (debrisoquine in ESI positive and 4, nitro benzoic acid in ESI negative mode) spiked in the extraction buffer. The data are log transformed and Pareto scaled. Targeted metabolomic analysis of plasma/serum samples was performed using the Biocrates Absolute-IDQ P180 (BIOCRATES, Life Science AG, Innsbruck, Austria). This validated targeted assay allows for simultaneous detection and quantification of metabolites in plasma samples (10 L) in a high-throughput manner. The methods have been described in detail [81,82]. The plasma samples were processed as per the instructions by the manufacturer and analyzed on a triple-quadrupole mass spectrometer (Xevo TQ-S, Waters Corporation, Milford, MA, USA) operating in the multiple reaction monitoring (MRM) mode. The measurements were made in a 96-well format for a total of 148 samples, and seven calibration standards and three quality control samples were integrated in the kit. Briefly, the flow injection analysis tandem mass spectrometry (MS/MS) method was used to quantify a panel of 144 lipids simultaneously by multiple reaction monitoring. The other metabolites are resolved on the UPLC and quantified using scheduled MRMs. The kit facilitates absolute quantitation of 21 amino acids, hexose, carnitine, 39 acylcarnitines (ACs), 15 sphingomyelins (SMs), 90 phosphatidylcholines (PCs) and 19 biogenic amines. Pre-analytical processing for the targeted metabolomic data was initially performed using the MetIQ software (BIOCRATES, Life Science AG, Innsbruck, Austria), followed by additional considerations [83], and developed into a similar Excel formatted targeted metabolomic data (see Supplementary Material) as the untargeted metabolomic data.

4.3. Metabolomic Biomarker Development

The goal of this analysis was to define a novel metabolomic classifier model that distinguished the subacute mTBI cases from TBI controls and to investigate whether biomarker similarities exist that may implicate TBI in the pathogenesis of PD. A similar untargeted plasma metabolomic biomarker development methodology for the subacute mTBI cohort as was described for our recent human acute mTBI investigation [12], taking advantage of the MetaboAnalyst 4.0 platform (www.metaboanalyst.ca) [84]. The primary steps involved in untargeted biomarker development for this portion of the analyses included running a preliminary annotation of normalized XCMS *m/z* features and their respective abundance data for each study subject using the preliminary annotation algorithm MSFmetabolomics (www.msfmetabolomics.com) [78], as previously described [12], and with a mass error stringency of 5 parts per million (ppm). Preliminarily annotated untargeted Excel (.csv) datasets (or similar targeted datasets) were uploaded into MetaboAnalyst 4.0 for biomarker development, utilizing the *Explorer* and *Tester modules*. After initial unbiased multivariate feature selection methods helped define potential biomarker panels, using LinSVM [85], PLS-DA [86], RandFor [87], and LASSO [88] algorithms, a customizable feature selection within the *Tester module* allowed optimization of model results and provided analytic outputs for comparing specific metabolite models via ROC AUC results and the LinSVM, PLS-DA, RandFor, logistic regression (LR) and logistic regression with 10-fold cross validation (LR + 10FCV) methods.

A second biomarker development approach, via the R package *mixOmics* [89], was used on both the untargeted subacute mTBI cohort metabolomic data as well as untargeted and targeted data from PD/PDD cohort. Complimentary to our prior analyses, we employed these methods to discover and evaluate biomarker signatures discriminating clinical cases (i.e., subacute mTBI or PD/PDD) from their respective healthy controls using both sparse PLS-DA (sPLS-DA) and non-spares PLS-DA methodologies, as appropriate. The sPLS-DA model offers an automated and integrated alternative to the manual selection of variables for inclusion into biomarker panels. Differing from our previous biomarker discovery and statistical work performed on the MetaboAnalyst 4.0 platform, all statistical computing in *mixOmics* was conducted using R. RVU measures initially underwent log base 2 (log_2) transformation. We also used *mixOmics* to provide ROC AUC results from the targeted metabolomic data derived from the subacute mTBI cohort using sPLS-DA, and to select specific metabolites for modeling using PLS-DA methodology.

4.4. Statistical Analyses

Numerical and categorical comparisons were performed using SPSS Statistics (version 24 for the Mac). Age distributions were plotted to assess normality for case and control groups in each cohort. The comparisons of the two independent group means for age were determined using parametric (*t*-test) and nonparametric (Mann-Whitney U Test) statistics based on normality of age distributions. Categorical analyses for diagnostic group and sex were performed using chi-square analyses. Statistical significance (with Bonferroni correction) was defined at the $p < 0.025$ level. Statistical algorithms within both MetaboAnalyst 4.0 and mixOmics platforms are detailed within their publications [84,89], as previously noted in the Metabolomic Biomarker Development section of the Methods. Both of these platforms utilize feature selection algorithms that account for multiple comparisons inherent in biomarker datasets, where multiple classification features are considered for a relatively small number of specimens ($p >> n$). We used ROC AUC results to compare classification of groups and specific biomarker panels in this investigation, with 1.0 indicating error-free classification and 0.5 indicating selection no better than by chance.

5. Conclusions

Based on this preliminary investigation, there appears to be a reciprocal relationship in blood-derived Glutamic Acid levels between cases and controls in our subacute mTBI and PD cohorts. Relatively elevated blood-derived Glutamic Acid was noted in the TBI cases compared to controls, where the opposite was defined in the PD cohort. Although unconfirmed, we propose such a blood biomarker difference may be associated with a central state of glutamate-specific pathobiology. We anticipate that such differences in blood Glutamic Acid levels would be relatively easy to document in a larger number of clinical specimens from similar subject groups and either reproduce or refute this study's findings. Under optimal conditions such comparisons would be performed on well-matched subject groups and via analysis of a single blood matrix (plasma OR serum) in both cohorts. Although we agree that an ultimate link between TBI and the pathogenesis of PD will require longitudinal assessments of a large number of subjects, future investigations utilizing blood biomarkers and appropriate animal models may provide additional correlative information that may lead to actionable clinical assessments and interventions. Finally, we anticipate that understanding the relationships between blood biomarkers and detailed clinical assessments derived from both TBI and PD subjects will provide additional focus for future investigations, including added neurobiological clues linking these distinct disorders.

Supplementary Materials: The following are available online at http://www.mdpi.com/2218-1989/8/3/50/s1, Supplementary READ ME file 1; S2, Supplementary targeted metabolomic Excel files (1–4); S3, Supplementary untargeted metabolomics Excel files (1,2); S4, Supplementary demographics Excel file.

Author Contributions: Conceptualization, M.S.F., T.J.G., M.T.H., R.W.-M., K. M.-B., J.B., M.M. and H.J.F.; Data curation, M.S.F., T.J.G., M.T.H., S.E., R.W.-M., K.M.-B., J.B. and A.K.C.; Formal analysis, M.S.F, T.G. and M.M.; Funding acquisition, M.S.F. and H.J.F.; Investigation, M.S.F., T.J.G., M.T.H., S.E., R.W.-M., K.M.-B., J.B. and H.J.F.; Methodology, M.S.F., S.E., A.K.C. and H.J.F.; Project administration, M.S.F., T.J.G. and H.J.F.; Resources, M.S.F., T.J.G., M.T.H., R.W.-M., K.M.-B., J.B., A.K.C. and H.J.F.; Supervision, M.S.F., T.J.G. and H.J.F.; Validation, M.S.F., M.M. and H.J.F.; Visualization, M.S.F. and T.J.G.; Writing—original draft, M.S.F.; Writing—review and editing, M.S.F., T.G., T.J.G., M.T.H., S.E., R.W.-M., K.M.-B., J.B., A.K.C., M.M. and H.J.F.

Funding: This research was funded through the U.S. Army Medical Research Acquisition Act (USAMRAA) and the U.S. Army Medical Research and Materiel Command (USAMRMC) grant numbers W81XWH-09-1-0103 and W81XWH-09-1-0107 to Howard Federoff, and W81XWH-16-1-0148 to Howard Federoff and Massimo Fiandaca.

Acknowledgments: The authors acknowledge the following institutional support: (1) the command and medical staff, as well as the sailors and marines from Naval Medical Center Camp Lejeune, for providing the subacute mTBI specimens evaluated in this study; (2) the athletes and staff participating in the Sports Related Concussion study at the University of Rochester for providing TBI control specimens used in this study; and, (3) the study participants and staff at the Oxford Parkinson's Disease Centre, and affiliated hospitals, for providing the PD/PDD and PD control specimens analyzed in this study. We thank Rond Malhas, Ricardo Miramontes, and Robert Padilla for their technical expertise in collecting, processing and storing the human blood specimens used for the metabolomic analyses, Tyrone Dowdy and Steven Payton for their technical assistance in developing the metabolomic data, and Amin Mahmoodi, Ron Sahyouni, and Nick Morris for their technical expertise in developing the MSFmetabolomics application and website.

Conflicts of Interest: T.M.J., M.T.H., S.E., and R.W.-M. declare no conflicts of interest. The other authors declare the filing of intellectual property related to blood-based biomarkers through Georgetown University and the University of California Irvine related to this work and prior human biomarker investigations related to memory and aging and TBI. The views expressed in this article reflect the results conducted by the authors and do not necessarily reflect the official policy or position of the Department of the Navy, Defense Department, nor the U.S. Government. Those providing funding for these investigations had no role in the design of the study; in the collection, analyses, or interpretation of data; in the writing of the manuscript, and in the decision to publish the results.

References

1. Jafari, S.; Etminan, M.; Aminzadeh, F.; Samii, A. Head injury and risk of parkinson disease: A systematic review and meta-analysis. *Mov. Disord.* **2013**, *28*, 1222–1229. [CrossRef] [PubMed]
2. Godbolt, A.K.; Cancelliere, C.; Hincapie, C.A.; Marras, C.; Boyle, E.; Kristman, V.L.; Coronado, V.G.; Cassidy, J.D. Systematic review of the risk of dementia and chronic cognitive impairment after mild traumatic brain injury: Results of the international collaboration on mild traumatic brain injury prognosis. *Arch. Phys. Med. Rehabil.* **2014**, *95*, S245–S256. [CrossRef] [PubMed]
3. Kristman, V.L.; Borg, J.; Godbolt, A.K.; Salmi, L.R.; Cancelliere, C.; Carroll, L.J.; Holm, L.W.; Nygren-de Boussard, C.; Hartvigsen, J.; Abara, U.; et al. Methodological issues and research recommendations for prognosis after mild traumatic brain injury: Results of the international collaboration on mild traumatic brain injury prognosis. *Arch. Phys. Med. Rehabil.* **2014**, *95*, S265–S277. [CrossRef] [PubMed]
4. Marras, C.; Hincapie, C.A.; Kristman, V.L.; Cancelliere, C.; Soklaridis, S.; Li, A.; Borg, J.; af Geijerstam, J.-L.; Cassidy, J.D. Systematic review of the risk of parkinson's disease after mild traumatic brain injury: Results of the international collaboration on mild traumatic brain injury prognosis. *Arch. Phys. Med. Rehabil.* **2014**, *95*, S238–S244. [CrossRef] [PubMed]
5. Wirdefeldt, K.; Adami, H.O.; Cole, P.; Trichopoulos, D.; Mandel, J. Epidemiology and etiology of parkinson's disease: A review of the evidence. *Eur. J. Epidemiol.* **2011**, *26* (Suppl. 1), S1–S58. [CrossRef]
6. Prince, C.; Bruhns, M.E. Evaluation and treatment of mild traumatic brain injury: The role of neuropsychology. *Brain Sci.* **2017**, *7*, 105. [CrossRef] [PubMed]
7. Gao, J.; Liu, R.; Zhao, E.; Huang, X.; Nalls, M.A.; Singleton, A.B.; Chen, H. Head injury, potential interaction with genes, and risk for parkinson's disease. *Parkinsonism Relat. Disord.* **2015**, *21*, 292–296. [CrossRef] [PubMed]
8. Gardner, R.C.; Burke, J.F.; Nettiksimmons, J.; Goldman, S.; Tanner, C.M.; Yaffe, K. Traumatic brain injury in later life increases risk for parkinson disease. *Ann. Neurol.* **2015**, *77*, 987–995. [CrossRef] [PubMed]
9. Gardner, R.C.; Yaffe, K. Epidemiology of mild traumatic brain injury and neurodegenerative disease. *Mol. Cell Neurosci.* **2015**, *66*, 75–80. [CrossRef] [PubMed]

10. Wong, J.C.; Hazrati, L.N. Parkinson's disease, parkinsonism, and traumatic brain injury. *Crit. Rev. Clin. Lab. Sci.* **2013**, *50*, 103–106. [CrossRef] [PubMed]
11. Faden, A.I.; Loane, D.J. Chronic neurodegeneration after traumatic brain injury: Alzheimer disease, chronic traumatic encephalopathy, or persistent neuroinflammation? *Neurotherapeutics* **2015**, *12*, 143–150. [CrossRef] [PubMed]
12. Fiandaca, M.S.; Mapstone, M.; Mahmoodi, A.; Gross, T.; Macciardi, F.; Cheema, A.K.; Merchant-Borna, K.; Bazarian, J.; Federoff, H.J. Plasma metabolomic biomarkers accurately classify acute mild traumatic brain injury from controls. *PLoS ONE* **2018**, *13*, e0195318. [CrossRef] [PubMed]
13. Shutter, L.; Tong, K.A.; Holshouser, B.A. Proton mrs in acute traumatic brain injury: Role for glutamate/glutamine and choline for outcome prediction. *J. Neurotrauma* **2004**, *21*, 1693–1705. [CrossRef] [PubMed]
14. Ashwal, S.; Holshouser, B.; Tong, K.; Serna, T.; Osterdock, R.; Gross, M.; Kido, D. Proton mr spectroscopy detected glutamate/glutamine is increased in children with traumatic brain injury. *J. Neurotrauma* **2004**, *21*, 1539–1552. [CrossRef] [PubMed]
15. Schaffer, S.; Takahashi, K.; Azuma, J. Role of osmoregulation in the actions of taurine. *Amino Acids* **2000**, *19*, 527–546. [CrossRef] [PubMed]
16. El Idrissi, A. Taurine increases mitochondrial buffering of calcium: Role in neuroprotection. *Amino Acids* **2008**, *34*, 321–328. [CrossRef] [PubMed]
17. Messina, S.A.; Dawson, R., Jr. Attenuation of oxidative damage to DNA by taurine and taurine analogs. *Adv. Exp. Med. Biol.* **2000**, *483*, 355–367. [PubMed]
18. Huxtable, R.J. Taurine in the central nervous system and the mammalian actions of taurine. *Prog. Neurobiol.* **1989**, *32*, 471–533. [CrossRef]
19. Hawkins, R.A. The blood-brain barrier and glutamate. *Am. J. Clin. Nutr.* **2009**, *90*, 867S–874S. [CrossRef] [PubMed]
20. Zhou, Y.; Danbolt, N.C. Glutamate as a neurotransmitter in the healthy brain. *J. Neural Transm.* **2014**, *121*, 799–817. [CrossRef] [PubMed]
21. Nakanishi, S.; Nakajima, Y.; Masu, M.; Ueda, Y.; Nakahara, K.; Watanabe, D.; Yamaguchi, S.; Kawabata, S.; Okada, M. Glutamate receptors: Brain function and signal transduction. *Brain Res. Rev.* **1998**, *26*, 230–235. [CrossRef]
22. Yudkoff, M. Interactions in the metabolism of glutamate and the branched-chain amino acids and ketoacids in the CNS. *Neurochem. Res.* **2017**, *42*, 10–18. [CrossRef] [PubMed]
23. Vandenberg, R.J.; Ryan, R.M. Mechanisms of glutamate transport. *Physiol. Rev.* **2013**, *93*, 1621–1657. [CrossRef] [PubMed]
24. Gottlieb, M.; Wang, Y.; Teichberg, V.I. Blood-mediated scavenging of cerebrospinal fluid glutamate. *J. Neurochem.* **2003**, *87*, 119–126. [CrossRef] [PubMed]
25. Helms, H.C.; Madelung, R.; Waagepetersen, H.S.; Nielsen, C.U.; Brodin, B. In vitro evidence for the brain glutamate efflux hypothesis: Brain endothelial cells cocultured with astrocytes display a polarized brain-to-blood transport of glutamate. *Glia* **2012**, *60*, 882–893. [CrossRef] [PubMed]
26. al-Sarraf, H.; Preston, J.E.; Segal, M.B. Changes in the kinetics of the acidic amino acid brain and csf uptake during development in the rat. *Dev. Brain Res.* **1997**, *102*, 127–134. [CrossRef]
27. al-Sarraf, H.; Preston, J.E.; Segal, M.B. Acidic amino acid accumulation by rat choroid plexus during development. *Dev. Brain Res.* **1997**, *102*, 47–52. [CrossRef]
28. Teichberg, V.I.; Cohen-Kashi-Malina, K.; Cooper, I.; Zlotnik, A. Homeostasis of glutamate in brain fluids: An accelerated brain-to-blood efflux of excess glutamate is produced by blood glutamate scavenging and offers protection from neuropathologies. *Neuroscience* **2009**, *158*, 301–308. [CrossRef] [PubMed]
29. Yi, J.H.; Hazell, A.S. Excitotoxic mechanisms and the role of astrocytic glutamate transporters in traumatic brain injury. *Neurochem. Int.* **2006**, *48*, 394–403. [CrossRef] [PubMed]
30. Faden, A.I.; Demediuk, P.; Panter, S.S.; Vink, R. The role of excitatory amino acids and nmda receptors in traumatic brain injury. *Science* **1989**, *244*, 798–800. [CrossRef] [PubMed]
31. Zauner, A.; Bullock, R. The role of excitatory amino acids in severe brain trauma: Opportunities for therapy: A review. *J. Neurotrauma* **1995**, *12*, 547–554. [CrossRef] [PubMed]
32. Olney, J.W.; Ho, O.L.; Rhee, V. Cytotoxic effects of acidic and sulphur containing amino acids on the infant mouse central nervous system. *Exp. Brain Res.* **1971**, *14*, 61–76. [CrossRef] [PubMed]

33. Dhillon, H.S.; Donaldson, D.; Dempsey, R.J.; Prasad, M.R. Regional levels of free fatty acids and evans blue extravasation after experimental brain injury. *J. Neurotrauma* **1994**, *11*, 405–415. [CrossRef] [PubMed]
34. Pilitsis, J.G.; Coplin, W.M.; O'Regan, M.H.; Wellwood, J.M.; Diaz, F.G.; Fairfax, M.R.; Michael, D.B.; Phillis, J.W. Free fatty acids in cerebrospinal fluids from patients with traumatic brain injury. *Neurosci. Lett.* **2003**, *349*, 136–138. [CrossRef]
35. Scheff, S.W.; Dhillon, H.S. Creatine-enhanced diet alters levels of lactate and free fatty acids after experimental brain injury. *Neurochem. Res.* **2004**, *29*, 469–479. [CrossRef] [PubMed]
36. Contreras, M.A.; Chang, M.C.; Kirkby, D.; Bell, J.M.; Rapoport, S.I. Reduced palmitate turnover in brain phospholipids of pentobarbital-anesthetized rats. *Neurochem. Res.* **1999**, *24*, 833–841. [CrossRef] [PubMed]
37. Newkirk, J.D.; Waite, M. Identification of a phospholipase a1 in plasma membranes of rat liver. *Biochim. Biophys. Acta* **1971**, *225*, 224–233. [CrossRef]
38. Su, X.; Chu, Y.; Kordower, J.H.; Li, B.; Cao, H.; Huang, L.; Nishida, M.; Song, L.; Wang, D.; Federoff, H.J. Pgc-1α promoter methylation in parkinson's disease. *PLoS ONE* **2015**, *10*, e0134087. [CrossRef] [PubMed]
39. Zheng, B.; Liao, Z.; Locascio, J.J.; Lesniak, K.A.; Roderick, S.S.; Watt, M.L.; Eklund, A.C.; Zhang-James, Y.; Kim, P.D.; Hauser, M.A.; et al. Pgc-1α, a potential therapeutic target for early intervention in parkinson's disease. *Sci. Transl. Med.* **2010**, *2*, 52ra73. [CrossRef] [PubMed]
40. Sheng, B.; Wang, X.; Su, B.; Lee, H.G.; Casadesus, G.; Perry, G.; Zhu, X. Impaired mitochondrial biogenesis contributes to mitochondrial dysfunction in alzheimer's disease. *J. Neurochem.* **2012**, *120*, 419–429. [CrossRef] [PubMed]
41. Agrawal, R.; Tyagi, E.; Vergnes, L.; Reue, K.; Gomez-Pinilla, F. Coupling energy homeostasis with a mechanism to support plasticity in brain trauma. *Biochim. Biophys. Acta* **2014**, *1842*, 535–546. [CrossRef] [PubMed]
42. Feng, Y.; Jankovic, J.; Wu, Y.C. Epigenetic mechanisms in parkinson's disease. *J. Neurol. Sci.* **2015**, *349*, 3–9. [CrossRef] [PubMed]
43. Salminen, A.; Haapasalo, A.; Kauppinen, A.; Kaarniranta, K.; Soininen, H.; Hiltunen, M. Impaired mitochondrial energy metabolism in alzheimer's disease: Impact on pathogenesis via disturbed epigenetic regulation of chromatin landscape. *Prog. Neurobiol.* **2015**, *131*, 1–20. [CrossRef] [PubMed]
44. Benveniste, H.; Drejer, J.; Schousboe, A.; Diemer, N.H. Elevation of the extracellular concentrations of glutamate and aspartate in rat hippocampus during transient cerebral ischemia monitored by intracerebral microdialysis. *J. Neurochem.* **1984**, *43*, 1369–1374. [CrossRef] [PubMed]
45. Rothman, S.M.; Olney, J.W. Glutamate and the pathophysiology of hypoxic-ischemic brain damage. *Ann. Neurol.* **1986**, *19*, 105–111. [CrossRef] [PubMed]
46. Blandini, F.; Greenamyre, J.T.; Nappi, G. The role of glutamate in the pathophysiology of parkinson's disease. *Funct. Neurol.* **1996**, *11*, 3–15. [PubMed]
47. Erecinska, M.; Dagani, F. Relationships between the neuronal sodium/potassium pump and energy metabolism. Effects of k+, na+, and adenosine triphosphate in isolated brain synaptosomes. *J. Gen. Physiol.* **1990**, *95*, 591–616. [CrossRef] [PubMed]
48. Albin, R.L.; Greenamyre, J.T. Alternative excitotoxic hypotheses. *Neurology* **1992**, *42*, 733–738. [CrossRef] [PubMed]
49. Beal, M.F.; Hyman, B.T.; Koroshetz, W. Do defects in mitochondrial energy-metabolism underlie the pathology of neurodegenerative diseases. *Trends Neurosci.* **1993**, *16*, 125–131. [CrossRef]
50. Blandini, F.; Porter, R.H.; Greenamyre, J.T. Glutamate and parkinson's disease. *Mol. Neurobiol.* **1996**, *12*, 73–94. [CrossRef] [PubMed]
51. Greene, J.G.; Greenamyre, J.T. Exacerbation of nmda, ampa, and l-glutamate excitotoxicity by the succinate dehydrogenase inhibitor malonate. *J. Neurochem.* **1995**, *64*, 2332–2338. [CrossRef] [PubMed]
52. Havelund, J.F.; Heegaard, N.H.H.; Faergeman, N.J.K.; Gramsbergen, J.B. Biomarker research in parkinson's disease using metabolite profiling. *Metabolites* **2017**, *7*, 42. [CrossRef] [PubMed]
53. Lei, S.; Zavala-Flores, L.; Garcia-Garcia, A.; Nandakumar, R.; Huang, Y.; Madayiputhiya, N.; Stanton, R.C.; Dodds, E.D.; Powers, R.; Franco, R. Alterations in energy/redox metabolism induced by mitochondrial and environmental toxins: A specific role for glucose-6-phosphate-dehydrogenase and the pentose phosphate pathway in paraquat toxicity. *ACS Chem. Biol.* **2014**, *9*, 2032–2048. [CrossRef] [PubMed]

54. Ahmed, S.S.; Santosh, W.; Kumar, S.; Christlet, H.T. Metabolic profiling of parkinson's disease: Evidence of biomarker from gene expression analysis and rapid neural network detection. *J. Biomed. Sci.* **2009**, *16*, 63. [CrossRef] [PubMed]
55. Wang, G.; Zhou, Y.; Huang, F.J.; Tang, H.D.; Xu, X.H.; Liu, J.J.; Wang, Y.; Deng, Y.L.; Ren, R.J.; Xu, W.; et al. Plasma metabolite profiles of alzheimer's disease and mild cognitive impairment. *J. Proteome Res.* **2014**, *13*, 2649–2658. [CrossRef] [PubMed]
56. Iwasaki, Y.; Ikeda, K.; Shiojima, T.; Kinoshita, M. Increased plasma concentrations of aspartate, glutamate and glycine in parkinson's disease. *Neurosci. Lett.* **1992**, *145*, 175–177. [CrossRef]
57. Mulder, C.; Wahlund, L.O.; Teerlink, T.; Blomberg, M.; Veerhuis, R.; van Kamp, G.J.; Scheltens, P.; Scheffer, P.G. Decreased lysophosphatidylcholine/phosphatidylcholine ratio in cerebrospinal fluid in alzheimer's disease. *J. Neural Transm.* **2003**, *110*, 949–955. [CrossRef] [PubMed]
58. Klavins, K.; Koal, T.; Dallmann, G.; Marksteiner, J.; Kemmler, G.; Humpel, C. The ratio of phosphatidylcholines to lysophosphatidylcholines in plasma differentiates healthy controls from patients with alzheimer's disease and mild cognitive impairment. *Alzheimers Dement.* **2015**, *1*, 295–302. [CrossRef] [PubMed]
59. Adachi, J.; Asano, M.; Yoshioka, N.; Nushida, H.; Ueno, Y. Analysis of phosphatidylcholine oxidation products in human plasma using quadrupole time-of-flight mass spectrometry. *Kobe J. Med. Sci.* **2006**, *52*, 127–140. [PubMed]
60. Mapstone, M.; Cheema, A.; Zhong, X.; Fiandaca, M.; Federoff, H. Biomarker validation: Methods and matrix matter (letter to the editor). *Alzheimers Dement.* **2017**, *13*, 608–609. [CrossRef] [PubMed]
61. Oresic, M.; Posti, J.P.; Kamstrup-Nielsen, M.H.; Takala, R.S.; Lingsma, H.F.; Mattila, I.; Jantti, S.; Katila, A.J.; Carpenter, K.L.; Ala-Seppala, H.; et al. Human serum metabolites associate with severity and patient outcomes in traumatic brain injury. *EBioMedicine* **2016**, *12*, 118–126. [CrossRef] [PubMed]
62. Postuma, R.B.; Gagnon, J.F.; Vendette, M.; Fantini, M.L.; Massicotte-Marquez, J.; Montplaisir, J. Quantifying the risk of neurodegenerative disease in idiopathic rem sleep behavior disorder. *Neurology* **2009**, *72*, 1296–1300. [CrossRef] [PubMed]
63. Postuma, R.B.; Bertrand, J.A.; Montplaisir, J.; Desjardins, C.; Vendette, M.; Rios Romenets, S.; Panisset, M.; Gagnon, J.F. Rapid eye movement sleep behavior disorder and risk of dementia in parkinson's disease: A prospective study. *Mov. Disord.* **2012**, *27*, 720–726. [CrossRef] [PubMed]
64. Holtbernd, F.; Gagnon, J.F.; Postuma, R.B.; Ma, Y.; Tang, C.C.; Feigin, A.; Dhawan, V.; Vendette, M.; Soucy, J.P.; Eidelberg, D.; et al. Abnormal metabolic network activity in rem sleep behavior disorder. *Neurology* **2014**, *82*, 620–627. [CrossRef] [PubMed]
65. Liu, L.; Aa, J.; Wang, G.; Yan, B.; Zhang, Y.; Wang, X.; Zhao, C.; Cao, B.; Shi, J.; Li, M.; et al. Differences in metabolite profile between blood plasma and serum. *Anal. Biochem.* **2010**, *406*, 105–112. [CrossRef] [PubMed]
66. Wedge, D.C.; Allwood, J.W.; Dunn, W.; Vaughan, A.A.; Simpson, K.; Brown, M.; Priest, L.; Blackhall, F.H.; Whetton, A.D.; Dive, C.; et al. Is serum or plasma more appropriate for intersubject comparisons in metabolomic studies? An assessment in patients with small-cell lung cancer. *Anal. Chem.* **2011**, *83*, 6689–6697. [CrossRef] [PubMed]
67. Gill, J.; Merchant-Borna, K.; Jeromin, A.; Livingston, W.; Bazarian, J. Acute plasma tau relates to prolonged return to play after concussion. *Neurology* **2017**, *88*, 595–602. [CrossRef] [PubMed]
68. Lawton, M.; Baig, F.; Rolinski, M.; Ruffman, C.; Nithi, K.; May, M.T.; Ben-Shlomo, Y.; Hu, M.T. Parkinson's disease subtypes in the oxford parkinson disease centre (OPDC) discovery cohort. *J. Parkinsons Dis.* **2015**, *5*, 269–279. [CrossRef] [PubMed]
69. Hughes, A.J.; Daniel, S.E.; Kilford, L.; Lees, A.J. Accuracy of clinical diagnosis of idiopathic parkinson's disease: A clinico-pathological study of 100 cases. *J. Neurol. Neurosurg. Psychiatry* **1992**, *55*, 181–184. [CrossRef] [PubMed]
70. Dubois, B.; Burn, D.; Goetz, C.; Aarsland, D.; Brown, R.G.; Broe, G.A.; Dickson, D.; Duyckaerts, C.; Cummings, J.; Gauthier, S.; et al. Diagnostic procedures for parkinson's disease dementia: Recommendations from the movement disorder society task force. *Mov. Disord.* **2007**, *22*, 2314–2324. [CrossRef] [PubMed]
71. Mapstone, M.; Cheema, A.K.; Fiandaca, M.S.; Zhong, X.; Mhyre, T.R.; MacArthur, L.H.; Hall, W.J.; Fisher, S.G.; Peterson, D.R.; Haley, J.M.; et al. Plasma phospholipids identify antecedent memory impairment in older adults. *Nat. Med.* **2014**, *20*, 415–418. [CrossRef] [PubMed]

72. Dromerick, A.W.; Edwardson, M.A.; Edwards, D.F.; Giannetti, M.L.; Barth, J.; Brady, K.P.; Chan, E.; Tan, M.T.; Tamboli, I.; Chia, R.; et al. Critical periods after stroke study: Translating animal stroke recovery experiments into a clinical trial. *Front. Hum. Neurosci.* **2015**, *9*, 231. [CrossRef] [PubMed]
73. Fiandaca, M.S.; Zhong, X.; Cheema, A.K.; Orquiza, M.H.; Chidambaram, S.; Tan, M.T.; Gresenz, C.R.; FitzGerald, K.T.; Nalls, M.A.; Singleton, A.B.; et al. Plasma 24-metabolite panel predicts preclinical transition to clinical stages of alzheimer's disease. *Front. Neurol.* **2015**, *6*, 237. [CrossRef] [PubMed]
74. Mapstone, M.; Lin, F.; Nalls, M.A.; Cheema, A.K.; Singleton, A.B.; Fiandaca, M.S.; Federoff, H.J. What success can teach us about failure: The plasma metabolome of older adults with superior memory and lessons for alzheimer's disease. *Neurobiol. Aging* **2017**, *51*, 148–155. [CrossRef] [PubMed]
75. Zhao, Z.; Xu, Y. An extremely simple method for extraction of lysophospholipids and phospholipids from blood samples. *J. Lipid Res.* **2010**, *51*, 652–659. [CrossRef] [PubMed]
76. Tautenhahn, R.; Patti, G.J.; Rinehart, D.; Siuzdak, G. Xcms online: A web-based platform to process untargeted metabolomic data. *Anal. Chem.* **2012**, *84*, 5035–5039. [CrossRef] [PubMed]
77. Huan, T.; Forsberg, E.M.; Rinehart, D.; Johnson, C.H.; Ivanisevic, J.; Benton, H.P.; Fang, M.; Aisporna, A.; Hilmers, B.; Poole, F.L.; et al. Systems biology guided by xcms online metabolomics. *Nat. Methods* **2017**, *14*, 461–462. [CrossRef] [PubMed]
78. MSFmetabolomics. Available online: https://www.mathworks.com/matlabcentral/fileexchange/60607-msfmetabolomics?s_tid=prof_contriblnk (accessed on 7 September 2018).
79. Xie, W.; Zhang, H.; Zeng, J.; Chen, H.; Zhao, Z.; Liang, Z. Tissues-based chemical profiling and semi-quantitative analysis of bioactive components in the root of salvia miltiorrhiza bunge by using laser microdissection system combined with uplc-q-tof-ms. *Chem. Cent. J.* **2016**, *10*, 42. [CrossRef] [PubMed]
80. Evans, A.M.; DeHaven, C.D.; Barrett, T.; Mitchell, M.; Milgram, E. Integrated, nontargeted ultrahigh performance liquid chromatography/electrospray ionization tandem mass spectrometry platform for the identification and relative quantification of the small-molecule complement of biological systems. *Anal. Chem.* **2009**, *81*, 6656–6667. [CrossRef] [PubMed]
81. Illig, T.; Gieger, C.; Zhai, G.; Romisch-Margl, W.; Wang-Sattler, R.; Prehn, C.; Altmaier, E.; Kastenmuller, G.; Kato, B.S.; Mewes, H.W.; et al. A genome-wide perspective of genetic variation in human metabolism. *Nat. Genet.* **2010**, *42*, 137–141. [CrossRef] [PubMed]
82. Romisch-Margl, W.; Prehn, C.; Bogumil, R.; Rohring, C.; Suhre, K.; Adamski, J. Procedure for tissue sample preparation and metabolite extraction for high-throughput targeted metabolomics. *Metabolomics* **2012**, *8*, 133–142. [CrossRef]
83. Gross, T.J.; Mapstone, M.; Miramontes, R.; Padilla, R.; Cheema, A.K.; Macciardi, F.; Federoff, H.J.; Fiandaca, M.S. Toward reproducible results from targeted metabolomic studies: Perspectives for data pre-processing and a basis for analytic pipeline development. *Curr. Top. Med. Chem.* **2018**, *18*, 883–895. [CrossRef] [PubMed]
84. Chong, J.; Soufan, O.; Li, C.; Caraus, I.; Li, S.; Bourque, G.; Wishart, D.S.; Xia, J. Metaboanalyst 4.0: Towards more transparent and integrative metabolomics analysis. *Nucleic Acids Res.* **2018**. [CrossRef] [PubMed]
85. Cortes, C.; Vapnik, V. Support-vector networks. *Mach. Learn.* **1995**, *20*, 273–297. [CrossRef]
86. Worley, B.; Powers, R. Multivariate analysis in metabolomics. *Curr. Metab.* **2013**, *1*, 92–107. [PubMed]
87. Chen, T.; Cao, Y.; Zhang, Y.; Liu, J.; Bao, Y.; Wang, C.; Jia, W.; Zhao, A. Random forest in clinical metabolomics for phenotypic discrimination and biomarker selection. *Evid. Based Complement. Altern. Med.* **2013**, *2013*, 298183. [CrossRef] [PubMed]
88. Tibshirani, R. Regression shrinkage and selection via the lasso. *J. R. Stat. Soc. Ser. B Methodol.* **1996**, *58*, 267–288.
89. Rohart, F.; Gautier, B.; Singh, A.; Le Cao, K.A. Mixomics: An R package for 'omics feature selection and multiple data integration. *PLoS Comput. Biol.* **2017**, *13*, e1005752. [CrossRef] [PubMed]

metabolites

MDPI

Review

High-Throughput Direct Mass Spectrometry-Based Metabolomics to Characterize Metabolite Fingerprints Associated with Alzheimer's Disease Pathogenesis

Raúl González-Domínguez [1,2,3,*], Ana Sayago [1,2] and Ángeles Fernández-Recamales [1,2]

1 Department of Chemistry, Faculty of Experimental Sciences, University of Huelva, 21007 Huelva, Spain; ana.sayago@dqcm.uhu.es (A.S.); recamale@dqcm.uhu.es (A.F.-R.)
2 International Campus of Excellence ceiA3, University of Huelva, 21007 Huelva, Spain
3 Biomarkers & Nutrimetabolomics Laboratory, Department of Nutrition, Food Sciences and Gastronomy, Faculty of Pharmacy and Food Sciences, University of Barcelona, 08028 Barcelona, Spain
* Correspondence: raul.gonzalez@dqcm.uhu.es or raul.gonzalez@ub.edu; Tel.: +34-959-219-975

Received: 23 August 2018; Accepted: 14 September 2018; Published: 18 September 2018

Abstract: Direct mass spectrometry-based metabolomics has been widely employed in recent years to characterize the metabolic alterations underlying Alzheimer's disease development and progression. This high-throughput approach presents great potential for fast and simultaneous fingerprinting of a vast number of metabolites, which can be applied to multiple biological matrices including serum/plasma, urine, cerebrospinal fluid and tissues. In this review article, we present the main advantages and drawbacks of metabolomics based on direct mass spectrometry compared with conventional analytical techniques, and provide a comprehensive revision of the literature on the use of these tools in the investigation of Alzheimer's disease.

Keywords: metabolomics; direct mass spectrometry; Alzheimer's disease; pathogenesis; biomarkers

1. The Potential of Direct Mass Spectrometry-Based Metabolomics

Metabolomics requires the use of powerful and versatile analytical techniques with the aim of covering the largest number of compounds comprising the great complexity of the metabolome, which is composed of metabolites with diverse molecular weights, polarities, acid-base properties, and other physicochemical characteristics. To this end, multiple metabolomic platforms have been proposed in the literature, including nuclear magnetic resonance (NMR), and mass spectrometry (MS) coupled to liquid chromatography (LC), to gas chromatography (GC), or to capillary electrophoresis (CE), each of them having their own strengths and weaknesses. For this reason, the combination of several of these complementary techniques is becoming a powerful workhorse to accomplish a global characterization of the metabolome [1–3]. Among these analytical tools, direct mass spectrometry (DMS)-based metabolomics has usually been relegated to the background due to its inherent drawbacks, such as the impossibility of resolving chemical isomers and problems associated with ion suppression due to the introduction of the whole sample into the mass spectrometry system without previous chromatographic or electrophoretic separation. However, some recently published review articles have also highlighted the great potential of this metabolomic approach, as illustrated in Figure 1 [4–7]. The most notable advantage of this tool is its high-throughput screening capability, due to the absence of a previous time-consuming separation step, which considerably reduces the total analysis time, thus allowing the analysis of hundreds of samples per day. The elimination of this chromatographic/electrophoretic separation also prevents the introduction of biased and selective retention mechanisms, so that DMS enables the simultaneous measurement of a huge number of metabolites, covering a wide physicochemical space. In this sense, it should also be noted that multiple instrumental configurations are available for performing DMS-based metabolomics, which can be combined to increase the metabolome coverage. For non-targeted metabolomics, direct infusion mass spectrometry (DIMS) is the simplest approach, since it only needs a syringe pump to introduce the sample extract into the mass spectrometer. Complementarily, the sample can also be delivered by flow injection (FIMS) using a LC pump. On the other hand, the multi-dimensional mass spectrometry-based shotgun lipidomic (MDMS-SL) approach developed by Han et al. allows the direct quantitation of hundreds of individual lipid species by means of a selective ionization of certain category of lipid classes at certain MS conditions [8]. In this context, simpler targeted metabolomic platforms are the AbsoluteIDQ^{TM} kits developed by Biocrates Life Sciences AG (Innsbruck, Austria), focused on the FI-MS/MS-based quantification of multiple metabolite classes, including lipids (phospholipids, sphingolipids, acyl-carnitines, glycerolipids), amino acids, hexoses and biogenic amines [9]. In turn, most of these DMS-based configurations can be coupled with various complementary atmospheric pressure ionization sources. Electrospray ionization (ESI) is the most commonly employed source in non-targeted metabolomics, which allows the simultaneous characterization of compounds with very diverse physico-chemical properties due to its sensitivity and versatility. Complementarily, atmospheric pressure chemical ionization (APCI) and atmospheric pressure photoionization (APPI) sources can also be employed for the ionization of less polar compounds. Thus, the combination of complementary ion sources and ionization modes (i.e., positive and negative polarities), is recommended to maximize the analytical coverage. To conclude, it is also worth noting that the lack of a separation step prior to MS detection facilitates the experimental design by avoiding common challenges associated with chromatography and electrophoresis, such as column/capillary clogging and deterioration, the need for complex data processing packages to align retention/migration times, as well as the minimization of the instrumental drift along batch analysis thanks to the reduced acquisition times usually employed in these approaches.

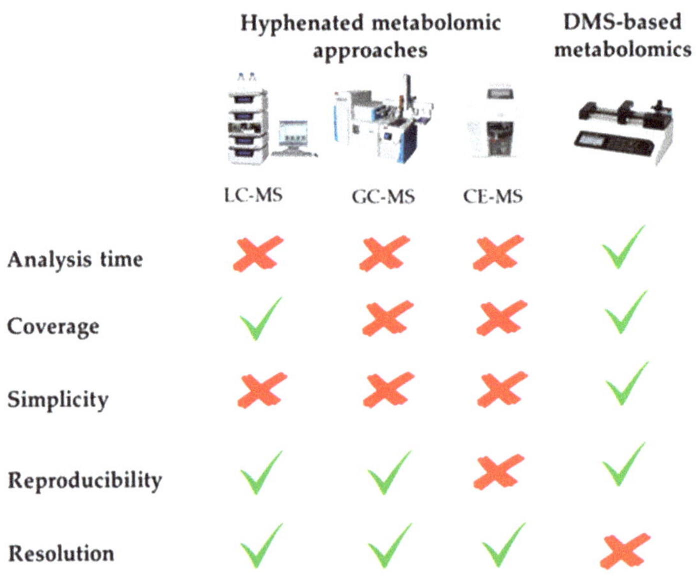

Figure 1. Advantages and drawbacks of DMS-based metabolomics compared with conventional hyphenated approaches.

2. Alzheimer's Disease, Mild Cognitive Impairment and Animal Models

Alzheimer's disease (AD) is the most prevalent neurodegenerative disorder worldwide in the elderly, and is primarily characterized by neuropathological alterations associated with the deposition of amyloid plaques and the formation of intra-neuronal neurofibrillary tangles. Furthermore, numerous authors have proposed that multiple other pathological processes can also play a pivotal role in the development of this disease, such as oxidative stress, abnormal mitochondrial functioning, neuroinflammatory mechanisms, impaired metal homeostasis and many others [10–12]. The investigation of AD etiology involves a great challenge to the scientific community due to its great complexity and the variability of clinical symptoms, its long pre-symptomatic period, and the impossibility of studying brain microscopic changes until the final stages of the disease. For these reasons, diagnosis of AD nowadays relies on the combination of various physical, neuropsychological and laboratory tests according to the clinical criteria of the National Institute of Neurological and Communicative Disorders and the Alzheimer's Disease and Related Disorders Association (NINCDS-ADRDA) [13]. However, this diagnostic method is only effective at advanced dementia, which hinders the application of pharmacological interventions, and in addition suffers from low specificity against other dementias as demonstrated after post mortem histopathological verification [14]. Thus, the discovery of novel biomarkers for accurate diagnosis of AD is mandatory, especially for predicting the development of disease from pre-dementia phases, also called mild cognitive impairment (MCI). MCI is a heterogeneous syndrome characterized by very mild symptoms of cognitive dysfunction, and is usually considered an intermediate pre-clinical stage of Alzheimer's disease. Although MCI has many common features with early AD, current data suggest that some MCI forms are part of the normal aging process [15]. Therefore, there is a great need to discover potential biomarkers for diagnosis and to investigate the pathological mechanisms associated with AD and MCI development and progression.

On the other hand, animal models are very useful tools for investigating the pathogenesis of AD and associated alterations in the central nervous system at different stages along the progression of disease [16], while studies in human cohorts are limited to post-mortem brain tissue, when the disease is in its final stage. Transgenic mice, obtained by the over-expression of mutated forms of human genes associated with AD such as the amyloid precursor protein (APP), presenilin 1 (PS1), presenilin 2 (PS2) or apolipoprotein E (ApoE), are the most useful models, since the neuropathology elicited by these animals is analogous to that observed in human AD, and furthermore, biochemical routes in humans and rodents are very similar [17]. The transgenic mice most commonly employed in AD research are based on the up-regulation of the APP, including the APP_{Tg2576}, APP_{V717F} and CRND8, transgenic lines, which usually show amyloid deposition in hippocampus and cortex and memory deficits, but not neuronal loss. In this vein, it has been demonstrated that the co-expression of mutated PS1, and to a lesser extent PS2, accelerates amyloid deposition, thus facilitating the appearance of the characteristic AD phenotype (APP $\times$ PS1, TASTPM). Taking into account the fact that the $\varepsilon 4$ allele of ApoE is one of the most important risk factors for AD, several knock-in mice in which this protein is expressed have been developed, which show significant cognitive and synaptic plasticity impairments. On the contrary, only a few transgenic models expressing tauopathy have been developed to date due to the lack of knowledge of genes involves in this process in AD (TAPP, 3 $\times$ Tg).

3. Application of Direct Mass Spectrometry-Based Metabolomics to AD Research

Considering the multifactorial nature of AD etiology, the application of holistic metabolomic approaches is emerging for the investigation of pathological hallmarks underlying this neurodegenerative disorder and for the discovery of potential diagnostic biomarkers [2,18,19]. In particular, DMS-based metabolomics has demonstrated great potential to characterize the AD metabotype in a comprehensive manner, as discussed in this section and summarized in Table 1.

Table 1. Summary of DMS-based metabolomics studies on Alzheimer's disease.

Cohort	Sample	Results	Ref.
AD (N = 22) HC (N = 18)	serum	imbalances in the PUFA/SFA composition of phospholipids; impairments in energy metabolism, neurotransmission, fatty acid homeostasis; hyperlipidemia	[20]
AD (N = 22) HC (N = 18)	serum	imbalances in the PUFA/SFA composition of phospholipids	[21]
AD (N = 30) HC (N = 30)	serum	up-regulated degradation of membrane phospholipids and sphingolipids (↑ diacylglycerols, ceramides); impairments in neurotransmission	[22]
AD (N = 22) HC (N = 18)	serum	impairments in membrane phospholipids (↓ PUFA, ↑diacylglycerols), homeostasis of neurotransmitter systems, nitrogen metabolism and oxidative stress	[23]
AD (N = 19) HC (N = 17)	serum	abnormal phospholipid homeostasis (imbalance of PUFA/SFA, over-activation of phospholipases, oxidative stress, peroxysomal dysfunction)	[24]
APP × PS1 (N = 30) WT (N = 30)	serum	impairments in phospholipid homeostasis, energy-related metabolism, oxidative stress, hyperlipidemia, hyperammonemia	[25]
APP × PS1 × IL4-KO (N = 7) APP × PS1 (N = 7) WT (N = 7)	serum	up-regulated production of eicosanoids, altered metabolism of amino acids and urea cycle	[26]
CRND8 (N = 6) WT (N = 6)	hippocampus	altered metabolism of arachidonic acid, carbohydrates and nucleotides	[27]
CRND8 (N = 6) WT (N = 6)	cerebellum	up-regulated production of eicosanoids; altered metabolism of amino acids and nucleotides	[28]
APP × PS1 (N = 30) WT (N = 30)	hippocampus, cortex, cerebellum, olfactory bulb	disturbances in the homeostasis of phospholipids, acyl-carnitines, fatty acids, nucleotides, amino acids, steroids, energy-related metabolites	[29]
AD young (N = 17) AD old (N = 17) MCI (N = 19) HC young (N = 20) HC old (N = 8)	CSF, frontal cortex grey and white matter	abnormal lipid homeostasis (plasmalogens, phosphatidylethanolamines, diacylglycerols)	[30]
APP × PS1 (N = 30) WT (N = 30)	liver, kidney, spleen, thymus	oxidative stress, lipid dyshomeostasis, imbalances in energy metabolism, homeostasis of amino acids and nucleotides	[31]
APP × PS1 (N = 10) WT (N = 10)	urine	unidentified discriminant signals	[32]
AD (N = 24) HC (N = 6) $APPV_{717F}$, APP_{sw}, WT	superior frontal cortex, superior temporal cortex, inferior parietal cortex, cerebellum	plasmalogen deficiency	[33]
AD (N = 17), HC (N = 5)	middle frontal gyrus, superior temporal gyrus, inferior parietal lobule, hippocampus, subiculum, entorhinal cortex	sulfatide deficiency	[34]
APP_{V717F}, APP_{sw}, WT	cortex, cerebellum	sulfatide deficiency	[35]
AD (N = 6) HC (N = 8)	superior frontal gyrus	sulfatide deficiency	[36]
AD (N = 26) HC (N = 26)	plasma	altered sphingolipidome	[37]
AD (N = 93) HC (N = 99)	serum	authors failed to replicate the 10-metabolite panel described by Mapstone et al. [38]	[39]

Table 1. *Cont.*

Cohort	Sample	Results	Ref.
MCI (N = 28) HC (N = 73)	plasma	discovery of a panel of 24 metabolites mainly phospholipids and acyl-carnitines)	[40]
AD (N = 143) MCI (N = 145) HC (N = 153)	plasma	impairments in phospholipid homeostasis	[41]
AD (N = 53) MCI (N = 33) HC (N = 35)	plasma	impairments in phospholipid homeostasis	[42]
AD, MCI, HC	brain, serum	impairments in the homeostasis of phospholipids and sphingolipids	[43]
APP × PS1 (N = 9) WT (N = 9)	brain, plasma	impairments in the homeostasis of phospholipids, acyl-carnitines, amino acids and polyamines	[44]

Numerous non-targeted DMS-based metabolomic studies have been conducted in serum samples, which is a very useful biofluid in clinical practice for the identification of diagnostic biomarkers in a non-invasive manner. González-Domínguez et al. employed a DIMS platform based on a two-step treatment of serum samples from AD patients to obtain a holistic snapshot of metabolite alterations associated with the early development of this neurodegenerative disorder [20,21]. The most notable findings could be associated with an abnormal homeostasis of neural membrane lipids, evidenced by reduced levels of circulating phospholipids containing polyunsaturated fatty acids (PUFAs) and increased content of lipid species composed of saturated fatty acids (SFAs) and some breakdown products (e.g., choline, glycerophosphocholine). Furthermore, significant impairments were also observed in biological pathways related to energy metabolism, neurotransmitter levels and fatty acid homeostasis. To complement this study, a FI-APPI-MS approach was subsequently applied to focus on the less polar metabolome, non-readily detectable by ESI-based metabolomics [22]. Increased serum levels of diacylglycerols and ceramides were detected in AD patients, indicative of up-regulated degradation of membrane phospholipids and sphingolipids by the action of phospholipases and sphingomyelinases, in line with results from DIMS analysis. Due to the central role that lipid dyshomeostasis seems to play in AD pathogenesis, serum samples from the same cohort of AD patients were subjected to DIMS-based lipidomics using a modification of the Bligh-Dyer extraction method [23]. Again, a reduced content of PUFA-containing phospholipids and increased levels of diacylglycerols were observed, corroborating previous hypotheses. Furthermore, changes in other low molecular weight metabolites also evidenced severe impairments in the homeostasis of various neurotransmitter systems, nitrogen metabolism and oxidative stress. Taking into account this evidence about the major role that phospholipids play in AD etiology, a metabolomic multiplatform based on the combination of DIMS and LC-MS, this later coupled to both molecular (ESI) and elemental (inductively coupled plasma, ICP) mass spectrometry was employed to get a deeper understanding of the AD-associated phospholipidome [24]. Thus, results evidenced that multiple factors are involved in this abnormal phospholipid homeostasis, including the imbalance of PUFA/SFA contained in their structure, the up-regulation of phospholipases, the implication of oxidative stress and peroxysomal malfunctioning, among others. Complementarily, González-Domínguez et al. also employed the DIMS and FI-APPI-MS approaches previously described to investigate the AD-like pathology in various transgenic mice models compared with wild type (WT) littermates. The analysis of serum samples from APP × PS1 mice revealed analogous metabolomic disturbances to those detected in previous studies with human cohorts, demonstrating the potential of these transgenic animals to model AD [25]. Additionally, DIMS-based fingerprinting has also been applied to the APP × PS1 × IL4-KO transgenic model with the aim of investigating the role of inflammation induced by means of interleukin-4 depletion in AD pathology [26]. Alterations in serum levels of eicosanoids, amino acids and related compounds, and metabolites involved in the urea cycle demonstrated that depletion of interleukin-4 exacerbates AD pathology in this transgenic line. It should be noted that all these results obtained by DMS analysis were subsequently validated by applying various orthogonal metabolomic techniques, including LC-MS, GC-MS and CE-MS [45–48], thus demonstrating the potential of MS-fingerprinting approaches to carry out fast and accurate screening of complex metabolic networks.

Other published studies on DMS-based metabolomics have focused on the characterization of metabolic impairments observed in brain from various transgenic mice models, a tissue of great interest in AD research, since it enables the in situ investigation of neuropathological processes related to this neurodegenerative disorder. Lin et al. applied an optimized DIMS platform to look for characteristic metabolic impairments in the hippocampus [27] and cerebellum [28] of the CRND8 mouse model. Major findings were observed with regard to an abnormal metabolism of amino acids and nucleotides, as well as the over-production of eicosanoids. In this vein, DIMS-based analysis of various brain regions from the APP × PS1 mouse model (i.e., hippocampus, cortex, cerebellum, striatum, and olfactory bulbs) evidenced that hippocampus and cortex are the most perturbed regions in AD pathology [29]. Similarly to previous studies, significant differences were observed in levels of phospholipids, acyl-carnitines, fatty acids, nucleotides, amino acids and many other metabolites, results which were then confirmed by LC/GC-MS metabolomic analysis [49]. Recently, Wood et al. also employed a lipidomic approach based on DIMS to define potential biomarkers with the aim of distinguishing healthy controls (HC) from MCI and AD patients [30]. They analyzed frontal cortex grey, white matter and cerebrospinal fluid (CSF), and detected abnormal levels of various lipid classes (e.g., plasmalogens, phosphatidylethanolamines, diglycerides), in agreement with previous studies. Alternatively, other peripheral organs from the APP × PS1 model have also been investigated to assess the possible systemic nature of AD, including the liver, kidneys, spleen and thymus [31]. In this work, authors found significant impairments associated with oxidative stress, lipid dyshomeostasis and imbalances in energy metabolism, among other processes, results which were subsequently validated by using a metabolomic multiplatform based on the combination of LC and GC coupled to MS [50,51]. Moreover, urine can also serve as a valid biological sample to study metabolomic perturbations associated with AD by using DIMS-based approaches, as demonstrated by González-Domínguez et al. [32]. For this purpose, various sample preparation methods and normalization strategies were tested, evidencing that ten-fold dilution of urine prior to MS-fingerprinting and subsequent statistical data normalization is enough to minimize ion suppression and to correct the inherent inter-individual variability of this matrix, respectively.

From a targeted perspective, the MDMS-SL platform optimized by Han et al. is a very interesting alternative for the comprehensive investigation of lipidomic alterations associated with AD, in samples coming from both human and animal models. The application of this tool to blood and brain samples showed significant changes in the content of plasmalogens [33], sulfatides [34–36], ceramides [34,37] and sphingomyelins [37], thus corroborating the pivotal role of lipid metabolism in the pathogenesis of AD. On the other hand, other authors proposed the use of AbsoluteIDQ^{TM} kits to analyze blood, brain and CSF samples from AD and MCI patients, observing major changes in the content of phospholipids and acyl-carnitines [39–44]. However, it should be noted that this tool presents a great drawback in the form of its low metabolome coverage.

4. Conclusions

Metabolomic approaches based on DMS analysis have been gaining great importance in recent years because of their high-throughput screening potential, reduced analysis time and wide metabolome coverage. In particular, these platforms have been widely applied for characterizing multifactorial disorders such as Alzheimer's disease, with the aim of elucidating the pathological mechanisms underlying disease development and progression and discovering potential diagnostic biomarkers. The analysis of multiple biological samples, including serum/plasma, urine, brain (hippocampus, cortex, cerebellum, etc.), cerebrospinal fluid and other organs (liver, kidney, spleen, thymus), has enabled obtaining a comprehensive snapshot of the major metabolic hallmarks associated with this neurodegenerative disorder, such as impairments in the homeostasis of membrane lipids, oxidative stress, inflammatory processes, imbalance in energy metabolism and neurotransmitter metabolism, among many others.

Author Contributions: Conceptualization and Original Draft Preparation, R.G.-D.; Review & Editing, R.G.-D., A.S. and Á.F.-R.

Funding: This research received no external funding. The APC was partially funded by University of Barcelona.

Conflicts of Interest: The authors declare no conflict of interest.

References

1. González-Domínguez, Á.; Durán-Guerrero, E.; Fernández-Recamales, Á.; Lechuga-Sancho, A.M.; Sayago, A.; Schwarz, M.; Segundo, C.; González-Domínguez, R. An overview on the importance of combining complementary analytical platforms in metabolomic research. *Curr. Top. Med. Chem.* **2017**, *17*, 3289–3295. [CrossRef] [PubMed]
2. González-Domínguez, R.; Sayago, A.; Fernández-Recamales, Á. Metabolomics in Alzheimer's disease: The need of complementary analytical platforms for the identification of biomarkers to unravel the underlying pathology. *J. Chromatogr. B Anal. Technol. Biomed. Life Sci.* **2017**, *1071*, 75–92. [CrossRef] [PubMed]
3. González-Domínguez, R.; González-Domínguez, Á.; Sayago, A.; Fernández-Recamales, Á. Mass spectrometry-based metabolomic multiplatform for Alzheimer's disease research. In *Biomarkers for Alzheimer's Disease Drug Development*; Perneczky, R., Ed.; Humana Press: New York, NY, USA, 2018; pp. 125–137. ISBN 978-1-4939-7703-1. [CrossRef]
4. Draper, J.; Lloyd, A.J.; Goodacre, R.; Beckmann, M. Flow infusion electrospray ionisation mass spectrometry for high throughput, non-targeted metabolite fingerprinting: A review. *Metabolomics* **2013**, *9*, 4–29. [CrossRef]
5. González-Domínguez, R.; Sayago, A.; Fernández-Recamales, Á. Direct infusion mass spectrometry for metabolomic phenotyping of diseases. *Bioanalysis* **2017**, *9*, 131–148. [CrossRef] [PubMed]
6. Habchi, B.; Alves, S.; Paris, A.; Rutledge, D.N.; Rathahao-Paris, E. How to really perform high throughput metabolomic analyses efficiently? *TrAC Trends Anal. Chem.* **2016**, *85*, 128–139. [CrossRef]
7. Fuhrer, T.; Zamboni, N. High-throughput discovery metabolomics. *Curr. Opin. Biotechnol.* **2015**, *31*, 73–78. [CrossRef] [PubMed]
8. Han, X.; Yang, J.; Cheng, H.; Ye, H.; Gross, R.W. Toward fingerprinting cellular lipidomes directly from biological samples by two-dimensional electrospray ionization mass spectrometry. *Anal. Biochem.* **2004**, *330*, 317–331. [CrossRef] [PubMed]
9. Römisch-Margl, W.; Prehn, C.; Bogumil, R.; Röhring, C.; Suhre, K.; Adamski, J. Procedure for tissue sample preparation and metabolite extraction for high-throughput targeted metabolomics. *Metabolomics* **2012**, *8*, 133–142. [CrossRef]
10. Maccioni, R.B.; Muñoz, J.P.; Barbeito, L. The molecular bases of Alzheimer's disease and other neurodegenerative disorders. *Arch. Med. Res.* **2001**, *32*, 367–381. [CrossRef]
11. Blennow, K.; de Leon, M.J.; Zetterberg, H. Alzheimer's disease. *Lancet* **2006**, *368*, 387–403. [CrossRef]
12. González-Domínguez, R.; García-Barrera, T.; Gómez-Ariza, J.L. Characterization of metal profiles in serum during the progression of Alzheimer's disease. *Metallomics* **2014**, *6*, 292–300. [CrossRef] [PubMed]
13. McKhann, G.; Knopman, D.S.; Chertkow, H.; Hymann, B.; Jack, C.R.; Kawas, C.; Klunk, W.; Koroshetz, W.; Manly, J.; Mayeux, R.; et al. The diagnosis of dementia due to Alzheimer's disease: Recommendations from the National Institute on Aging-Alzheimer's Association workgroups on diagnostic guidelines for Alzheimer's disease. *Alzheimers Dement.* **2011**, *7*, 263–269. [CrossRef] [PubMed]
14. Dubois, B.; Feldman, H.H.; Jacova, C.; DeKosky, S.T.; Barberger-Gateau, P.; Cummings, J.; Delacourte, A.; Galasko, D.; Gauthier, S.; Jicha, G.; et al. Research criteria for the diagnosis of Alzheimer's disease: Revising the NINCDS-ADRDA criteria. *Lancet Neurol.* **2007**, *6*, 734–746. [CrossRef]
15. Petersen, R.C.; Smith, G.E.; Waring, S.C.; Ivnik, R.J.; Tangalos, E.G.; Kokmen, E. Mild cognitive impairment: Clinical characterization and outcome. *Arch. Neurol.* **1999**, *56*, 303–308. [CrossRef] [PubMed]
16. Hall, A.M.; Roberson, E.D. Mouse models of Alzheimer's disease. *Brain Res. Bull.* **2012**, *88*, 3–12. [CrossRef] [PubMed]
17. Trushina, E.; Mielke, M.M. Recent advances in the application of metabolomics to Alzheimer's Disease. *Biochim. Biophys. Acta Mol. Basis Dis.* **2014**, *1842*, 1232–1239. [CrossRef] [PubMed]

18. Enche Ady, C.N.A.; Lim, S.M.; The, L.K.; Salleh, M.Z.; Chin, A.V.; Tan, M.P.; Poi, P.J.H.; Kamaruzzaman, S.B.; Abdul Majeed, A.B.; Ramasamy, K. Metabolomic-guided discovery of Alzheimer's disease biomarkers from body fluid. *J. Neurosci. Res.* **2017**, *95*, 2005–2024. [CrossRef] [PubMed]
19. Wilkins, J.M.; Trushina, E. Application of Metabolomics in Alzheimer's Disease. *Front. Neurol.* **2017**, *8*, 719. [CrossRef] [PubMed]
20. González-Domínguez, R.; García-Barrera, T.; Gómez-Ariza, J.L. Using direct infusion mass spectrometry for serum metabolomics in Alzheimer's disease. *Anal. Bioanal. Chem.* **2014**, *406*, 7137–7148. [CrossRef] [PubMed]
21. González-Domínguez, R.; García-Barrera, T.; Gómez-Ariza, J.-L. Metabolomic approach to Alzheimer's disease diagnosis based on mass spectrometry. *Chem. Pap.* **2012**, *66*, 829–835. [CrossRef]
22. González-Domínguez, R.; García-Barrera, T.; Gómez-Ariza, J.L. Application of a novel metabolomic approach based on atmospheric pressure photoionization mass spectrometry using flow injection analysis for the study of Alzheimer's disease. *Talanta* **2015**, *131*, 480–489. [CrossRef] [PubMed]
23. González-Domínguez, R.; García-Barrera, T.; Gómez-Ariza, J.L. Metabolomic study of lipids in serum for biomarker discovery in Alzheimer's disease using direct infusion mass spectrometry. *J. Pharm. Biomed. Anal.* **2014**, *98*, 321–326. [CrossRef] [PubMed]
24. González-Domínguez, R.; García-Barrera, T.; Gómez-Ariza, J.L. Combination of metabolomic and phospholipid-profiling approaches for the study of Alzheimer's disease. *J. Proteom.* **2014**, *104*, 37–47. [CrossRef] [PubMed]
25. González-Domínguez, R.; García-Barrera, T.; Vitorica, J.; Gómez-Ariza, J.L. Application of metabolomics based on direct mass spectrometry analysis for the elucidation of altered metabolic pathways in serum from the APP/PS1 transgenic model of Alzheimer's disease. *J. Pharm. Biomed. Anal.* **2015**, *107*, 378–385. [CrossRef] [PubMed]
26. González-Domínguez, R.; García-Barrera, T.; Vitorica, J.; Gómez-Ariza, J.L. Metabolomic research on the role of interleukin-4 in Alzheimer's disease. *Metabolomics* **2015**, *11*, 1175–1183. [CrossRef]
27. Lin, S.; Liu, H.; Kanawati, B.; Liu, L.; Dong, J.; Li, M.; Huang, J.; Schmitt-Kopplin, P.; Cai, Z. Hippocampal metabolomics using ultrahigh-resolution mass spectrometry reveals neuroinflammation from Alzheimer's disease in CRND8 mice. *Anal. Bioanal. Chem.* **2013**, *405*, 5105–5117. [CrossRef] [PubMed]
28. Lin, S.; Kanawati, B.; Liu, L.; Witting, M.; Li, M.; Huang, J.; Schmitt-Kopplin, P.; Cai, Z. Ultrahigh resolution mass spectrometry-based metabolic characterization reveals cerebellum as a disturbed region in two animal models. *Talanta* **2014**, *118*, 45–53. [CrossRef] [PubMed]
29. González-Domínguez, R.; García-Barrera, T.; Vitorica, J.; Gómez-Ariza, J.L. Metabolomic screening of regional brain alterations in the APP/PS1 transgenic model of Alzheimer's disease by direct infusion mass spectrometry. *J. Pharm. Biomed. Anal.* **2015**, *102*, 425–435. [CrossRef] [PubMed]
30. Wood, P.L.; Barnette, B.L.; Kaye, J.A.; Quinn, J.F.; Woltjer, R.L. Non-targeted lipidomics of CSF and frontal cortex grey and white matter in control, mild cognitive impairment, and Alzheimer's disease subjects. *Acta Neuropsychiatr.* **2015**, *18*, 270–278. [CrossRef] [PubMed]
31. González-Domínguez, R.; García-Barrera, T.; Vitorica, J.; Gómez-Ariza, J.L. High throughput multiorgan metabolomics in the APP/PS1 mouse model of Alzheimer's disease. *Electrophoresis* **2015**, *36*, 2237–2249. [CrossRef] [PubMed]
32. González-Domínguez, R.; Castilla-Quintero, R.; García-Barrera, T.; Gómez-Ariza, J.L. Development of a metabolomic approach based on urine samples and direct infusion mass spectrometry. *Anal. Biochem.* **2014**, *465*, 20–27. [CrossRef] [PubMed]
33. Han, X.; Holtzman, D.M.; McKeel, D.W. Plasmalogen deficiency in early Alzheimer's disease subjects and in animal models: Molecular characterization using electrospray ionization mass spectrometry. *J. Neurochem.* **2001**, *77*, 1168–1180. [CrossRef] [PubMed]
34. Han, X.; Holtzman, D.; McKeel, D.W.; Kelley, J.; Morris, J.C. Substantial sulfatide deficiency and ceramide elevation in very early Alzheimer's disease: Potential role in disease pathogenesis. *J. Neurochem.* **2002**, *82*, 809–818. [CrossRef] [PubMed]
35. Cheng, H.; Zhou, Y.; Holtzman, D.M.; Han, X. Apolipoprotein E mediates sulfatide depletion in animal models of Alzheimer's disease. *Neurobiol. Aging* **2010**, *31*, 1188–1196. [CrossRef] [PubMed]
36. Cheng, H.; Wang, M.; Li, J.L.; Cairns, N.J.; Han, X. Specific changes of sulfatide levels in individuals with pre-clinical Alzheimer's disease: An early event in disease pathogenesis. *J. Neurochem.* **2013**, *127*, 733–738. [CrossRef] [PubMed]

37. Han, X.; Rozen, S.; Boyle, S.H.; Hellegers, C.; Cheng, H.; Burke, J.R.; Welsh-Bohmer, K.A.; Doraiswamy, P.M.; Kaddurah-Daouk, R. Metabolomics in early Alzheimer's disease: Identification of altered plasma sphingolipidome using shotgun lipidomics. *PLoS ONE* **2011**, *6*, e21643. [CrossRef] [PubMed]
38. Mapstone, M.; Cheema, A.K.; Fiandaca, M.S.; Zhong, X.; Mhyre, T.R.; MacArthur, L.H.; Hall, W.J.; Fisher, S.G.; Peterson, D.R.; Haley, J.M.; et al. Plasma phospholipids identify antecedent memory impairment in older adults. *Nat. Med.* **2014**, *20*, 415–418. [CrossRef] [PubMed]
39. Casanova, R.; Varma, S.; Simpson, B.; Kim, M.; An, Y.; Saldana, S.; Riveros, C.; Moscato, P.; Griswold, M.; Sonntag, D.; et al. Blood metabolite markers of preclinical Alzheimer's disease in two longitudinally followed cohorts of older individuals. *Alzheimer's Dement.* **2016**, *12*, 815–822. [CrossRef] [PubMed]
40. Fiandaca, M.S.; Zhong, X.; Cheema, A.K.; Orquiza, M.H.; Chidambaram, S.; Tan, M.T.; Gresenz, C.R.; FitzGerald, K.T.; Nalls, M.A.; Singleton, A.B.; et al. Plasma 24-metabolite panel predicts preclinical transition to clinical stages of Alzheimer's disease. *Front. Neurol.* **2015**, *6*, 1–13. [CrossRef] [PubMed]
41. Li, D.; Misialek, J.R.; Boerwinkle, E.; Gottesman, R.F.; Sharrett, A.R.; Mosley, T.H.; Coresh, J.; Wruck, L.M.; Knopman, D.S.; Alonso, A. Plasma phospholipids and prevalence of mild cognitive impairment and/or dementia in the ARIC Neurocognitive Study (ARIC-NCS). *Alzheimer's Dement. Diagn. Assess. Dis. Monit.* **2016**, *3*, 73–82. [CrossRef] [PubMed]
42. Klavins, K.; Koal, T.; Dallmann, G.; Marksteiner, J.; Kemmler, G.; Humpel, C. The ratio of phosphatidylcholines to lysophosphatidylcholines in plasma differentiates healthy controls from patients with Alzheimer's disease and mild cognitive impairment. *Alzheimer's Dement.* **2015**, *1*, 295–302. [CrossRef] [PubMed]
43. Varma, V.R.; Oommen, A.M.; Varma, S.; Casanova, R.; An, Y.; Andrews, R.M.; O'Brien, R.; Pletnikova, O.; Troncoso, J.C.; Toledo, J.; et al. Brain and blood metabolite signatures of pathology and progression in Alzheimer disease: A targeted metabolomics study. *PLoS Med.* **2018**, *15*, e1002482. [CrossRef] [PubMed]
44. Pan, X.; Nasaruddin, M. Bin; Elliott, C.T.; McGuinness, B.; Passmore, A.P.; Kehoe, P.G.; Hölscher, C.; McClean, P.L.; Graham, S.F.; Green, B.D. Alzheimer's disease-like pathology has transient effects on the brain and blood metabolome. *Neurobiol. Aging* **2016**, *38*, 151–163. [CrossRef] [PubMed]
45. González-Domínguez, R.; Rupérez, F.J.; García-Barrera, T.; Barbas, C.; Gómez-Ariza, J.L. Metabolomic-driven elucidation of serum disturbances associated with Alzheimer's disease and mild cognitive impairment. *Curr. Alzheimer Res.* **2016**, *13*, 641–653. [CrossRef] [PubMed]
46. González-Domínguez, R.; García-Barrera, T.; Gómez-Ariza, J.L. Metabolite profiling for the identification of altered metabolic pathways in Alzheimer's disease. *J. Pharm. Biomed. Anal.* **2015**, *107*, 75–81. [CrossRef] [PubMed]
47. González-Domínguez, R.; García, A.; García-Barrera, T.; Barbas, C.; Gómez-Ariza, J.L. Metabolomic profiling of serum in the progression of Alzheimer's disease by capillary electrophoresis-mass spectrometry. *Electrophoresis* **2014**, *35*, 3321–3330. [CrossRef] [PubMed]
48. González-Domínguez, R.; García-Barrera, T.; Vitorica, J.; Gómez-Ariza, J.L. Deciphering metabolic abnormalities associated with Alzheimer's disease in the APP/PS1 mouse model using integrated metabolomic approaches. *Biochimie* **2015**, *110*, 119–128. [CrossRef] [PubMed]
49. González-Domínguez, R.; García-Barrera, T.; Vitorica, J.; Gómez-Ariza, J.L. Region-specific metabolic alterations in the brain of the APP/PS1 transgenic mice of Alzheimer's disease. *Biochim. Biophys. Acta Mol. Basis Dis.* **2014**, *1842*, 2395–2402. [CrossRef] [PubMed]
50. González-Domínguez, R.; García-Barrera, T.; Vitorica, J.; Gómez-Ariza, J.L. Metabolomic investigation of systemic manifestations associated with Alzheimer's disease in the APP/PS1 transgenic mouse model. *Mol. Biosyst.* **2015**, *11*, 2429–2440. [CrossRef] [PubMed]
51. González-Domínguez, R.; García-Barrera, T.; Vitorica, J.; Gómez-Ariza, J.L. Metabolomics reveals significant impairments in the immune system of the APP/PS1 transgenic mice of Alzheimer's disease. *Electrophoresis* **2015**, *36*, 577–587. [CrossRef] [PubMed]

metabolites

MDPI

Article

Evidence That Parietal Lobe Fatty Acids May Be More Profoundly Affected in Moderate Alzheimer's Disease (AD) Pathology Than in Severe AD Pathology

Muhammad L. Nasaruddin [1], Xiaobei Pan [1], Bernadette McGuinness [2], Peter Passmore [2], Patrick G. Kehoe [3], Christian Hölscher [4], Stewart F. Graham [5,6] and Brian D. Green [1,*]

[1] Institute for Global Food Security (IGFS), Queen's University Belfast, Stranmillis Road, Belfast BT9 6AG, Ireland; mbinnasaruddin01@qub.ac.uk (M.L.N.); x.pan@qub.ac.uk (X.P.)
[2] Centre for Public Health, School of Medicine, Dentistry and Biomedical Sciences, Queen's University, Belfast BT12 6BA, Ireland; B.McGuinness@qub.ac.uk (B.M.); P.Passmore@qub.ac.uk (P.P.)
[3] Dementia Research Group, Institute of Clinical Neurosciences, School of Clinical Sciences, University of Bristol, Bristol BS10 5NB, UK; Patrick.Kehoe@bristol.ac.uk
[4] Research and Experimental Center, Henan University of Traditional Chinese Medicine, Longzihu University Campus, 156 Jinshui Dong Road, Zhengzhou 450000, China; christian_holscher@me.com
[5] Metabolomics Research, Beaumont Research Institute 3811 W. 13 Mile Road, Royal Oak, MI 48073, USA; Stewart.Graham@beaumont.org
[6] Metabolomics Research, Oakland University-William Beaumont School of Medicine, Rochester, MI 48309, USA
* Correspondence: b.green@qub.ac.uk; Tel.: +44-2890-976541

Received: 10 September 2018; Accepted: 25 October 2018; Published: 26 October 2018

Abstract: Brain is a lipid-rich tissue, and fatty acids (FAs) play a crucial role in brain function, including neuronal cell growth and development. This study used GC-MS to survey all detectable FAs in the human parietal cortex (Brodmann area 7). These FAs were accurately quantified in 27 cognitively normal age-matched controls, 16 cases of moderate Alzheimer's disease (AD), 30 severe AD, and 14 dementia with Lewy bodies (DLB). A total of 24 FA species were identified. Multiple comparison procedures, using stepdown permutation tests, noted higher levels of 13 FAs but the majority of changes were in moderate AD and DLB, rather than severe AD. Subjects with moderate AD and DLB pathology exhibited significantly higher levels of a number of FAs (13 FAs and 12 FAs, respectively). These included nervonic, lignoceric, *cis*-13,16-docosadienoic, arachidonic, *cis*-11,14,17-eicosatrienoic, erucic, behenic, α-linolenic, stearic, oleic, *cis*-10-heptanoic, and palmitic acids. The similarities between moderate AD and DLB were quite striking—arachidic acid was the only FA which was higher in moderate AD than control, and was not similarly affected in DLB. Furthermore, there were no significant differences between moderate AD and DLB. The associations between each FA and a number of variables, including diagnosis, age, gender, Aβ plaque load, tau load, and frontal tissue pH, were also investigated. To conclude, the development of AD or DLB pathology affects brain FA composition but, intriguingly, moderate AD neuropathology impacts this to a much greater extent. Post-mortem delay is a potential confounding factor, but the findings here suggest that there could be a more dynamic metabolic response in the earlier stages of the disease pathology.

Keywords: fatty acid; GC-MS; Alzheimer's disease; dementia with Lewy bodies; metabolomics; lipidomics

1. Introduction

Alzheimer's disease (AD) is a neurodegenerative disorder characterized by progressive memory loss and deteriorating cognitive abilities in older populations. Dementia currently affects around 820,000 people in the United Kingdom, and the estimated cost of the condition to the economy is £24 billion per annum [1]. Global projections suggest that the number of people affected by AD will triple to 115 million by the year 2050 [1]. The main pathological hallmarks of AD are the accumulation of β-amyloid plaques and neurofibrillary tangles [2,3]. There has been a concerted effort to investigate these two histological features, but the exact causes and pathological consequences of AD remain to be elucidated.

Metabolomics is the investigation of small, often chemically diverse molecules, including lipids, saccharides, steroids, bile acids, and amino acids [4,5]. Lipidomics is a subdiscipline of metabolomics focusing entirely on the measurement of lipid and lipid-like molecules. Profiling of the lipidome has revealed alterations in (i) lipid metabolism [4,6,7]; (ii) lipid-mediated signaling processes [4,8,9]; and (iii) biochemical interactions with other lipids, proteins, and metabolites [10,11]. Ultimately, these techniques could lead to new discoveries in terms of disease pathogenesis and pharmacological targets [12,13]. Lipid profiling techniques are increasingly being employed in AD in the hope that they might provide new mechanistic insights or novel diagnostic biomarkers [4]. The most common approach has been the profiling of fatty acids (FAs) by either GC-MS or LC-MS. There are clear reasons for applying precise lipid profiling techniques in an AD context. The brain is one of the most lipid-rich tissues in the human body [14] and, also, the earliest reports of AD pathology noted the occurrence of "fat inclusions" or "lipid granules" in the brain. Around 50% of neuronal cell membranes are composed of fatty acids (FA) that are polyunsaturated [15]. These FAs appear to be incorporated into membrane phospholipids and secondary signaling messengers, which modulate oxidative stress and inflammatory processes in neurons [15].

It is surprising that FA profiling techniques have shown relatively subtle changes in brain FA composition [16]. This, in part, can be attributed to current technical limitations, which makes it difficult to detect changes in the subcellular distribution of FAs. However, it is clear that the brain concentrations of a small number of FAs are affected by the development of AD pathology [16]. Altered levels of stearic acid, arachidonic acid, and oleic acid have been observed in the frontal and temporal cortex of human cases of AD [16]. This investigation also showed that the parietal cortex is comparably much less affected [16]. Currently, it is a major challenge to distinguish between FA alterations that are secondary to the development of AD, and those that which may contribute to the disease process. Progress in this area is hampered by the fact that most studies (including our own) have not specifically determined how the severity of AD pathology affects the FA profile [12,13,16,17]. There is a need for more research in this area. Recent metabolomic studies profiling APP/PS1 mice found that metabolites, including many lipid species, are affected by the development of AD pathology. However, the disturbances are transient in nature, not occurring in either a persistent or progressive manner [18]. Lipid disturbances in both brain and plasma appear to be highly dependent on the disease stage/severity, and many lipids differ much less in the later stages of the pathology [18].

This study investigated how exactly human brain FA profiles are affected by the extent of neuropathological change. The parietal cortex was selected because it is not the primary site of AD pathology and, also, because FA composition is largely unaffected in AD. We used a fatty acid methyl ester (FAME) methodology to survey all detectable FA species present in this brain region. A quantitative method was established for 24 of these. A full range of analytical FAME standards and internal standards were employed to quantify FAs in pathologically and clinically confirmed cases of AD and dementia with Lewy bodies (DLB), in addition to normal age-matched control cases. To assess the influence of AD severity on FAs, the cases of AD were classified as either moderate ("intermediate" AD neuropathological change, Braak tangle stage III–IV) or severe ("high" AD neuropathological change, Braak tangle stage V–VI).

2. Results

Initial survey of the FA composition of human post-mortem parietal cortex identified 24 quantifiable (calibration curves for FAMEs with R^2 values of <0.98 were deemed inadmissible) individual FAME peaks that were consistently present in all subject groups. The potential for there to be differences between subject groups was immediately evident from relative peak intensity data. FAME peak identities were assigned by mass spectrometry and subsequently confirmed using FAME analytical standards. Typical total ion chromatograms (TICs) for specimens from cases of moderate and severe AD, DLB, and age-matched control subjects, are shown in Figure 1. Initially the Kruskal–Wallis test was used to shortlist those FAs presenting at least one statistically significant difference between any of the 4 subgroups ($p < 0.05$; FDR < 0.05) (Table 1). A total of 13 FAs differed, including (i) 4 saturated FAs: lignoceric acid (C24:0), behenic acid (C22:0), arachidic acid (C20:0), and stearic acid (C18:0); (ii) 5 monounsaturated FAs: nervonic acid (C24:1Δ15), oleic acid (c-C18:1Δ9), *cis*-10-heptadecanoic acid (C17:1Δ10), palmitic (c-C18:1Δ9), and erucic acid (C22:1Δ130); and (iii) 4 polyunsaturated FAs; *cis*-13,16-docosadienoic acid (C22:2Δ13,16), arachidonic acid (C20:4Δ5,8,11,14), *cis*-11,14,17-eicosatrienoic acid (C20:3Δ11,14,17), and α-linolenic acid (C18:3Δ9,12,15) (Table 1). Multiple comparison procedures using stepdown permutation *p*-values, on the ranked data, were used to investigate the nature of group differences for variables, with a false discovery rate of 0.05 or less (Table 2). This revealed that only one FA (oleic acid (c-C18:1Δ9)) significantly differed between severe AD and control samples (Table 2). Contrastingly, a total of 12 and 13 FAs were significantly ($p < 0.05$) elevated in DLB and moderate AD, respectively, compared with control. The levels of 5 FAs differed between severe AD and DLB: arachidonic acid (C20:4Δ5,8,11,14), *cis*-11,14,17-eicosatrienoic acid (C20:3Δ11,14,17), erucic acid (C22:1Δ130), behenic acid (C22:0), and linolenic acid (C18:3Δ9,12,15) (Table 2). The profile of moderate and severe AD samples differed markedly, with 9 FAs being significantly lower in severe AD (Figure 2). Significant differences were found for Total FA, Total SFA, Total monounsaturated FA (MUFA), Total polyunsaturated FA (PUFA), but not for either Omega 3/Omega 6 ratio or 16:1/16:0 ratio (Table S2). No significant associations were detected between any FA and the subject age, frontal tissue pH, and beta-amyloid levels (Table S3). Only linoleic acid differed between male and female subjects (41% higher in females). Six FAs (lignoceric acid, arachidonic acid, *cis*-11,14,17-eicosatrienoic acid, linolenic acid, arachidic acid, and stearic acid) negatively correlated with post-mortem delay. Three FAs (*cis*-11,14,17-eicosatrienoic acid, erucic acid, and linoelaidic acid) negatively correlated with levels of tau protein.

The scores plot generated by unsupervised principal component analysis (PCA) showed only very weak separation between the 4 groups (Figure S1). The PCA results showed that the first component (PC1) explained 36.3% of the variation. Although there was overlap among the 4 groups, it was noteworthy that moderate AD and DLB cases were distributed at the extreme end of PC1.

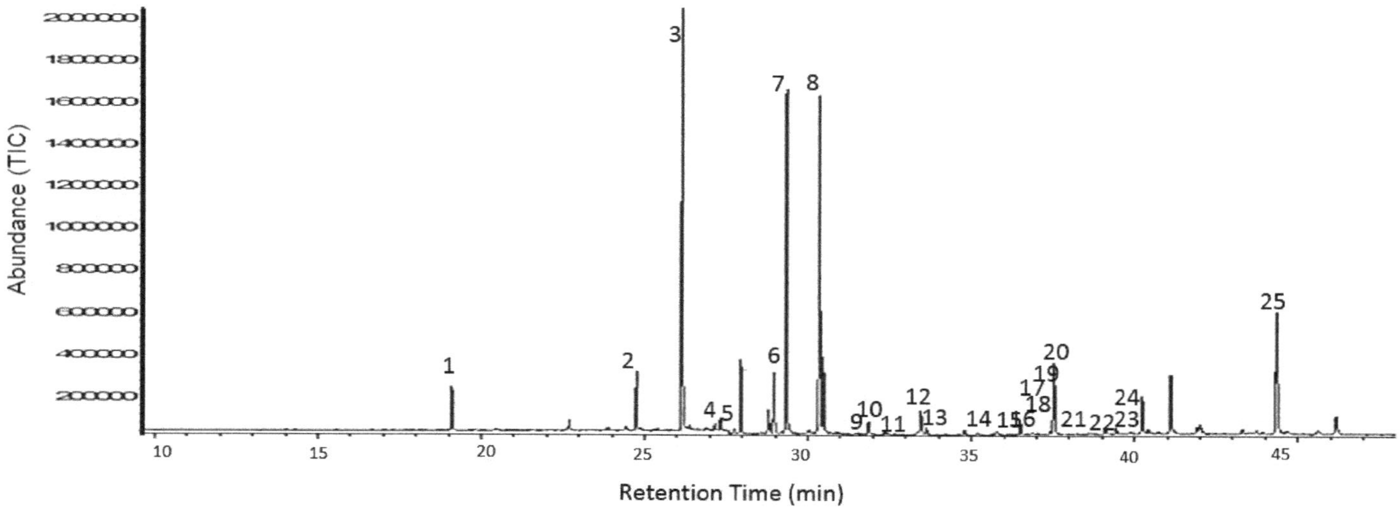

Figure 1. GC-MS detection of brain fatty acids. A typical total ion chromatogram (TIC) of fatty acid methyl esters (FAMEs) from post-mortem brain tissue. The peaks were identified as 1. lauric acid, 2. pentadecanoic acid, 3. palmitic acid, 4. palmitoleic acid, 5. heptadecanoic acid, 6. *cis*-10-heptadecanoic acid, 7. stearic acid, 8. oleic acid, 9. linolelaidic acid, 10. linoleic acid, 11. arachidic acid, 12. *cis*-11-eicosanoic acid, 13. linolenic acid, 14. heneicosanoic acid, 15. *cis*-8,11,14-eicosatrienoic acid, 16. erucic acid, 17. *cis*-11,14,17-eicosatrienoic acid, 18. *cis*-11,14-eicosadienoic acid, 19. behenic acid, 20. arachidonic acid, 21. tricosanoic acid, 22. *cis*-13,16-docosadienoic acid, 23. lignoceric acid, 24. nervonic acid, 25. docosahexaenoic acid.

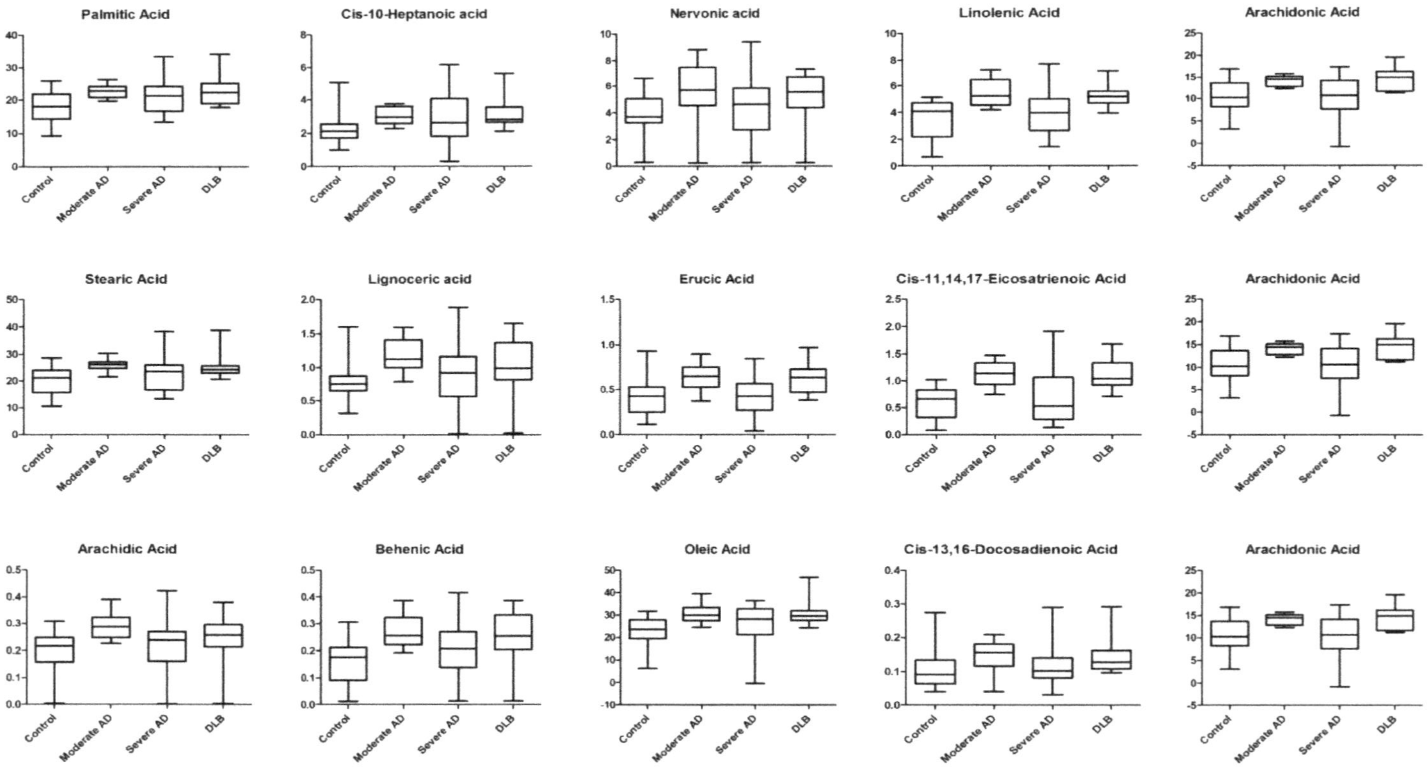

Figure 2. Brain fatty acids (g/kg) significantly altered across moderate Alzheimer's disease (AD), severe AD, and dementia with Lewy bodies (DLB) groups. A total of 13 brain fatty acid (FA) species were found to be significant across the 4 groups. FA species concentrations are displayed as box-and-whisker plots with maximum and minimum values.

Table 1. Fatty acid measurements in post-mortem brain samples of Brodmann 7 region from control, moderate stage AD, full/late stage AD, and DLB subjects. FA concentrations (g/kg dry weight) displayed as mean ± standard deviation (SD), 1st Quartile (Q1), median, and 3rd Quartile (Q3). The adjusted false discovery rate (FDR) value of $p < 0.05$ generated from the Kruskal–Wallis (KW) tests was deemed to be significant. Total FA was calculated based on all 24 measured FAs.

Fatty Acid				Control				Moderate AD			
Name	**Lipid No.**	**RT**	**Mass**	**Mean ± SD**	**Q1**	**Median**	**Q3**	**Mean ± SD**	**Q1**	**Median**	**Q3**
Docosahexanoic Acid	C22:6Δ4,7,10,13,16,19	45.01	342.51	24.77 ± 6.695	18.830	24.640	30.200	30.184 ± 4.215	27.600	28.540	33.130
Nervonic Acid	C24:1Δ15	40.68	380.65	3.925 ± 1.527	3.331	3.686	5.014	5.517 ± 2.549	4.795	5.706	7.311
Lignoceric Acid	C24:0	39.55	382.66	0.767 ± 0.317	0.657	0.756	0.851	1.187 ± 0.256	1.003	1.126	1.404
***cis*-13,16-Docosadienoic acid**	C22:2Δ13,16	38.89	350.58	0.103 ± 0.052	0.065	0.091	0.125	0.146 ± 0.044	0.122	0.156	0.178
Arachidonic acid	C20:4Δ5,8,11,14	38.08	318.49	10.81 ± 3.309	8.280	10.270	13.600	14.04 ± 1.244	12.950	14.480	15.050
Tricosanoic Acid	C23:0	37.42	368.64	0.165 ± 0.140	0.024	0.138	0.248	0.218 ± 0.207	0.014	0.285	0.417
***cis*-11,14,17-Eicosatrienoic acid**	C20:3Δ11,14,17	28.93	320.51	0.565 ± 0.292	0.322	0.656	0.807	1.098 ± 0.242	0.937	1.139	1.294
Erucic acid	C22:1Δ13	35.60	352.59	0.409 ± 0.188	0.260	0.427	0.514	0.627 ± 0.145	0.539	0.653	0.720
***cis*-8,11,14-Eicosatreinoic acid**	C20:3Δ8,11,14	37.05	320.51	1.010 ± 0.389	0.729	1.031	1.245	1.253 ± 0.523	1.262	1.295	1.519
Behenic acid	C22:0	35.18	354.61	0.160 ± 0.085	0.095	0.177	0.211	0.273 ± 0.060	0.222	0.258	0.323
***cis*-11,14-Eicosadienoic acid**	C20:2 Δ11,14	35.18	308.50	0.549 ± 0.482	0.253	0.357	0.629	0.534 ± 0.411	0.242	0.449	0.648
Heneicosanoic acid	C21:0	34.39	340.58	0.018 ± 0.007	0.011	0.020	0.024	0.026 ± 0.009	0.022	0.026	0.030
Linolenic acid	C18:3Δ9,12,15	34.20	294.26	3.514 ± 1.315	2.469	4.067	4.613	5.401 ± 1.048	4.564	5.209	6.059
***cis*-11-Eicosanoic acid**	C20:1 Δ11	33.45	310.50	2.010 ± 0.803	1.539	2.016	2.597	2.616 ± 1.266	2.075	2.905	3.513
Arachidic acid	C20:0	32.85	326.56	0.200 ± 0.070	0.165	0.216	0.247	0.292 ± 0.047	0.259	0.288	0.315
Linoleic acid	c-C18:2Δ9,12	32.38	294.47	0.702 ± 0.478	0.386	0.777	1.012	0.941 ± 0.593	0.751	0.989	1.327
Linolelaidic acid	t-C18:2Δ9,12	31.29	294.47	0.019 ± 0.009	0.012	0.016	0.024	0.023 ± 0.011	0.014	0.023	0.029
Oleic acid	c-C18:1Δ9	30.82	297.49	22.92 ± 6.249	19.400	23.560	27.370	30.664 ± 4.219	27.490	29.960	33.460
Stearic acid	C18:0	29.79	298.50	20.37 ± 5.037	15.730	21.290	23.950	25.985 ± 2.407	24.770	26.170	26.770
***cis*-10-Heptadecanoic acid**	C17:1Δ10	29.30	284.48	2.275 ± 0.918	1.720	2.144	2.537	3.054 ± 0.485	2.631	2.965	3.532
Heptadecanoic acid	C17:0	28.21	283.48	0.308 ± 0.104	0.254	0.300	0.366	0.366 ± 0.035	0.344	0.372	0.396
Palmitoleic acid	C16:1Δ9	27.77	270.45	0.842 ± 0.322	0.653	0.790	1.062	1.046 ± 0.221	0.928	1.113	1.186
Palmitic acid	C16:0	26.60	270.45	18.09 ± 4.308	14.450	18.110	21.500	22.670 ± 1.848	21.080	22.680	23.820
Pentadecanoic acid	C15:0	24.91	256.42	0.159 ± 0.061	0.118	0.165	0.184	0.174 ± 0.028	0.154	0.184	0.192

Fatty Acid				Severe AD				DLB				*p*-Value
Name	**Lipid No.**	**RT**	**Mass**	**Mean ± SD**	**Q1**	**Median**	**Q3**	**Mean ± SD**	**Q1**	**Median**	**Q3**	**KW FDR**
Docosahexanoic Acid	C22:6Δ4,7,10,13,16,19	45.01	342.51	27.694 ± 9.311	20.090	26.970	34.740	32.329 ± 9.102	26.010	31.960	36.140	0.0[illegible]525
Nervonic Acid	C24:1Δ15	40.68	380.65	4.352 ± 2.385	2.747	4.633	5.775	5.348 ± 2.763	4.417	5.547	6.530	DL[illegible]0251
Lignoceric Acid	C24:0	39.55	382.66	0.942 ± 0.442	0.576	0.924	1.125	1.023 ± 2.384	0.846	0.990	1.243	0.0[illegible]027
***cis*-13,16-Docosadienoic acid**	C22:2Δ13,16	38.89	350.58	0.116 ± 0.059	0.084	0.102	0.140	0.144 ± 0.049	0.114	0.129	0.161	0.0[illegible]034
Arachidonic acid	C20:4Δ5,8,11,14	38.08	318.49	10.83 ± 4.169	7.929	10.660	13.790	14.43 ± 5.755	12.040	14.930	16.070	0.0[illegible]027
Tricosanoic Acid	C23:0	37.42	368.64	0.219 ± 0.154	0.139	0.209	0.336	0.238 ± 3.177	0.011	0.292	0.391	0.56[illegible]651
***cis*-11,14,17-Eicosatrienoic acid**	C20:3Δ11,14,17	28.93	320.51	0.684 ± 0.508	0.299	0.527	1.008	1.105 ± 7.267	0.942	1.042	1.267	0.00[illegible]006
Erucic acid	C22:1Δ13	35.60	352.59	0.421 ± 0.217	0.276	0.429	0.559	0.633 ± 5.170	0.517	0.636	0.707	0.00[illegible]006
***cis*-8,11,14-eicosatreinoic acid**	C20:3Δ8,11,14	37.05	320.51	1.193 ± 0.371	0.990	1.227	1.419	1.211 ± 4.636	0.969	1.176	1.531	0.05[illegible]755
Behenic acid	C22:0	35.18	354.61	0.202 ± 0.098	0.142	0.207	0.266	0.258 ± 8.088	0.213	0.255	0.325	0.00[illegible]006
***cis*-11,14-eicosadienoic acid**	C20:2 Δ11,14	35.18	308.50	0.687 ± 0.629	0.222	0.429	1.083	0.856 ± 9.508	0.435	0.859	1.369	0.20[illegible]431
Heneicosanoic acid	C21:0	34.39	340.58	0.021 ± 0.008	0.015	0.021	0.026	0.023 ± 8.009	0.017	0.024	0.028	0.03[illegible]525
Linolenic acid	C18:3Δ9,12,15	34.20	294.26	3.954 ± 1.650	2.722	3.979	4.928	5.195 ± 9.834	4.679	5.519	5.528	0.00[illegible]006
***cis*-11-Eicosanoic acid**	C20:1 Δ11	33.45	310.50	1.842 ± 1.316	0.377	1.975	2.640	2.137 ± 2.357	1.123	2.669	2.845	0.14[illegible]919
Arachidic acid	C20:0	32.85	326.56	0.224 ± 0.079	0.169	0.239	0.266	0.250 ± 9.088	0.216	0.257	0.287	0.00[illegible]020
Linoleic acid	c-C18:2Δ9,12	32.38	294.47	0.641 ± 0.576	0.016	0.743	0.988	0.720 ± 9.747	0.020	0.730	1.283	0.3[illegible]903
Linolelaidic acid	t-C18:2Δ9,12	31.29	294.47	0.017 ± 0.010	0.011	0.041	0.019	0.021 ± 1.009	0.013	0.020	0.028	0.21[illegible]413
Oleic acid	c-C18:1Δ9	30.82	297.49	26.416 ± 8.524	21.350	28.260	32.480	30.818 ± 8.451	28.110	29.720	31.870	0.00[illegible]014
Stearic acid	C18:0	29.79	298.50	22.679 ± 5.695	17.820	23.630	25.710	25.708 ± 6.692	23.420	24.240	25.580	0.00[illegible]029
***cis*-10-Heptadecanoic acid**	C17:1Δ10	29.30	284.48	2.932 ± 1.470	1.874	2.621	3.898	3.177 ± 5.925	2.671	2.826	3.444	0.00[illegible]046
Heptadecanoic acid	C17:0	28.21	283.48	0.320 ± 0.094	0.248	0.321	0.397	0.375 ± 5.073	0.333	0.366	0.386	0.04[illegible]672
Palmitoleic acid	C16:1Δ9	27.77	270.45	0.855 ± 0.409	0.647	0.895	1.098	0.760 ± 1.489	0.289	0.722	1.282	0.17[illegible]193
Palmitic acid	C16:0	26.60	270.45	20.953 ± 5.023	17.150	21.220	23.920	22.929 ± 7.392	20.150	22.230	24.660	0.00[illegible]072
Pentadecanoic acid	C15:0	24.91	256.42	0.165 ± 0.051	0.133	0.163	0.199	0.187 ± 3.048	0.152	0.173	0.223	0.2[illegible]558

Table 2. Fatty acids significantly higher in post-mortem brain tissue from subjects with moderate AD, severe AD, or DLB pathology. Following the Kruskal–Wallis tests, multiple comparison procedures using stepdown permutation *p*-values on the ranked data were used to investigate the nature of group differences for variables with a false discovery rate of 0.05 or less.

Fatty Acid	*p*-Values					
	Control vs. Moderate AD	**Control vs. Severe AD**	**Control vs. DLB**	**Moderate AD vs. Severe AD**	**Moderate AD vs. DLB**	**Severe AD vs. DLB**
Nervonic acid (mono)	0.0453	0.7397	0.0453	0.1811	0.9578	0.1811
Erucic acid (mono)	0.0011	1	0.0011	0.0022	1	0.0022
Oleic acid (mono)	0.0006	0.0337	0.0011	0.1912	0.8664	0.1917
cis-10-Heptanoic acid (mono)	0.0049	0.0899	0.0068	0.3476	0.8887	0.3476
cis-13,16-Docosadienoic acid (poly)	0.0036	0.5416	0.0104	0.0331	0.7307	0.0707
Arachidonic acid (poly)	0.0094	1	0.0094	0.0128	1	0.0128
cis-11,14,17-Eicosatrienoic acid (poly)	0.0001	0.4203	0.0001	0.0004	0.9930	0.0004
Linolenic acid (poly)	0.0003	0.6126	0.0004	0.0029	0.8786	0.0039
Lignoceric acid (saturated)	0.0004	0.1263	0.0333	0.0514	0.4317	0.4317
Behenic acid (saturated)	0.0002	0.1151	0.0006	0.0186	0.7104	0.0455
Arachidic acid (saturated)	0.0001	0.3774	0.0783	0.0046	0.2069	0.3774
Stearic acid (saturated)	0.0006	0.2969	0.0344	0.0344	0.3416	0.3416
Palmitic acid (saturated)	0.0065	0.0902	0.0143	0.4114	0.7813	0.4832
Number of FAs significantly different	13	1	12	9	0	5

3. Discussion

This study accurately quantified the FA composition of post-mortem human brain tissue from 87 subjects: control subjects (n = 27); moderate AD (n = 16), severe AD (n = 30), and DLB (n = 14). A key strength of this study is its focused approach to brain FA measurements. It represents the widest brain FA coverage measured, thus far, in AD or other forms of dementia. Previous studies have found gender-related differences in brain FA content in AD [12], but this was not evident in the present study. Other studies typically measure a much narrower range of FAs, of perhaps 14 or 16 FAs at most [16,19]. We compared the measured brain FA levels as percentage % of total FA content, and found them to be broadly comparable with a number of other studies [16,17,20–26]. In general, the findings indicate that, in parietal cortex, there is a trend for higher FA levels in moderate AD, but not in cases of severe AD. Several FAs where significantly affected, including lignoceric acid, *cis*-13,16-docosadienoic, arachidonic acid, *cis*-11,14,17-eicosatrienoic acid, erucic acid, linolenic acid, and stearic acid. This is not the first time that FA concentrations have been shown to be higher in AD [20], however, these earlier changes suggest mechanisms involving lipid metabolism in response to the development of disease pathology. It has previously been suggested that the relatively higher brain FA content may be due to ceramide accumulation due to the activation of sphingomyelinase in oligodendrocytes, induced by increases in amyloid beta peptide levels [20].

A particular limitation of the present study was that post-mortem delay was significantly longer in cases of severe AD, compared with cases of moderate AD. In general, the post-mortem delay durations for the brain specimens obtained were relatively long. In our tissue requests, we did our best to obtain samples with as short a delay as possible, to minimize tissue degradation/oxidation which, in turn, could affect the FA composition. Where possible, we attempted to match groups as much as possible with respect to post-mortem delay. This was performed within the constraints of the available tissue with the appropriate pathological characteristics. It is possible that post-mortem delay is a confounding factor which explains significant differences for some FAs between Severe and Moderate AD groups. An in-depth analysis finds that there is some support for this—there were modest but significant associations between the duration of post-mortem delay and concentrations of six individual FAs. All of these were negative associations, suggesting that longer post-mortem delay may decrease these FAs. Five of the six FAs where among the 13 shortlisted as differing between groups. Furthermore, it is also clear that there were group differences as far as aggregated FA levels are concerned (total FA, total saturated, total monounsaturated, or total polyunsaturated). For this reason, the findings here should be interpreted with a degree of care because post-mortem delay is an uncontrolled variable.

Nervonic acid is a product of the desaturation and elongation processes of several fatty acids, including palmitic acid, stearic acid, and oleic acid. Through the action of stearoyl CoA desaturase (SCD) enzyme, these fatty acids undergo a series of elongation steps prior to the production of nervonic acid. Higher concentrations of these fatty acids, in moderate AD pathology, could be attributed to the increased activity of SCD itself. A recent study has demonstrated that nervonic acid and several mono-unsaturated fatty acids produced by SCD are markedly upregulated in brain samples of AD patients [27]. Higher FA levels and SCD activity have been shown to correlate closely with cognitive impairment [27]. Although our findings contrast with that of Astarita et al. [27], SCD activity does provide a potential pathogenic mechanism.

Higher palmitic acid concentrations in moderate AD pathology indicate damaging effects on the brain. Palmitic acid has been shown to induce tau hyperphosphorylation and to elevate β-secretase activity in embryonic rat cortex cultures [28]. Furthermore, the trend for palmitic acid is in keeping with the plasma levels observed in another study [19]. The case is also similar for oleic and stearic acid plasma levels in AD patients recently reported in a longitudinal population-based study by Ronnema et al. (2012) [29]. It is interesting to note that only oleic acid was found to be significantly elevated when Control was paired with Severe AD (Figure 2). Another study reported similar finding, but uniquely in the white brain matter of AD patients [17].

Omega-3 and -6 (n-3/n-6) fatty acids are referred to as essential fatty acids. synthesized from dietary linolenic or linoleic acids through series of saturation and elongation reactions [30]. Both n-3 and n-6 fatty acids have been implicated in the modulation of brain inflammatory processes [30–32]. Fatty acids, such as arachidonic acid (AA), eicosapentanoic acid (EPA; C20:5), and DHA are thought to modulate the severity and duration of AD inflammatory processes [33,34]. These fatty acids exert their inflammatory function through their conversion to potent eicosanoids, such as prostaglandins, thromboxanes, and leukotrienes (by AA), or resolvins and docosatrienes (by EPA and DHA) by means of cyclooxygenase (COX) and/or lipoxygenase (LOX) enzymes [29,35].

In the present study, DHA did not differ between any of the 4 groups, a finding which is consistent with several previous studies [16,17,36]. Other reports, on the other hand, showed lower levels in AD brain subjects [13,19,21,22,37]. Snowden and colleagues found DHA levels in individuals with significant AD neuropathology to be unchanged in both the inferior temporal gyrus and the cerebellum, but it was elevated in middle frontal gyrus [38]. For linolenic acid (a precursor of DHA as well as EPA) however, we did detect higher levels in moderate AD. Linolenic acid is an omega-3 FA with anti-inflammatory activity [37,38]. The elevated linolenic acid (in moderate AD) observed, here, may constitute an anti-inflammatory to the development of early AD disease pathology. Interestingly, arachidonic acid was higher in moderate AD. Eicosanoids derived from arachidonic acid are typically pro-inflammatory agents. It has been demonstrated that arachidonic acid is converted by the enzymatic activity of COX and LOX on increased levels of pro-inflammatory cytokines and activation of neutrophils, resulting from eicosanoid biosynthesis from enzymatic activity of COX and LOX on arachidonic acid [39]. Our study showed decreased levels of both linolenic acid and arachidonic acid (Figure 2). While the reason for such occurrences in the brain are yet to be fully determined, the trend, however, was found to be in keeping with plasma levels of recently studied AD patients [40]. The decreasing trend of AA and LA seen in plasma AD patients was indicative of a deficiency in neuroprotective elements against the pathology of the disease, and the excessive inflammation that was occurring [40]. Furthermore, the significantly higher *cis*-13,16-docosadienoic concentrations in moderate AD group could form part of an anti-inflammatory response to the development of AD. Higher levels of *cis*-13,16-docosadienoic concentrations could provide neuroprotection by blockading COX enzyme activity [34]. The overlapping levels of fatty acids seen and described earlier in age-matched Controls with moderate and severe AD subjects, provide a unique profile which may serve as indicating potential biomarkers of the disease.

Dementia with Lewy bodies (DLB) is the second major type of senile, degenerative dementia after AD. It is characterized by the presence of cytoplasmic inclusions of highly conserved amyloidogenic α-synuclein [41] protein that mirrors tau proteins in AD. It is localized, in part, to presynaptic terminals where it loosely associates with synaptic vesicles [42,43]. Numerous studies have been conducted on its structural organization at various stages of fibrillation and inclusion formation.

Fatty acids have been found to play an important role in the conversion of normal, soluble α-synuclein to insoluble and potentially cytotoxic forms. In vitro studies of mouse embryonic stem cells (MES) neurons showed PUFA promotes the appearance of oligomeric forms of α-synuclein in a time-dependent manner [44], while increasing the degree of unsaturation of fatty acids strikingly enhanced the amount of soluble forms [44]. It is further noted that higher levels of α-synuclein in DLB brains may be linked to the changes in the composition of endogenous brain fatty acid species. Sharon et al. (2003) reported a consistently lower linolenic acid and higher DHA levels in frontal cortex of DLB brains [45]. In contrast to their findings, our results indicate higher concentrations of linolenic acid when DLB are either paired with Control or severe AD groups (Figure 2; Table 2).

In addition, the study also found no changes or significant alterations in arachidonic acid levels, as well as other MUFAs [45]. We, however, saw higher concentrations of such species in both Control, and few in severe AD-paired groups. Furthermore, the higher abundance of saturated fatty acids, including palmitic acid, stearic acid, behenic acid, and lignoceric acid, observed in our study, may suggest a protective mechanism by counterbalancing the oligomeric formation of α-synuclein by eicosatrienoic, linolenic, and arachidonic acid in DLB brains (Figure 2; Table 2). One study demonstrated the impact of DHA and arachidonic acids imposing conformational changes in the structure of α-synuclein, whereas the action of palmitic and arachidic acids remained unchanged [46]. To the best of the author's knowledge, this is the first study that directly compares brain fatty acid profiles of Alzheimer's disease and DLB. Despite the fact that the overlapping FA species between severe AD and Control with DLB are relatively unknown, with regards to their individual function, it still highlights the probability of such fatty acids being used as potential biomarkers to distinguish the differences between these groups. Principal component acids in both healthy control, moderate, and severe AD groups could serve as potential biomarkers that warranted further functional studies. Particular attention, in future, should be paid to controlling the controlling of post-mortem delay. While most FAs mentioned above played intrinsic roles in the inflammatory/anti-inflammatory processes of AD, some remained to be understood, given the discrepancies and limited data available in the current literature. The FA profiles of AD and DLB are very similar, perhaps reflecting common progression of the disease, but conclusions such as this require additional studies. This paper provides an insight into the changes of FA metabolism in the development of AD, as well differences that can be observed in the profiles of DLB subjects.

4. Materials and Methods

4.1. Reagents and Analytical Standards

The following fatty acid and fatty acid methyl ester standards were purchased from Sigma Aldrich (Gillingham, Dorset, UK): tricosanoic acid (C23:0), behenic acid (C22:0), *cis*-11,14-eicosadienoic acid (C20:2 Δ11,14), heptadecanoic acid (C17:0), *cis*-10-pentadecanoic acid (C15:1Δ10), pentadecanoic acid methyl ester (C15:0) (PDA), palmitic acid methyl ester (C16:0), palmitoleic acid methyl ester (C16:1Δ9), *cis*-10-heptadecanoic acid methyl ester (C17:1Δ10), stearic acid methyl ester (C18:0), elaidic acid methyl ester (t-C18:1Δ9), oleic acid methyl ester (C18:1Δ9), linolelaidic acid methyl ester (t-C18:2 Δ9, 12) linoleic acid methyl ester (C18:2 Δ9,12), arachidic acid methyl ester (C20:0), linolenic acid methyl ester (C18:3 Δ9,12,15), heneicosanoic acid methyl ester (C21:0), *all-cis*-11,14,17-eicosatrienoic acid methyl ester (C20:3 Δ11,14,17), arachidonic acid methyl ester (C20:4 Δ5,8,11,14), *all-cis*-13,16-docosadienoic acid methyl ester (C22:2 Δ13,16), lignoceric acid methyl ester (C24:0), nervonic acid methyl ester (C21:1 Δ15), docosahexanoic acid methyl ester (C22:6 Δ4,7,10,13,16,19) (DHA), *all-cis*-8,11,14-eicosaterinoic acid methyl ester (C20:3 Δ8,11,14), erucic acid methyl ester (C22:1 Δ13). All solvents used (methanol *n*-hexane and dichloromethane) were CHROMASOLV HPLC grade (Sigma Aldrich, UK). Hydrogen chloride (1.25 M) in methanol was purchased from Fluka Analytical (UK).

4.2. Human Post-Mortem Tissue

As previously described by Graham et al. (2014) [47], post-mortem tissue samples (parietal neocortex, Brodmann area 7) were obtained from pathologically and clinically confirmed cases of AD ($n = 46$), DLB ($n = 14$), and normal age-matched controls ($n = 27$) [47] (Table S1). The parietal cortex (Brodmann 7) region was selected as it not the primary site of AD pathology. Studies have shown that FA composition of the parietal cortex is less affected than other brain regions, such as the frontal or temporal cortex [16]. The AD was classified as moderate ("intermediate" AD neuropathological change [48], and also Braak tangle stage III–IV; $n = 16$) or severe ("high" AD neuropathological change [48], and also Braak tangle stage V–VI; $n = 30$). Mixed pathology cases were excluded from the study. Cases were geographically spread across the United Kingdom (Bristol, Newcastle, and London) and were obtained through the Brains for Dementia Research (BDR; see acknowledgements). Sample selection did not control for post-mortem delay, and there was a significantly difference between Severe AD and Moderate AD ($p = 0.03$). The neuropathological diagnoses were made using widely accepted criteria [48,49], uniformly applied according to a standardized protocol by members of the BDR Neuropathology Group. DLB was diagnosed on the basis that (i) there was a clinical diagnosis of dementia, (ii) some Lewy bodies were present in the neocortex [50]. Consent and ethical approval for the use of tissue were obtained by individual brain banks, all of which are licensed by the Human Tissue Authority (UK). Consent and ethical approval for the use of tissue was obtained by individual brain banks, all of which are licensed by the Human Tissue Authority. The study was conducted in accordance with the Declaration of Helsinki, and the protocol was reviewed by the School Research Ethics Committee (Biological Sciences, Queen's University Belfast) (0512-GrahamS).

4.3. Sample Preparation and GC-MS Analysis

Frozen tissue samples stored at −80 °C were initially lyophilized by placing them in a freeze-drier at −50 °C. Following the complete removal of moisture tissue specimens, they were placed into a cryogenic grinder (Model 6850, SPEX SamplePREP, UK), which uses liquid nitrogen as a coolant throughout the milling process and, thus, avoids heat generation which could affect the concentration of metabolites in the brain tissue samples. Once milled, all samples were stored at −80 °C. Lipids were extracted from tissue by the Folch extraction method [51]. Briefly, 50 mg of powder (weighed out into sterile Eppendorf tubes (2 mL) chilled on ice). Methanol/water (50% v/v; 1 mL) was added to each sample and shaken for 10 min using a Minimix Standard Shaker (Merris Engineering, Maidenhead, Berkshire). Samples were then sonicated for 15 min. Protein was removed by centrifugation at 16,000× *g* at 4 °C for 20 min. To assess the recoveries of FA species, lauric acid (C12:0) was selected as an internal standard, as it was found to be completely absent in all the post-mortem human brain samples. Lauric acid was added to each sample pellet (100 ng/μL) and FAs were extracted in 1 mL of DCM, transferred to sterile test tubes, and evaporated to dryness under nitrogen. FAs were derivatized to fatty acid methyl esters (FAMEs) by reconstituting the dried extracts in 2 mL hydrogen chloride in methanol. Samples were subsequently cooled, and 1 mL of water was added. The FAMEs were extracted in 1 mL of hexane and subsequently analyzed using an Agilent GC (model 7890, Wilmington, DE, USA) coupled to an MS detector (Agilent model 5975C, Wilmington, DE, USA). The FAMEs procedure creates methyl esters from free fatty acids, but also from esterified fatty acids and, therefore, measurements reflect the total levels of unesterified and esterified fatty acids.

Samples were injected (inlet temp 220 °C, split-mode ration of 15:1) onto a CP-Sil88 fused silica capillary column (100 m × 0.25 mm × 0.25 5 μm) (Agilent, UK) with helium as the carrier gas at a constant flow of 1 mL/min. The initial temperature gradient began at 100 °C, increasing at 4 °C/min to 220 °C/min, and held for 5 min. Following this, the gradient increased at 4 °C/min to 240 °C/min, and was held for 8 min. The mass selective detector (MSD) operated at 70 eV in dual scan/single ion monitoring (SIM) mode; source temp 230 °C, quad temp 150 °C and the interface temp 225 °C. The full scan ranged from *m/z* 50 to 550, whilst SIM mode targeted the molecular ion and another appropriate ion selected from the fragmentation pattern, each ion having a dwell time of 100 ms. All FAMEs were confirmed using purchased analytical-grade standards. Quantification was based on a linear regression model formed from a five-point calibration curve from the individual FAMEs, which were acquired at a low (0–20 ng/mL) or high concentration range (0–300 ng/mL). FFA concentrations are reported as g/kg post-mortem human brain dry weight, and corrected to the internal standard. Fatty acids with calibration curves ($n = 3$) of poor linearity of ($R^2 < 0.9$) were not quantified

4.4. Data Analysis

Normality of distribution was assessed using the Anderson–Darling test. Nonparametric methods were used since the assumption that FA concentrations were normally distributed was not satisfied for many variables. The four groups were compared on metabolite levels using Kruskal–Wallis tests. Elaidic and cis-10-pentadeconaic acid were excluded, given the large prevalence of values of 0. Given the large number of variables tested, false discovery rates were computed based on the *p*-values from the Kruskal–Wallis tests on the remaining 24 variables. Multiple comparison procedures using stepdown permutation *p*-values, on the ranked data, were used to investigate group differences for variables with a false discovery rate of 0.05 or less. Statistical analysis used the SAS System for Windows version 9.3 (Cary, NC, USA). All tests were two-sided, and $p < 0.05$ was considered statistically significant. Graphical representations of the data were produced using Prism (GraphPad 5.0, La Jolla, CA, USA). Group comparisons of patient characteristics (age, % female, and post-mortem delay) were compared by parametric one-way ANOVA. Associations between FAs and continuous variables (age, post-mortem delay, beta-amyloid, or tau) were tested by Spearman's rank correlation coefficient. The relationship between FA levels and gender were carried out by dividing subjects into male and female and conducting a Mann–Whitney *U t*-test for each FA. PCA was completed using Simca P (v14.1; Umetrics, Umea, Sweden) using mean centered and log transformed data.

Supplementary Materials: The following are available online at http://www.mdpi.com/2218-1989/8/4/69/s1, Table S1: Descriptive characteristics of the study population by group, Table S2: Aggregated fatty acid concentrations and ratios, Table S3: Associations between fatty acids and clinical variables (age, gender, post-mortem delay, frontal tissue pH, beta-amyloid, tau); Figure S1. Principal component analysis (PCA) of fatty acid concentrations in post-mortem human brain specimens.

Author Contributions: Conceptualization, B.M., P.P., C.H., S.F.G., P.G.K. and B.D.G.; Formal analysis, M.L.N., X.P. and S.F.G.; Funding acquisition, B.M., P.P., C.H., S.F.G., P.G.K. and B.D.G.; Methodology, M.L.N., X.P. and S.F.G.; Project administration, B.D.G.; Software, M.L.N. and X.P.; Supervision, S.F.G. and B.D.G.; Writing—original draft, M.L.N.; Writing—review & editing, X.P., B.M., P.P., C.H., S.F.G., P.G.K. and B.D.G.

Funding: Studies into Alzheimer's disease have been supported by grants from Alzheimer's Research UK [ARUK-NCH2012B-5; ARUK-PPG2011B-8 and ARUK-Network2012-11]. The authors' work is also supported by a Proof of Concept grant from Invest Northern Ireland [INI-PoC406].

Acknowledgments: We would like to thank Rachael Hill for her invaluable assistance, both for her help with the sample preparation and mass spectral analyses. We would like to thank all the staff of the regional brain banks of Brains for Dementia Research for their advice, assistance and co-ordination; in particular we thank Laura E. Palmer and Hannah Tayler (Bristol), Maria Monteiro (Newcastle) and Richard Hudspith (London). Brain tissue was provided by the South West Dementia Brain Bank (SWDBB), Newcastle Brain Tissue Resource (NBTR) and King's College London (KCL) who are members of the Brains for Dementia Research (BDR) Network.

Conflicts of Interest: The authors have no conflict of interest to declare.

References

1. Prince, M.; Bryce, R.; Albanese, E.; Wimo, A.; Ribeiro, W.; Ferri, C.P. The global prevalence of dementia: A systematic review and metaanalysis. *Alzheimers Dement.* **2013**, *9*, 63–75. [CrossRef] [PubMed]
2. Karran, E.; Mercken, M.; De Strooper, B. The amyloid cascade hypothesis for Alzheimer's disease: An appraisal for the development of therapeutics. *Nat. Rev. Drug Discov.* **2011**, *10*, 698–712. [CrossRef] [PubMed]
3. Ittner, L.M.; Gotz, J. Amyloid-beta and tau-a toxic pas de deux in Alzheimer's disease. *Nat. Rev. Neurosci.* **2011**, *12*, 65–72. [CrossRef] [PubMed]
4. Touboul, D.; Gaudin, M. Lipidomics of Alzheimer's disease. *Bioanalysis* **2014**, *6*, 541–561. [CrossRef] [PubMed]
5. Abu-Saad, K.; Shahar, D.R.; Vardi, H.; Fraser, D. Importance of ethnic foods as predictors of and contributors to nutrient intake levels in a minority population. *Eur. J. Clin. Nutr.* **2010**, *64*, S88–S94. [CrossRef] [PubMed]
6. Cutler, R.G.; Kelly, J.; Storie, K.; Pedersen, W.A.; Tammara, A.; Hatanpaa, K.; Troncoso, J.C.; Mattson, M.P. Involvement of oxidative stress-induced abnormalities in ceramide and cholesterol metabolism in brain aging and Alzheimer's disease. *Proc. Natl. Acad. Sci. USA* **2004**, *101*, 2070–2075. [CrossRef] [PubMed]
7. Yanagisawa, K.; Odaka, A.; Suzuki, N.; Ihara, Y. GM1 ganglioside-bound amyloid beta-protein (A beta): A possible form of preamyloid in Alzheimer's disease. *Nat. Med.* **1995**, *1*, 1062–1066. [CrossRef] [PubMed]
8. Berridge, M.J. Inositol trisphosphate and calcium signaling. *Ann. N. Y. Acad. Sci.* **1995**, *766*, 31–43. [CrossRef] [PubMed]
9. Yuan, Q.; Zhao, S.; Wang, F.; Zhang, H.; Chen, Z.J.; Wang, J.; Wang, Z.; Du, Z.; Ling, E.A.; Liu, Q.; et al. Palmitic acid increases apoptosis of neural stem cells via activating c-Jun N-terminal kinase. *Stem Cell Res.* **2013**, *10*, 257–266. [CrossRef] [PubMed]
10. Takenawa, T.; Itoh, T. Phosphoinositides, key molecules for regulation of actin cytoskeletal organization and membrane traffic from the plasma membrane. *Biochim. Biophys. Acta* **2001**, *1533*, 190–206. [CrossRef]

11. Wenk, M.R.; De Camilli, P. Protein-lipid interactions and phosphoinositide metabolism in membrane traffic: Insights from vesicle recycling in nerve terminals. *Proc. Natl. Acad. Sci. USA* **2004**, *101*, 8262–8269. [CrossRef] [PubMed]
12. Nasaruddin, M.L.; Holscher, C.; Kehoe, P.; Graham, S.F.; Green, B.D. Wide-ranging alterations in the brain fatty acid complement of subjects with late Alzheimer's disease as detected by GC-MS. *Am. J. Transl. Res.* **2016**, *8*, 154–165. [PubMed]
13. Martin, V.; Fabelo, N.; Santpere, G.; Puig, B.; Marin, R.; Ferrer, I.; Diaz, M. Lipid alterations in lipid rafts from Alzheimer's disease human brain cortex. *J. Alzheimers Dis.* **2010**, *19*, 489–502. [CrossRef] [PubMed]
14. Wood, P.L. Lipidomics of Alzheimer's disease: Current status. *Alzheimers Res. Ther.* **2012**, *4*, 5. [CrossRef] [PubMed]
15. Wang, D.C.; Sun, C.H.; Liu, L.Y.; Sun, X.H.; Jin, X.W.; Song, W.L.; Liu, X.Q.; Wan, X.L. Serum fatty acid profiles using GC-MS and multivariate statistical analysis: Potential biomarkers of Alzheimer's disease. *Neurobiol. Aging* **2012**, *33*, 1057–1066. [CrossRef] [PubMed]
16. Fraser, T.; Tayler, H.; Love, S. Fatty Acid Composition of Frontal, Temporal and Parietal Neocortex in the Normal Human Brain and in Alzheimer's Disease. *Neurochem. Res.* **2010**, *35*, 503–513. [CrossRef] [PubMed]
17. Igarashi, M.; Ma, K.; Gao, F.; Kim, H.W.; Rapoport, S.I.; Rao, J.S. Disturbed choline plasmalogen and phospholipid fatty acid concentrations in Alzheimer's disease prefrontal cortex. *J. Alzheimers Dis.* **2011**, *24*, 507–517. [CrossRef] [PubMed]
18. Pan, X.; Nasaruddin, M.B.; Elliott, C.T.; McGuinness, B.; Passmore, A.P.; Kehoe, P.G.; Holscher, C.; McClean, P.L.; Graham, S.F.; Green, B.D. Alzheimer's disease-like pathology has transient effects on the brain and blood metabolome. *Neurobiol. Aging* **2016**, *38*, 151–163. [CrossRef] [PubMed]
19. Cunnane, S.C.; Schneider, J.A.; Tangney, C.; Tremblay-Mercier, J.; Fortier, M.; Bennett, D.A.; Morris, M.C. Plasma and brain fatty acid profiles in mild cognitive impairment and Alzheimer's disease. *J. Alzheimers Dis.* **2012**, *29*, 691–697. [CrossRef] [PubMed]
20. Roher, A.E.; Weiss, N.; Kokjohn, T.A.; Kuo, Y.M.; Kalback, W.; Anthony, J.; Watson, D.; Luehrs, D.C.; Sue, L.; Walker, D.; et al. Increased A beta peptides and reduced cholesterol and myelin proteins characterize white matter degeneration in Alzheimer's disease. *Biochemistry* **2002**, *41*, 11080–11090. [CrossRef] [PubMed]
21. Soderberg, M.; Edlund, C.; Kristensson, K.; Dallner, G. Fatty-Acid Composition of Brain Phospholipids in Aging and in Alzheimers-Disease. *Lipids* **1991**, *26*, 421–425. [CrossRef] [PubMed]
22. Wang, G.; Zhou, Y.; Huang, F.J.; Tang, H.D.; Xu, X.H.; Liu, J.J.; Wang, Y.; Deng, Y.L.; Ren, R.J.; Xu, W.; et al. Plasma metabolite profiles of Alzheimer's disease and mild cognitive impairment. *J. Proteome Res.* **2014**, *13*, 2649–2658. [CrossRef] [PubMed]
23. Skinner, E.R.; Watt, C.; Besson, J.A.; Best, P.V. Differences in the fatty acid composition of the grey and white matter of different regions of the brains of patients with Alzheimer's disease and control subjects. *Brain* **1993**, *116*, 717–725. [CrossRef] [PubMed]
24. Corrigan, F.M.; Horrobin, D.F.; Skinner, E.R.; Besson, J.A.; Cooper, M.B. Abnormal content of n-6 and n-3 long-chain unsaturated fatty acids in the phosphoglycerides and cholesterol esters of parahippocampal cortex from Alzheimer's disease patients and its relationship to acetyl CoA content. *Int. J. Biochem. Cell Biol.* **1998**, *30*, 197–207. [CrossRef]
25. Pamplona, R.; Dalfo, E.; Ayala, V.; Bellmunt, M.J.; Prat, J.; Ferrer, I.; Portero-Otin, M. Proteins in human brain cortex are modified by oxidation, glycoxidation, and lipoxidation: Effects of Alzheimer disease and identification of lipoxidation targets. *J. Biol. Chem.* **2005**, *280*, 21522–21530. [CrossRef] [PubMed]
26. Guan, Z.Z.; Soderberg, M.; Sindelar, P.; Edlund, C. Content and fatty acid composition of cardiolipin in the brain of patients with Alzheimer's disease. *Neurochem. Int.* **1994**, *25*, 295–300. [CrossRef]
27. Astarita, G.; Jung, K.M.; Vasilevko, V.; Dipatrizio, N.V.; Martin, S.K.; Cribbs, D.H.; Head, E.; Cotman, C.W.; Piomelli, D. Elevated stearoyl-CoA desaturase in brains of patients with Alzheimer's disease. *PLoS ONE* **2011**, *6*, e24777. [CrossRef] [PubMed]
28. Patil, S.; Chan, C. Palmitic and stearic fatty acids induce Alzheimer-like hyperphosphorylation of tau in primary rat cortical neurons. *Neurosci. Lett.* **2005**, *384*, 288–293. [CrossRef] [PubMed]
29. Ronnemaa, E.; Zethelius, B.; Vessby, B.; Lannfelt, L.; Byberg, L.; Kilander, L. Serum fatty-acid composition and the risk of Alzheimer's disease: A longitudinal population-based study. *Eur. J. Clin. Nutr.* **2012**, *66*, 885–890. [CrossRef] [PubMed]

30. Schmitz, G.; Ecker, J. The opposing effects of n-3 and n-6 fatty acids. *Prog. Lipid Res.* **2008**, *47*, 147–155. [CrossRef] [PubMed]
31. Wall, R.; Ross, R.P.; Fitzgerald, G.F.; Stanton, C. Fatty acids from fish: The anti-inflammatory potential of long-chain omega-3 fatty acids. *Nutr. Rev.* **2010**, *68*, 280–289. [CrossRef] [PubMed]
32. Patterson, E.; Wall, R.; Fitzgerald, G.F.; Ross, R.P.; Stanton, C. Health implications of high dietary omega-6 polyunsaturated Fatty acids. *J. Nutr. Metab.* **2012**, *2012*, 539426. [CrossRef] [PubMed]
33. Sanchez-Mejia, R.O.; Mucke, L. Phospholipase A2 and arachidonic acid in Alzheimer's disease. *Biochim. Biophys. Acta* **2010**, *1801*, 784–790. [CrossRef] [PubMed]
34. Serhan, C.N.; Arita, M.; Hong, S.; Gotlinger, K. Resolvins, docosatrienes, and neuroprotectins, novel omega-3-derived mediators, and their endogenous aspirin-triggered epimers. *Lipids* **2004**, *39*, 1125–1132. [CrossRef] [PubMed]
35. Youdim, K.A.; Martin, A.; Joseph, J.A. Essential fatty acids and the brain: Possible health implications. *Int. J. Dev. Neurosci.* **2000**, *18*, 383–399. [CrossRef]
36. Brooksbank, B.W.; Martinez, M. Lipid abnormalities in the brain in adult Down's syndrome and Alzheimer's disease. *Mol. Chem. Neuropathol.* **1989**, *11*, 157–185. [CrossRef] [PubMed]
37. Prasad, M.R.; Lovell, M.A.; Yatin, M.; Dhillon, H.; Markesbery, W.R. Regional membrane phospholipid alterations in Alzheimer's disease. *Neurochem. Res.* **1998**, *23*, 81–88. [CrossRef] [PubMed]
38. Lukiw, W.J.; Cui, J.G.; Marcheselli, V.L.; Bodker, M.; Botkjaer, A.; Gotlinger, K.; Serhan, C.N.; Bazan, N.G. A role for docosahexaenoic acid-derived neuroprotectin D1 in neural cell survival and Alzheimer disease. *J. Clin. Investig.* **2005**, *115*, 2774–2783. [CrossRef] [PubMed]
39. Hong, S.; Gronert, K.; Devchand, P.R.; Moussignac, R.L.; Serhan, C.N. Novel docosatrienes and 17S-resolvins generated from docosahexaenoic acid in murine brain, human blood, and glial cells. Autacoids in anti-inflammation. *J. Biol. Chem.* **2003**, *278*, 14677–14687. [CrossRef] [PubMed]
40. Bagga, D.; Wang, L.; Farias-Eisner, R.; Glaspy, J.A.; Reddy, S.T. Differential effects of prostaglandin derived from omega-6 and omega-3 polyunsaturated fatty acids on COX-2 expression and IL-6 secretion. *Proc. Natl. Acad. Sci. USA* **2003**, *100*, 1751–1756. [CrossRef] [PubMed]
41. Lucke, C.; Gantz, D.L.; Klimtchuk, E.; Hamilton, J.A. Interactions between fatty acids and alpha-synuclein. *J. Lipid Res.* **2006**, *47*, 1714–1724. [CrossRef] [PubMed]
42. George, J.M.; Jin, H.; Woods, W.S.; Clayton, D.F. Characterization of a novel protein regulated during the critical period for song learning in the zebra finch. *Neuron* **1995**, *15*, 361–372. [CrossRef]
43. Hashimoto, M.; Masliah, E. Alpha-synuclein in Lewy body disease and Alzheimer's disease. *Brain Pathol.* **1999**, *9*, 707–720. [CrossRef] [PubMed]
44. Sharon, R.; Bar-Joseph, I.; Frosch, M.P.; Walsh, D.M.; Hamilton, J.A.; Selkoe, D.J. The formation of highly soluble oligomers of alpha-synuclein is regulated by fatty acids and enhanced in Parkinson's disease. *Neuron* **2003**, *37*, 583–595. [CrossRef]
45. Sharon, R.; Bar-Joseph, I.; Mirick, G.E.; Serhan, C.N.; Selkoe, D.J. Altered fatty acid composition of dopaminergic neurons expressing alpha-synuclein and human brains with alpha-synucleinopathies. *J. Biol. Chem.* **2003**, *278*, 49874–49881. [CrossRef] [PubMed]
46. Broersen, K.; van den Brink, D.; Fraser, G.; Goedert, M.; Davletov, B. Alpha-synuclein adopts an alpha-helical conformation in the presence of polyunsaturated fatty acids to hinder micelle formation. *Biochemistry* **2006**, *45*, 15610–15616. [CrossRef] [PubMed]
47. Graham, S.F.; Nasaruddin, M.B.; Carey, M.; Holscher, C.; McGuinness, B.; Kehoe, P.G.; Love, S.; Passmore, P.; Elliott, C.T.; Meharg, A.A.; et al. Age-associated changes of brain copper, iron, and zinc in Alzheimer's disease and dementia with Lewy bodies. *J. Alzheimers Dis.* **2014**, *42*, 1407–1413. [CrossRef] [PubMed]
48. Montine, T.J.; Phelps, C.H.; Beach, T.G.; Bigio, E.H.; Cairns, N.J.; Dickson, D.W.; Duyckaerts, C.; Frosch, M.P.; Masliah, E.; Mirra, S.S.; et al. National Institute on A and Alzheimer's A. National Institute on Aging-Alzheimer's Association guidelines for the neuropathologic assessment of Alzheimer's disease: A practical approach. *Acta Neuropathol.* **2012**, *123*, 1–11. [CrossRef] [PubMed]
49. McKeith, I.G.; Dickson, D.W.; Lowe, J.; Emre, M.; O'Brien, J.T.; Feldman, H.; Cummings, J.; Duda, J.E.; Lippa, C.; Perry, E.K.; et al. Diagnosis and management of dementia with Lewy bodies: Third report of the DLB Consortium. *Neurology* **2005**, *65*, 1863–1872. [CrossRef] [PubMed]

50. Hansen, L.A.; Daniel, S.E.; Wilcock, G.K.; Love, S. Frontal cortical synaptophysin in Lewy body diseases: Relation to Alzheimer's disease and dementia. *J. Neurol. Neurosurg. Psychiatry* **1998**, *64*, 653–656. [CrossRef] [PubMed]
51. Folch, J.; Lees, M.; Sloane-Stanley, G.H. A simple method for the isolation and purification of total lipides from animal tissues. *J. Biol. Chem.* **1957**, *226*, 497–509. [PubMed]

metabolites

MDPI

Review

No Country for Old Worms: A Systematic Review of the Application of *C. elegans* to Investigate a Bacterial Source of Environmental Neurotoxicity in Parkinson's Disease

Kim A. Caldwell [1,2,*], **Jennifer L. Thies** [1] **and Guy A. Caldwell** [1,2]

1 Department of Biological Sciences, The University of Alabama, Box 870344, Tuscaloosa, AL 35487, USA; Jthies@crimson.ua.edu (J.L.T.); gcaldwel@ua.edu (G.A.C.)
2 Departments of Neurology and Neurobiology, Center for Neurodegeneration and Experimental Therapeutics, Nathan Shock Center for Research on the Basic Biology of Aging, University of Alabama at Birmingham School of Medicine, Birmingham, AL 35294, USA
* Correspondence: kcaldwel@ua.edu

Received: 7 October 2018; Accepted: 26 October 2018; Published: 29 October 2018

Abstract: While progress has been made in discerning genetic associations with Parkinson's disease (PD), identifying elusive environmental contributors necessitates the application of unconventional hypotheses and experimental strategies. Here, we provide an overview of studies that we conducted on a neurotoxic metabolite produced by a species of common soil bacteria, *Streptomyces venezuelae (S. ven)*, indicating that the toxicity displayed by this bacterium causes stress in diverse cellular mechanisms, such as the ubiquitin proteasome system and mitochondrial homeostasis. This dysfunction eventually leads to age and dose-dependent neurodegeneration in the nematode *Caenorhabditis elegans*. Notably, dopaminergic neurons have heightened susceptibility, but all of the neuronal classes eventually degenerate following exposure. Toxicity further extends to human SH-SY5Y cells, which also degenerate following exposure. Additionally, the neurons of nematodes expressing heterologous aggregation-prone proteins display enhanced metabolite vulnerability. These mechanistic analyses collectively reveal a unique metabolomic fingerprint for this bacterially-derived neurotoxin. In considering that epidemiological distinctions in locales influence the incidence of PD, we surveyed soils from diverse regions of Alabama, and found that exposure to ~30% of isolated *Streptomyces* species caused worm dopaminergic neurons to die. In addition to aging, one of the few established contributors to PD appears to be a rural lifestyle, where exposure to soil on a regular basis might increase the risk of interaction with bacteria producing such toxins. Taken together, these data suggest that a novel toxicant within the *Streptomyces* genus might represent an environmental contributor to the progressive neurodegeneration that is associated with PD.

Keywords: neurodegeneration; Parkinson's disease; *C. elegans*; *Streptomyces venezuelae*; natural product

1. Genetics of Parkinson's Disease

As the second most common neurodegenerative disorder, Parkinson's Disease (PD) is considered a disease of aging, since it primarily affects individuals over the age of 65. It is characterized by a progressive loss of dopaminergic (DA) neurons in the substantia nigra pars compacta and results in resting tremors, muscle rigidity, and impaired balance. Current treatments provide limited, and purely symptomatic, relief.

Several cellular processes have been associated with PD, including DA chemistry imbalances, abnormal vesicular trafficking, proteasome dysfunction, disrupted protein homeostasis, mitochondrial impairment, and impaired autophagy. While these are often examined as separate processes, it is likely that once a single inciting molecular event occurs, the pathogenic process encompasses overlapping molecular mechanisms. Advances have been made in understanding the processes that are associated with neurodegeneration through the analysis of gene mutations that are more commonly associated with autosomal-dominant [α-synuclein (PARK1/4) and LRRK2 (PARK8)] and recessive forms of familial PD [parkin (PARK2), PINK1 (PARK6), DJ-1 (PARK7), and ATP13A2 (PARK9)]. The gene products encoding many of these familial mutations have been implicated in cellular pathways involved in PD pathological disease mechanisms. However, even the most common gene mutation, LRRK2 (G2019S), is associated with <2% of identified PD cases [1]. In total, genetics account for ~10% of diagnosed PD, and it has become apparent that investigation into purely genetic factors will not elucidate all or even most PD incidence.

2. Environmental Factors Impacting Neurodegeneration

Recent research suggests that since familial forms of PD are fairly rare, environmental determinants may significantly contribute to the onset of neurodegenerative pathology. In addition to aging, one of the few established and reproducible epidemiological contributors to PD appears to be a rural lifestyle, where drinking well water, farming, and exposure to pesticides or herbicides may all be risk factors [2–8]. The herbicides (paraquat) and pesticides (rotenone) that are used in farming result in the formation of excessive reactive oxygen species (ROS). These both induce Parkinsonian phenotypes in animals [9], and rotenone also inhibits mitochondrial complex I [10]. Mitochondrial defects also are a common theme in PD pathogenesis. Mutations in the autosomal recessive familial genes, PINK1 and parkin, result in mitochondrial dysfunction [11]. However, it is important to note that the levels of herbicide and pesticide exposures that are often encountered in farming cannot completely account for the increased PD odds ratio for those living in rural areas [12]. With this information, we sought to identify an alternative environmental exposure that could partially account for the enhanced PD risk associated with rural living.

3. Soil Bacteria and Neurodegeneration

The gap in our understanding between environmental and inherited causes of PD remains long unresolved. Our ability to successfully reduce PD among susceptible individuals is dependent upon knowledge about factors that render certain populations at risk. The higher incidence of PD in rural areas, where the disease may actually be underreported due to health care disparities, remains a rare clue to a potential environmental contribution to the disease. Lifestyle and occupational distinctions among individuals from rural areas may present a more consistent exposure to terrestrial environmental factors that are simply less common in more developed lands and cities. For example, living on dirt floors, drinking well water, farming, and general interaction with soil microbes may represent a source of risk to certain individuals. Approximately one million distinct microbial species are estimated to comprise the approximately one billion microorganisms in a single gram of soil [13,14]. *Streptomyces*, a bacterial genus within the order Actinomycetales, is ubiquitously prevalent in soil samples (~6% of the microbial population) [15]. Notably, *Streptomyces* are renowned as a source of secondary metabolites; the genus includes over 70% of known antibiotics [16]. Therefore, we surmised that a putative source of the undefined environmental contributors to PD onset and progression could come from exposure to these common soil bacteria. We further hypothesized that exposure to a potentially neurotoxic compound of bacterial origin could be exacerbated by factors influencing genetic pre-susceptibility to neurodegeneration (or PD).

Proteasome inhibitors, many of which are isolated from Actinomycetes, have been shown to induce neurodegeneration in animal model systems. At least four characterized proteasome inhibitors are products of *Streptomycetes* isolated from soil, including lactacystin [17]. The selective loss of DA neurons after the systemic administration of epoxomicin, which is a naturally occurring proteasome inhibitor, or PSI ((Z-Ile-Glu(OtBu)-Ala-Leu-al), which is a synthetic proteasome inhibitor, to rodents was reported [18]. While promising, these data proved to be difficult to reproduce by other research groups [19–21]. Nevertheless, the reliable progressive degeneration of dopaminergic neurons was, in fact, achieved via the direct administration of either of these agents into rodent brains [22]. Another intriguing study demonstrated that mice injected with a pathogenic actinomycete that is found commonly in soil and water, *Nocardia asteroides*, developed symptoms that phenocopied PD; moreover, these infected animals responded positively to the administration of levodopa [23]. Although the relevant mechanism of action remains unknown, subsequent in vitro and in vivo studies showed that infection with GUH-2, a specific strain of *N. asteroides*, resulted in the apoptotic death of substantia nigra DA neurons [24]. Further studies with GUH-2 revealed that it was neuroinvasive in both mice and monkeys, and that their brain infections resulted in the apoptotic death of DA neurons due to proteasome inhibition [24–28]. While these data have heightened speculation that bacterial exposure/infection could be a potential risk factor for PD, progress in this area of research has been limited. Taken together, we decided to ask if common soil bacteria from the genus *Streptomyces* could produce neurodegenerative secondary metabolites.

4. Using *C. elegans* to Model Neurodegenerative Phenotypes

Our lab has focused on the application of the nematode, *C. elegans*, as model system whereby genetic or external factors influencing DA neuron survival can be rapidly evaluated [29]. Importantly, this model has proven to be predictive of downstream effects that have been observed in mammalian neurons, as well as genetic modifiers of neurodegeneration that have emerged in human genomic studies [30–35]. While evolutionarily distant from humans, *C. elegans* neurons retain many of the hallmarks of mammalian neuronal function. Among these, neuropeptides and neurotransmitters (dopamine, serotonin, GABA, glutamate, acetylcholine), as well as ion channel families, vesicular transporters, receptors, synaptic components, and axonal guidance molecules are highly conserved [36]. The *C. elegans* nervous system is comprised of exactly 302 neurons, eight of which produce DA (Figure 1A,B,D).

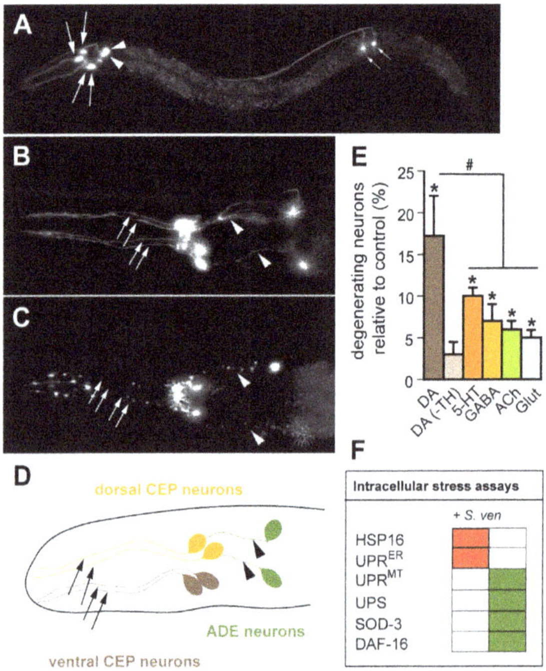

Figure 1. Exposure to the *S. venezuelae* metabolite causes dopaminergic neurodegeneration and intracellular stress in *C. elegans*. (**A**) Dopaminergic cell bodies and processes are illuminated using green fluorescent protein (GFP) driven from the dopaminergic (DA) transporter promoter (P_{dat-1}::GFP). The six anterior DA neurons that are shown on the left side of this image include two pairs of cephalic neurons (CEPs, large arrows indicating cell bodies) and one pair of anterior deirid neurons (ADEs, arrowheads indicating cell bodies). There is also one pair of posterior deirid neurons (PDEs, small arrows indicating cell bodies); (**B**) Close-up of the anterior region of *C. elegans* where the six most anterior DA neurons are highlighted by GFP (P_{dat-1}::GFP) with the four CEP neurons (arrows indicating neuronal processes) and two ADE neurons (arrowheads indicating neuronal processes) highlighted; (**C**) A worm expressing GFP in DA neurons displays neurodegenerative changes in all six anterior DA neurons following nine days of exposure to the *S. venezualae* (*S. ven*) metabolite; (**D**) Drawing of the *C. elegans* anterior DA neurons. Precisely six DA neurons in the anterior of *C. elegans* are found in pairs defined as two dorsal CEPs, two ventral CEPs, and two ADE neurons. The dorsal CEPs are post-synaptic to the ADEs, and are connected within this circuit, whereas the ventral CEPs and the ADEs do not display connectivity; (**E**) Separate neuronal subtypes within isogenic populations of worms were scored for neurodegeneration in animals where GFP was exclusively expressed to illuminate either the dopaminergic (DA) [+ and − tyrosine hydroxylase (TH) expression], serotonergic (5-HT), GABAergic (GABA), cholinergic (ACh), and glutamatergic (Glut) neuronal subclasses. All of the neuronal classes that were examined exhibited significant neurodegeneration following eight days of exposure to an *S. ven* conditioned medium, except animals wherein the DA neurons were devoid of TH through a genetic mutation (*cat-2*) (* $p < 0.01$; ANOVA). Significantly, the DA neurons displayed increased degeneration compared to all other neuronal classes (# $p < 0.05$; one-way ANOVA). The amount of neurodegeneration that was observed in animals exposed to an *E. coli* control conditioned medium was used as a baseline for standardization. To compensate for distinct neuronal classes containing different numbers of neurons, the percentage of total degenerating neurons (not worms with degeneration) was used for comparisons; (**F**) Intracellular stress response summary following exposure to *S. ven* metabolite. Gene reporters for stress assays tested are shown here. HSP16 is a homolog of the hsp16/hsp20/alphaB-crystallin family of heat shock proteins. The endoplasmic reticulum (ER) unfolded protein response (UPR^{ER}) was assessed by measuring P_{hsp-4}::GFP in *C. elegans*; HSP-4 is homologous to the mammalian ER chaperone, BiP. The mitochondrial unfolded protein response (UPR^{mt}) was measured via P_{hsp-6}::GFP; HSP-6 in *C. elegans* is a transcriptional reporter for mitochondrial stress and is a member of the DnaK/Hsp70 superfamily. The UPS assay examined a ubiquitin-related degradation signal (a "degron") fused to CFP (P_{dat-1}::CFP::CL-1). SOD-3 encodes mitochondrial superoxide dismutase. We examined oxidative stress using the transgenic line P_{sod-3}::GFP as an inducible assay system. The activity of DAF-16, which is homologous to the FOXO transcription factor, was monitored using a transgenic line, P_{daf-16}::DAF-16::GFP, where upregulation in response to metabolite was determined by the nuclear localization of DAF-16::GFP. For all of the assays described, upregulation is indicated with a green box, while a red box indicates no response to the metabolite using these reporter strains.

In considering the numerous advantageous attributes of *C. elegans* as a model, we employed this system to determine whether or not *Streptomyces* could cause neurodegeneration. This rapidly cultured organism (three days from egg to adult) has an experimentally accommodating lifespan (two to three weeks), and is well-suited to studies that are designed to take more exploratory concepts to mechanistic fruition rapidly and inexpensively. *C. elegans* has also been used for toxicology studies on a variety of agents that are relevant to PD, including heavy metals [37,38], pharmaceuticals [39–41], and ROS-inducing chemicals [42,43].

Worms eat bacteria. Indeed, *E. coli* is the standard food source that is used to maintain *C. elegans* cultures in research laboratories. Thus, we examined three common soil bacteria from the genus *Streptomyces* (*S. venezuelae, S. griseus,* and *S. coelicolor*). We initially attempted the direct exposure of *Streptomyces* spp. to the nematodes through feeding. Unfortunately, the worms displayed substantial aversion behavior in response to the *Streptomyces* spp. This was not too surprising, as *C. elegans* display chemosensory avoidance of unfamiliar bacteria [44]. We subsequently grew *Streptomyces* spp. (or *E. coli*, as a control) in liquid cultures to saturation and tested the resulting conditioned media for *C. elegans* DA neurodegeneration [45]. Actinomycete metabolites are typically produced during the stationary phase; therefore, *Streptomyces* spp. were sporulated and grown for two weeks in SYZ (starch, yeast extract, NZ amine) media, which is commonly used for metabolite production. Following filtration, the conditioned medium was incorporated into the nematode growth media, and animals were transferred to fresh petri dishes every two days. Strikingly, the progressive degeneration of DA neurons in worm populations was observed following exposure to *S. venezuelae* (*S. ven*) conditioned media, but not following exposure to *E. coli*, *S. griseus*, or *S. coelicolor* conditioned media (Figure 1B,C).

A notable attribute of *C. elegans* includes the ability to quantify the precise cellular complement of specific neuronal classes in genetically invariant populations. Therefore, this facilitated our capacity to evaluate sensitivity to the *S. ven* metabolite among four other neuronal subclasses, including serotonergic (5-HT), GABAergic (GABA), cholinergic (ACh), and glutamatergic (Glut) neurons [45]. Strains expressing green fluorescent protein (GFP) exclusively within these defined neuronal subclasses were scored over the course of time, specifically at four, six, and eight days of age. While DA neurons showed significant degenerative changes after four days of exposure to *S. ven* conditioned media, other neuronal classes did not exhibit degenerative changes until eight days of exposure and even at this time point, significantly more DA neurons were degenerated compared with other neuronal classes (Figure 1E). Since DA neurons exhibited enhanced vulnerability compared to other neuronal classes, we hypothesized that the presence of DA might enhance the neurodegeneration that is associated with exposure to the *S. ven* metabolite. To examine this, we exposed *cat-2* mutant worms [46] to the conditioned medium. *cat-2* worms express reduced levels of tyrosine hydroxylase (TH), the rate-limiting enzyme in the production of DA, and as a result, they contain only 40% of normal DA levels [47]. After six and eight days of exposure, *cat-2* worms displayed significantly less degeneration than wild-type worms (Figure 1E) [45]. Therefore, the presence of L-DOPA or DA might provide a sensitized cellular milieu for the *S. ven* metabolite that exacerbates neurodegeneration.

To examine if the *S. ven* metabolite causes degeneration in human cells, we used SH-SY5Y neuroblastoma cells, which is a line that is widely used in cellular models of PD. The cells were exposed to *S. ven* medium for 48 h, and cell viability was measured by the release of the intracellular enzyme, lactate dehydrogenase (LDH), and compared to the amount of LDH released by exposure to *S. coelicolor* conditioned media (which was negative for neurodegeneration in the *C. elegans* assays). Notably, dose-related toxicity was observed for *S. ven* conditioned media, but not for *S. coelicolor* [45].

We proceeded to characterize several mechanisms that are known to have a causal relationship with intracellular stress and/or neurodegeneration using *C. elegans* transgenic strains. As described in Figure 1F, we learned that the metabolite does not elicit a generalized cytoplasmic chaperone response via the hsp16/hsp20.alphaB-crystallin family of heat shock proteins [45]. Additionally, the metabolite does not upregulate the endoplasmic reticulum unfolded protein response (UPR^{ER}) pathway [45]. However, it does trigger the mitochondrial unfolded protein response (UPR^{MT}) pathway, and it blocks ubiquitin-related degron degradation, indicating that the ubiquitin proteasome system (UPS) is functionally impaired following metabolite exposure [43,45].

5. Gene-by-Environment (GxE) Interactions Modeled in *C. elegans*

PD-associated dopaminergic neuropathology is characterized by the accumulation of α-synuclein (α-syn) in Lewy bodies. Significantly, α-syn itself can induce neurodegeneration when overexpressed or mutated [48]. The *C. elegans* genome encodes homologs of all of the familial parkinsonism genes except α-syn, which, as an autosomal dominant modulator of PD, can be overexpressed to recapitulate time-dependent degenerative pathology in vivo, including progressive DA neuron loss (Figure 2A,D) and the accumulation of misfolded protein. We were interested in determining if exposure to *S. ven* in α-syn expressing *C. elegans* would result in enhanced neurodegeneration, because there is precedent for this in the literature. For example, paraquat increases α-syn expression in mice [49]. Similarly, paraquat and rotenone also accelerate the formation of α-syn inclusions, causing a conformational change to α-syn itself, which in turn effects fibril formation [50].

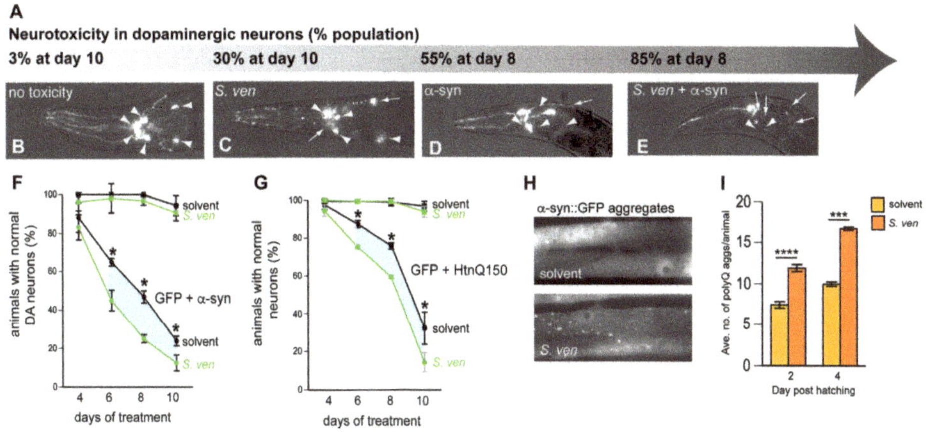

Figure 2. Gene-by-environment interactions exacerbate *C. elegans* phenotypes. (**A**–**E**) Genetic or environmental factors can be rapidly quantified for degenerative phenotypes by examining the six anterior dopaminergic neurons of *C. elegans*. Since the majority of cases of sporadic Parkinson's disease (PD) are idiopathic, undefined factors from the environment and innate genetics that sensitize individuals to PD, or combinations thereof, could lower the threshold for neurodegeneration. Using *C. elegans*, the impact of both genetic and environmental exposure can be evaluated for neurotoxicity in the isogenic lines of animals for additive or synergistic effects, depending on the dosage or age of animals. (**A**) Animals displaying dopaminergic neurodegeneration can be modulated based on a causative factor; (**B**) Normal *C. elegans* rarely show dopaminergic neurodegeneration, even as animals reach old age (Day 10). Arrowheads indicate intact DA neurons; (**C**) The addition of the *S. ven* metabolite induces the age-dependent accumulation of degenerative phenotypes (arrows) and can be visualized as neuronal loss and blebbing; *S. ven* metabolite treatment alone results in ~30% neurodegeneration in contrast to animals overexpressing α-syn in the absence of *S. ven* exposure; (**D**) *S. ven* exposure in combination with a PD genetic factor, human α-synuclein (α-syn) overexpression is additive; (**E**) Whereby ~85% of the population displays neurodegeneration. Exposure to other reactive oxygen species (ROS)-inducing chemicals (i.e., 6-OHDA, rotenone, paraquat) can produce similar phenotypes; (**F**,**G**) *S. ven* exposure has been examined in combination with genetic susceptibility factors to uncover gene and environment interactions. In these scenarios, *C. elegans* with and without the expression of different heterologous aggregation-prone proteins were examined for neurotoxicity following exposure to metabolites. It should be noted that the concentration of metabolites that was used was much lower, since we wanted to ensure that, on its own, metabolite would not cause neurodegeneration, yet would reveal potential neurodegeneration in combination with α-syn [51]. The results shown here displayed this concept with either α-syn expressed in dopaminergic neurons; (**F**) or mutant huntingtin (Htn_{Q150}) expressed in the *C. elegans* ASH-sensory neuron; (**G**) As shown; (**H**) The metabolite also induces α-syn-dependent proteostasis disruption in the readily visualized body wall muscle cells that express α-syn::GFP under control of a body wall muscle promoter ($P_{unc\text{-}54}$), and were treated with *S. ven* metabolite continuously since hatching; (**I**) *C. elegans* expressing a nucleotide repeat encoding 35 polyglutatmines (Q35) in body wall muscles ($P_{unc\text{-}54}$::polyQ35::GFP) that were exposed to the metabolite display an increase in aggregate number compared to solvent [51].

Since both *S. ven* and α-syn can cause DA neurodegeneration independently, we wanted to establish chronic supplementation conditions using lower metabolite dosages in *C. elegans* with a GFP (only) marker in DA neurons so that it would no longer elicit a neurodegenerative response (Figure 2A,F). Using this sub-toxic dosing regimen, we uncovered a GxE interaction in *C. elegans* expressing α-syn in DA neurons whereby the percentage of animals with normal DA neurons was significantly decreased following six to 10 days of exposure [51] (Figure 2A–F). At day four, neurodegeneration was not enhanced, suggesting that the accumulation of α-syn is not extensive enough to manifest this neurotoxic phenotype. We performed analogous experiments with two other neuronally-expressed pathogenic proteins in vivo. One of these models expressed Aβ in the glutamatergic neurons, while the other expressed mutant huntingtin (Htn-Q_{150}) in the ASH-type sensory neuron. We found that treatment with the *S. ven* metabolite similarly enhanced neurotoxicity at lower dosages (Figure 2G). These pathogenic proteins served as surrogate indictor markers of disease progression, where the dysregulation of normal homeostatic pathway function accumulates over time [52,53]. In this regard, in our three models of protein misfolding, neurodegenerative phenotypes were not observed in young animals in the absence of pathogenic protein expression, suggesting that the metabolite might synergize with threshold state animals [51]. Notably, GxE effects were not limited to misfolded proteins, as the *S. ven* metabolite similarly enhanced the toxicity of another autosomal dominant form of PD that we modeled in *C. elegans*. The LRRK2(G2019S) mutation has been shown to decrease both mitochondrial membrane potential and ATP production, resulting in neuronal toxicity [54,55]. We overexpressed LRRK2(G2019S) in DA neurons and found that *S. ven* significantly enhanced neuronal degeneration compared to solvent control (from 20% to 43% of the population with degenerating DA neurons when exposed to *S. ven* metabolite [43]). This GxE interaction was similar to a study in Drosophila where the overexpression of mutant LRRK2 (G2019S or G2385R), in combination with rotenone, caused an increase in neurodegeneration [56].

6. A Metabolic Fingerprint is Revealed in Response to the *S. ven* Metabolite

We wanted to assess if the metabolite-induced enhancement of neurodegeneration was correlated with alterations in protein handling. Therefore, we monitored established *C. elegans* muscle models of the overexpression of α-syn, Aβ, or polyglutamine for changes in aggregate density and/or for behavioral phenotypes following exposure to the metabolite. With all three pathogenic proteins, exposure to the *S. ven* metabolite induced phenotypic changes (Figure 2H,I, [51]).

We also have evidence that *S. ven* metabolite exposure increases reactive oxygen species (ROS) in *C. elegans* lysates using both in vivo assays and an ex vivo DCF assay from whole animal extracts [43]. An additional study examined worms expressing an oxidative stress-inducible reporter that is known to be upregulated against endogenous ROS, *sod-3*::GFP. ROS was significantly increased in these worms (Figure 1F) [43]. Since the upregulation of SOD-3 is associated with defense against oxidative stress, we treated animals expressing α-syn in either DA neurons or muscle cells with antioxidants to determine if the protein mishandling we observed was associated with oxidative damage. The four antioxidants that we tested did not attenuate phenotypes in either muscle cells or DA neurons; however, glutathione (GSH) was the only that antioxidant that suppressed both α-syn aggregate formation in muscle cells and attenuated DA neurodegeneration [51]. Through a systematic series of pharmacological and genetic studies, we determined that ubiquitin proteasome system (UPS)-linked protein homeostasis defects may result from *S. ven* exposure and that glutathione homeostasis is a regulator of metabolite-induced proteotoxicity [51].

Additional studies are required to mechanistically associate the glutathione metabolic changes with the alterations in proteostasis that we observed following *S. ven* exposure. For example, it is hypothesized that GSH attenuation occurs through the repair of damaged cysteine residues [57]. Alternatively, the glutathione couple, GSH/GSSG, might operate as a surveillance mechanism within the cellular proteome to identify redox changes within the thiol–disulfide balance of the cell [58]. Regardless of which hypothesis is correct, we suggest that shifting the GSH couple to a more reduced state might beneficially alter the neurodegenerative threshold state of *C. elegans* cells [59].

We also hypothesized that it was possible that the UPS perturbations and ROS induction we identified from metabolite exposure might be the result of modulating the PINK1 and/or parkin pathways. These two gene products are associated with autosomal recessive forms of PD and have been identified to regulate both protein and mitochondrial homeostasis, as well as autophagy, in many organisms [11,60–64]. We first depleted *pdr-1* (the *C. elegans* homolog of parkin) cell autonomously in α-syn-expressing DA neurons. Animals were treated with *S. ven* metabolite and/or the proteasome inhibitor MG132. A combination of all three stressors (α-syn, metabolite, and MG132), along with *pdr-1*(RNAi) resulted in a more severe DA neurodegenerative phenotype than any two stressors alone (Figure 3A). In contrast, when we performed a comparable experiment in α-syn animals where *pink-1* was knocked down by RNAi, increased neurodegeneration was not observed; instead, these stressors behaved in a similar manner (Figure 3B). Therefore, these data suggest that proteasome inhibition and metabolite-induced protein misfolding are epistatically regulated via *pink-1*, but are not regulated in this manner when considering *pdr-1* [51].

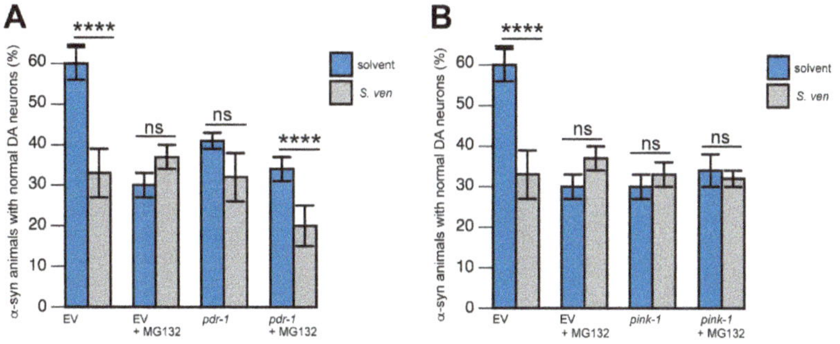

Figure 3. Epistatic regulation of enhanced α-synuclein toxicity by *pink-1*. (**A**,**B**) Combinations of dopaminergic stressors (such as MG132 and/or *Streptomyces venezuelae*, or *S. ven*) were applied to transgenic *C. elegans*, which was also expressed human α-syn in DA neurons. (**A**) Along with knockdown of *pdr-1* (RNAi), a more severe DA neurodegenerative phenotype was observed if compared to combinations of just two stressors at a time; (**B**) In animals where *pink-1*(RNAi) was knocked down, enhanced neurodegeneration was not observed. Instead, all of the combinations of stressors yielded similar levels of neurodegeneration with the depletion of *pink-1*, which was thereby indicative of a common mechanism revealed by putative epistatic relationships.

To further understand the mechanisms that elicit an oxidative stress response following *S. ven* metabolite treatment, we asked if the FOXO transcription factor protein, DAF-16, which is directly inhibited by the insulin signaling pathway, translocated to the nucleus following metabolite exposure. Notably, DAF-16 was translocated to the nucleus (Figure 1F) [43]; this was similar to what had been reported for paraquat exposure [65]. It is known that the nuclear accumulation of DAF-16 correlates with increased ROS. In response to this stress, the activation of genetic factors associated with pathogen defense, mitochondrial stress mechanisms, and/or cell death pathways occurs [66–69]. In this regard, we are interested in determining the targets of DAF-16 that are upregulated by metabolite exposure.

These mechanistic studies collectively reveal an emerging metabolomic fingerprint in response to the *S. ven* metabolite that share features with known toxins and the other metabolic effectors that are associated with PD, yet appear distinct in its emerging signature. As detailed below, a major component of this metabolic profile involves altered mitochondrial homeostasis.

7. The *S. ven* Metabolite Causes Mitochondrial Dysfunction

As previously described, following metabolite exposure the cytoprotective DAF-16 transcription factor accumulates in the nucleus and *sod-3* expression levels significantly increase. Since dysfunctional mitochondria can result in ROS and misfolded protein accumulation, we decided to explore whether a stress-response pathway referred to as the mitochondrial unfolded protein response (UPR^{mt}) was triggered in response to *S. ven* metabolite exposure. The UPR^{mt} activates the transcription of mitochondrial chaperone genes to promote protein homeostasis. In *C. elegans*, *S. ven* metabolite exposure resulted in a significant upregulation of the UPR^{mt}, as assayed via monitoring a nuclear-encoded mitochondrial chaperone, *hsp-6* (Figure 1F) [43]. These data are suggestive of a disturbance of mitochondrial homeostasis, especially considering that the *S. ven* metabolite does not activate the unfolded protein response signaling pathways (UPR) in the cytosol or ER (Figure 1F) [45].

The mitochondrial protein-folding environment is sensitive to alterations in organelle structure, the excess production of free radicals, and/or the improper function of the electron transport chain [70]. Therefore, while we had determined that the metabolite increased ROS, it was important to further characterize the mitochondrial phenotype associated with *S. ven* exposure. We also examined ATP levels by using an ex vivo luciferase assay with *C. elegans* extracts [43]. We determined that worms exposed to the *S. ven* metabolite displayed significantly lower overall levels of ATP compared to the solvent control; these data indicate that metabolite exposure caused the impairment of mitochondrial function. Since the structure and bioenergetics of mammalian and nematode mitochondrial respiratory chains are very similar [71], we further investigated whether the metabolite inhibited mitochondrial complex I in a manner similar to rotenone, an environmental toxin that also causes DA neurodegeneration in *C. elegans* [10,72]. Furthermore, we elected to evaluate the activity of two specific compounds for their capacity to protect against the DA neurodegeneration that was caused by *S. ven* exposure. In this regard, we examined the treatment of *C. elegans* with riboflavin, which is a mitochondrial complex I (NADH dehydrogenase) activator, and D-beta-hydroxybutyrate (DβHB), which is an activator of mitochondrial complex II (succinate dehydrogenase) that can rescue complex I deficiencies via a mechanism dependent on complex II function [73,74]. Both riboflavin and DβHB treatment significantly rescued *S. ven* neurotoxicity [43].

We also wanted to determine if the metabolite altered mitochondrial membrane potential ($\Delta\Psi_m$); therefore, we exposed worms to *S. ven*, and then measured the relative mitochondrial uptake of the fluorescent dye tetramethylrhodamine ethyl ester (TMRE), which accumulates in active mitochondria. Live animals exposed to metabolite displayed significant decreases in TMRE fluorescence, thus demonstrating that the metabolite is associated with $\Delta\Psi_m$ collapse.

8. The *S. ven* Metabolite Disrupts Mitochondrial Homeostasis

Since abnormal mitochondrial fission/fusion is associated with $\Delta\Psi_m$ collapse [75,76], we decided to explore whether the metabolite altered the regular fission/fusion cycles that are regulated by the GTPases (Drp1 and Opa1) and mitofusins (Mfn1/Mfn2) that are located on the outer and inner mitochondrial membranes. Furthermore, it is well-established that other environmental toxins alter mitochondrial homeostasis [77–81]. We determined that the *S. ven* metabolite increases mitochondrial fragmentation/fission, as visualized in the large body wall muscle cells of *C. elegans*, in a time-dependent manner, whereby animals are more sensitive to the metabolite as they age (Figure 4A,B). Additionally, there is a concomitant decrease in *fzo-1* gene expression and an increase in *drp-1* gene expression (outer mitochondrial fusion and fission genes, respectively) (Figure 4C). Comparable with results from studies that have been conducted using human cells in culture, our worm data revealed that mitochondrial fragmentation resulted from mitochondrial oxidative stress due to an imbalance in Mfn2 and Drp1 activity [82]. In *C. elegans*, *drp-1* and *fzo-1*, the outer mitochondrial membrane genes have changes in expression following treatment with the *S. ven* metabolite.

We proceeded to compare the gene expression profiles obtained from *S. ven* exposure to those obtained from other environmental PD toxins. While *S. ven* enhances mitochondrial fragmentation, rotenone decreases fission and induces fusion with an associated decrease in *Drp1* gene expression [77,78,81]. Conversely, MPP^+ increases mitochondrial fragmentation and increases Drp1 protein levels, which is similar to *S. ven* [79,80]. However, when human *Drp1* is genetically inactivated in SH-SY5Y cells, this blocks MPP^+ mitochondrial fragmentation [79]. Whereas in an analogous experiment in *C. elegans*, when *drp-1* is knocked down, fragmentation still occurs following treatment with *S. ven* [83]. Thus, the mitotoxic mechanisms of action are clearly distinguishable among these substances, even though all three inhibit mitochondrial complex I.

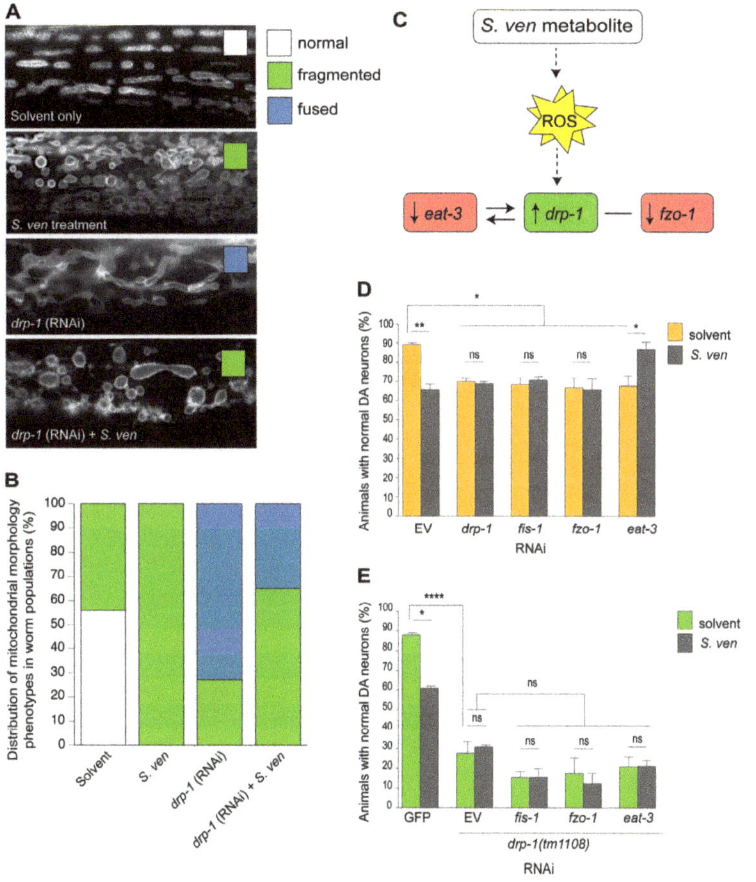

Figure 4. *S. ven* metabolite increases mitochondrial fragmentation. The impact of the *S. ven* metabolite on the mitochondrial outer membrane in *C. elegans* body wall muscle cells in animals expressing an outer mitochondrial protein targeted RFP (P_{myo3}::TOM20::mRFP). (**A**) Compared with solvent treatment, *S. ven* metabolite-treated animals exhibit significantly more mitochondrial fragmentation following exposure (green box). RNAi knockdown of *drp-1*, an outer mitochondrial membrane fission gene, reveals significantly increased fusion in these animals (blue box). Notably, in *drp-1* (RNAi) worms treated with *S. ven*, many animals return to a fragmented mitochondrial phenotype. Mitochondrial morphology is defined as normal (tubular—white box), fused (elongated—blue), or fragmented (circular and irregular—green); (**B**) Quantitation of mitochondrial morphology phenotypes in *C. elegans* populations. The distribution of fragmented mitochondria is different between all of the samples. In *S. ven* and *drp-1* RNAi + *S. ven* treated populations, increased fragmentation is indicative of damaged mitochondria that cannot be turned over by mitophagy [83]. The color scheme (white, blue, green) is the same as shown in (**A**); (**C**) Schematic representation of qPCR data showing that *S. ven* metabolite exposure in *C. elegans* leads to increased *drp-1* gene expression, as well as lowered *fzo-1* and *eat-3* gene expression [83]. DRP-1 is an outer mitochondrial membrane fission protein, while FZO-1 and EAT-3 are fusion proteins that are located at the outer and inner mitochondrial membranes, respectively; (**D**) *eat-3*(RNAi) knockdown suppresses dopaminergic neurotoxicity caused by the metabolite. Notice that the knockdown of all of the other mitochondrial fission and fusion genes still causes toxicity in the presence of metabolite, except for *eat-3* (RNAi), which now exhibits neuroprotection; (**E**) The RNAi sensitive strain used in part D (above) was crossed to *drp-1* loss-of-function mutant (allele *tm1108*) animals. In this background, all of the fission and fusion genes that were examined display enhanced sensitivity to neurodegeneration, notably, *eat-3* (RNAi*); drp-1(tm1108)* no longer shows resistance in the presence of the metabolite.

Since metabolite exposure resulted in pronounced mitochondrial morphological changes, we also examined how modulating fission/fusion impacted DA neurodegeneration. We utilized a RNAi strain that allows for selective RNAi knockdown exclusively in DA neurons to examine mitochondrial fission/fusion components following the exposure of the metabolite [84]. There was significant DA neurodegeneration following RNAi depletion for fission (*drp-1* and *fis-1*) and fusion (*fzo-1* and *eat-3*) genes when compared to solvent-only empty vector (EV) control (Figure 4D). Following the addition of metabolite, neurodegeneration was enhanced in EV, but no further degeneration was observed in *drp-1*, *fis-1*, or *fzo-1* RNAi knockdown (Figure 4D) [83]. In contrast, the OPA1 homolog, *eat-3*(RNAi) and metabolite revealed a resistance to DA neurotoxicity (Figure 4D). The *eat-3* data seemed curious until we found a previous publication reporting that *drp-1*, an outer mitochondrial fission component, can act genetically upstream of *eat-3*, which is an inner mitochondrial fusion component [85]. The report further suggested that there is mutual compensation for physiological defects. In this regard, we considered our own data where *S. ven* metabolite increases *drp-1* gene expression levels (Figure 4C) and hypothesized that we could reverse the resistance to DA neurodegeneration that occurs in an *eat-3* background by depleting both *eat-3* and *drp-1* in the same animals. Therefore, using our DA neuron-selective RNAi strain to test the putative interaction between *eat-3* and metabolite-induced *drp-1*, the effect of *eat-3* (RNAi) on DA neurodegeneration in a *drp-1 (tm1108)* null mutant background was evaluated. We observed no neuroprotection against the *S. ven* metabolite in the *drp-1* null mutant background. Thus, we surmised that there is an epistatic regulatory relationship between the metabolite-induced *drp-1* activity depletion of *eat-3* [83] Figure 4E). When *fis-1* or *fzo-1* were knocked down in the *drp-1* mutant background, further neurodegeneration did not occur (Figure 4E). The interdependence between *drp-1* and *eat-3* was further confirmed with qPCR (data not shown; Figure 4C; [83]).

9. PINK-1/DRP-1-Dependent Fission Induced by *S. ven* Metabolite

PINK1 and parkin often function together to promote DRP-1-dependent mitochondrial fission [86,87]. In exploring this functional relationship in the context of the *S. ven* metabolite, we postulated that if metabolite-induced fission is independent of PINK-1 activity, it would suppress DA neurodegeneration in *drp-1* mutants. Conversely, when *pink-1* was depleted, we predicted that treatment with the *S. ven* metabolite would not enhance neurodegeneration if *drp-1* was dependent on *pink-1* function. Indeed, we discerned that the DA neurodegeneration that was caused by metabolite exposure was not further enhanced by *dpr-1* RNAi in the absence of *pink-1*. Therefore, DRP-1 activity in mitochondrial fission appears to be required for metabolite-induced DA neurodegeneration in a *pink-1* mutant background. We also discerned that the reduction of *eat-3* (RNAi) significantly suppressed *pink-1*-induced neurodegeneration with or without metabolite exposure. Similar results were obtained when examining a *pdr-1* loss-of-function mutant. These data indicate that downregulation of *eat-3* is neuroprotective in *pink-1* or *pdr-1* mutant conditions [83].

Following mitochondrial dysfunction, an increase in DRP-1 activity can occur through AMP-activated protein kinase (AMPK), which is a key regulator of energy metabolism [88,89]. AMPK can be activated by rotenone and antimycin A and can promote mitochondrial division [88]. In this regard, we wanted to determine if the *S. ven* metabolite was also associated with this type of response. We performed both genetic and pharmacological studies, and assayed DA neurodegeneration in *C. elegans.* AMPK suppressed metabolite-induced DA neurotoxicity in N2 wildtype, *eat-3* (RNAi), and *pink-1; drp-1*(RNAi) animals. From these results we concluded that AMPK plays a mechanistic role in *S. ven* metabolite-induced DA neurodegeneration, although more research will be required to determine how AMPK activity is modulated [83].

10. Toward the Identification of the Neurotoxic Metabolite

In most of our previously described assays, the *S. ven* metabolite was partitioned sequentially through DCM, water, and chloroform to provide an enriched form, which we know from thin-layer chromatography and bioassay testing contains six fractions, two of which cause neurodegeneration.

This neutral-lipid fraction is dried down, resuspended in ethyl acetate, and tested for DA neuronal death in *C. elegans* assays. When the activity is confirmed, a concentration that is appropriate for neuronal death equivalents/mL of metabolite is established. The metabolite is then incorporated into Petri dishes for use in culturing/exposure to *C. elegans*.

Our goal is to identify the chemical structure of the neurotoxic molecule and then chemically synthesize it for long-term experimental use. Fractionation-guided purification was performed, starting from spent *S. ven* media, whereby it was subjected to reverse-phase semi-preparative HPLC. The resulting fractions were analyzed using the *C. elegans* DA neurodegeneration assay, and one fraction was found to be significantly active. High-resolution electrospray ionization mass spectrometry (HRESIMS) and nuclear magnetic resonance (NMR) analysis permitted elucidation of the structure, which is novel. Current efforts are underway for compound identification via proof-of-structure by chemical synthesis. The authors are not comfortable prematurely revealing structural information until this "gold standard" for biomolecule identity has been achieved.

11. Exposure to Other Soil *Streptomyces* Species also Causes Neurodegeneration

All of the research described within this review was performed with *Streptomyces venezuelae*, which is an American Type Culture Collection isolate. However, given the ubiquitous distribution of *Streptomyces* species (spp.) within the soil, either chronic or acute exposure to this metabolite could represent a previously unreported contributor to the onset of neurodegeneration, and may be exacerbated by factors influencing genetic pre-susceptibility to neurodegeneration. Therefore, we proceeded to conduct a regional microbial ecology survey of 1200 natural *Streptomyces* spp. from diverse land uses in the state of Alabama; 180 of these soil samples grew in laboratory conditions (Figure 5). From these, we learned that 51 of the species (28%) caused significant DA neuron death in *C. elegans* [90].

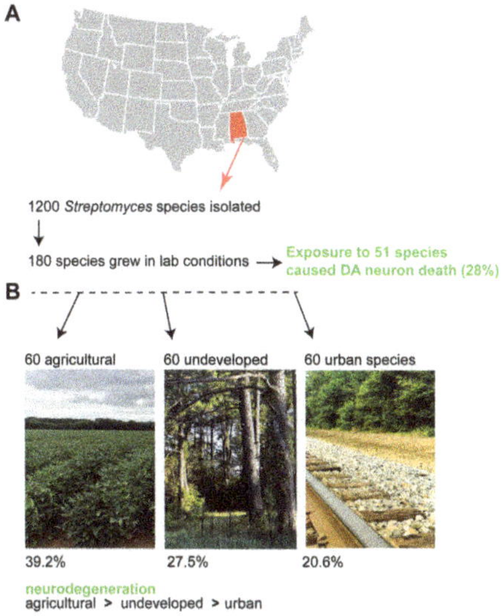

Figure 5. *Streptomyces* isolates from the environment cause DA neuron death in *C. elegans*. (**A**) We isolated 1200 *Streptomyces* from diverse regions of Alabama, of which 180 samples were grown in laboratory conditions [90]. Of these, 51/180 resulted in significant DA neuron death (28%); (**B**) These same 180 samples that grew in the lab represented isolates from three diverse locations: agricultural soils, urban soils, and pristine/undeveloped areas. We examined whether rates of neurodegeneration varied according to sample location, and learned that there were significant differences, with all three soil types exhibiting significant differences in neurodegeneration from each other. Agricultural soils caused the most degeneration, and urban soils caused the least amount of neurodegeneration [90]. Exemplar images of the areas that were sampled are displayed here.

The species that grew in lab conditions could be further subdivided into land-use patterns (agricultural, undeveloped, or urban soils; Figure 5). Notably, there was significant differences in neurodegeneration among all three soil types, with 39.2% of agricultural *Streptomyces* spp., 27.5% of undeveloped *Streptomyces* spp., and only 20.6% of urban spp. causing neurodegeneration [90]. These data suggest that there could be a common environmental toxicant(s) within the *Streptomyces* genus that causes neurotoxicity. In this regard, it is common for multiple species within a bacterial genus to produce related metabolites [91,92]. Understanding how distinctions in microbial ecology might intersect with the socio-economic disparities that impact health with respect to neurodegenerative diseases of aging represents a major unmet medical challenge of our time.

12. Summary and Future Studies

We have described the phenotypic consequences in *C. elegans* that are associated with exposure to bacterial lipophilic and amphipathic secondary metabolites by chemically extracting post-log bacteria cultures of the soil bacterium *S. venezuelae*. In a broad sense, *S. ven* exposure in *C. elegans* mirrors some of the pathological hallmarks of idiopathic PD, including ubiquitin proteasome system (UPS) disruption, glutathione homeostasis perturbation, general perturbation of proteostasis, mitochondrial dysfunction, and mitophagic alteration [93–95]. These observations are suggestive of the following two points about *S. ven* exposure.

First, *S. ven* metabolite toxicity mimics idiopathic PD in a way that is reminiscent to other environmental compounds that also cause stress in broad, diverse, cellular pathways. The cellular response to *S. ven* exemplifies the concept whereby genetic pathways function as interactome networks [96,97]. In these networks, distant genetic components (UPS, mitochondrial dysfunction, etc.) are eventually directed into large, central, organizing pathways (for example, mitophagy). Thus, peripheral pathway dysfunction will eventually lead to central pathway dysfunction, and widespread cellular failure will occur. As such, the pathology of idiopathic PD may arise from the collapse of interconnected pathways with time.

In the future, we plan to perform a metabolomic profile analysis in *C. elegans* to identify the small molecules that influence the cellular dysfunction associated with *S. ven* exposure. Our interest in profiling the metabolome is a direct result of data showing that glutathione directly modulates α-syn-induced neurodegeneration and misfolding. Metabolomic profiling has successfully identified metabolic differences in *C. elegans* through studies examining natural variation in populations, as well as in an analysis of longevity and another of transgenic amyloid-beta expression [98–100]. We also intend to profile the transcriptome following exposure to the *S. ven* metabolite. As described in Section 6, *C. elegans* DAF-16 (the FOXO transcription factor) is translocated to the nucleus following metabolite exposure [43]. Based on our prior results, we predict that the transcriptome will include gene products that are modulated by DAF-16 and/or the UPR^{mt} in addition to illuminating previously unattained regulators. After we obtain data from both the metabolomic and transcriptomic platforms, we plan to superimpose these data in a cross-omics approach, as is often performed to identify network interactions [98]. This dual analysis will allow us to capture changes occurring at two regulatory levels and pinpoint common pathways of cellular dysfunction that occur following *S. ven* exposure. We also have an interest in further exploring metabolite-induced cellular dysfunction as it pertains to mitochondrial biogenesis. Here, we will explore AMPK activity as a means to modulate *drp-1* and *eat-3* gene transcription levels and examine their impact on mitochondrial activity and dysfunction.

Second, the environment is replete with damaging toxins from either bacteria or other natural sources. For example, as we have shown from a recent collaborative study in our lab, many other *Streptomyces* spp. secondary metabolite products display neurodegenerative potential [90]. Prokaryotic organisms represent an overwhelming majority of life on earth [101]. Despite the enormous diversity of bacteria, their relationship with Eukarya is not yet well described. Therefore, their native environments, and biochemical and cellular interactions, may well provide us with a wealth of future information. Thus, it is intriguing to consider the secondary metabolites that have been excreted from these soil bacteria as a potentially large source of environmental stress. Future studies could include examining various environmental conditions, such as rotenone or paraquat exposure, on soil *Streptomyces* spp. for their impact on toxin production.

In summary, the *Streptomyces venezuelae* metabolite causes cellular stress in a manner that is similar yet distinct from other PD environmental toxins. It is this unique signature, and prevalence of this genus within the environment, that makes this metabolite an intriguing molecule for further investigation. It is tempting to envision chronic exposure to such a factor as an unforeseen environmental component impacting long-term neuronal survival during the human aging process. Given our increasingly aging global population, the potential ramifications of identifying a causal, or even contributory, environmental factor for neurodegeneration is substantial. This information would open a door toward the identification of factors controlling disease susceptibility, which could be used to gauge or reduce environmental risk and serve to accelerate the development of novel treatment strategies.

Author Contributions: Conceptualization, K.A.C. and G.A.C.; Writing—Original draft preparation, K.A.C.; Resources, J.L.T.; Writing—review and editing, J.L.T. and G.A.C.

Funding: The research described in this review article was primarily funded by a National Institutes of Health grant [R21ES01422] to G.A.C. and a National Institutes of Health grant [R15NS074197] to K.A.C.

Acknowledgments: Special thanks to Hanna Kim, Bryan Martinez, Julie Olson, Anna Watkins, and Arpita Ray for their contributions to this research. The authors wish to respectfully acknowledge this review does not strive to be inclusive of all the fine neurotoxicological research that has been reported using *C. elegans* models but is intended to focus and summarize on our own published research in this area, as requested by the Editor.

Conflicts of Interest: The authors declare no conflicts of interest.

References

1. Goldwurm, S.; Di Fonzo, A.; Simons, E.J.; Rohé, C.F.; Zini, M.; Canesi, M.; Tesei, S.; Zecchinelli, A.; Antonini, A.; Mariani, C.; et al. The G6055A (G2019S) mutation in LRRK2 is frequent in both early and late onset Parkinson's disease and originates from a common ancestor. *J. Med. Genet.* **2005**, *42*, e65. [CrossRef] [PubMed]
2. Tanner, C.M. Occupational and environmental causes of parkinsonism. *Occup. Med.* **1992**, *7*, 503–513. [PubMed]
3. Liou, H.H.; Tsai, M.C.; Chen, C.J.; Jeng, J.S.; Chang, Y.C.; Chen, S.Y.; Chen, R.C. Environmental risk factors and Parkinson's disease: A case-control study in Taiwan. *Neurology* **1997**, *48*, 1583–1588. [CrossRef] [PubMed]
4. Priyadarshi, A.; Khuder, S.A.; Schaub, E.A.; Priyadarshi, S.S. Environmental risk factors and Parkinson's disease: A metaanalysis. *Environ. Res.* **2001**, *86*, 122–127. [CrossRef] [PubMed]
5. Costello, S.; Cockburn, M.; Bronstein, J.; Zhang, X.; Ritz, B. Parkinson's disease and residential exposure to maneb and paraquat from applications in the central valley of California. *Am. J. Epidemiol.* **2009**, *169*, 919–926. [CrossRef] [PubMed]
6. Gatto, N.M.; Cockburn, M.; Bronstein, J.; Manthripragada, A.D.; Ritz, B. Well-water consumption and Parkinson's disease in rural California. *Environ. Health Perspect.* **2009**, *117*, 1912–1918. [CrossRef] [PubMed]
7. Tanner, C.M.; Ross, G.W.; Jewell, S.A.; Hauser, R.A.; Jankovic, J.; Factor, S.A.; Bressman, S.; Deligtisch, A.; Marras, C.; Lyons, K.E.; et al. Occupation and risk of Parkinsonism: A multicenter case-control study. *Arch. Neurol.* **2009**, *66*, 1106–1113. [CrossRef] [PubMed]
8. Freire, C.; Koifman, S. Pesticide exposure and Parkinson's disease: Epidemiological evidence of association. *Neurotoxicology* **2012**, *22*, 947–971. [CrossRef] [PubMed]
9. Blesa, J.; Phani, S.; Jackson-Lewis, V.; Przedborski, S. Classic and new animal models of Parkinson's disease. *J. Biomed. Biotechnol.* **2012**, *2012*, 845618. [CrossRef] [PubMed]
10. Betarbet, R.; Sherer, T.B.; MacKenzie, G.; Garcia-Osuna, M.; Panov, A.V.; Greenamyre, J.T. Chronic systemic pesticide exposure reproduces features of Parkinson's disease. *Nat. Neurosci.* **2000**, *3*, 1301–1306. [CrossRef] [PubMed]
11. Narendra, D.; Walker, J.E.; Youle, R. Mitochondrial quality control mediated by PINK1 and Parkin: Link to Parkinsonism. *Cold Spring Harb. Perspect. Biol.* **2012**, *4*, a011338. [CrossRef] [PubMed]
12. Gorell, J.M.; Johnson, C.C.; Rybicki, B.A.; Peterson, E.L.; Richardson, R.J. The risk of Parkinson's disease with exposure to pesticides, farming, well water, and rural living. *Neurology* **1998**, *50*, 1346–1350. [CrossRef] [PubMed]

13. Roesch, L.F.W.; Fulthorpe, R.R.; Riva, A.; Casella, G.; Hadwin, A.K.M.; Kent, A.D.; Daroub, S.H.; Camargo, F.A.O.; Farmerie, W.G.; Triplett, E.W. Pyrosequencing enumerates and contrasts soil microbial diversity. *ISME J.* **2007**, *1*, 283–290. [CrossRef] [PubMed]
14. Gans, J.; Wolinsky, M.; Dunbar, J. Computational improvements reveal great bacterial diversity and high metal toxicity in soil. *Science* **2005**, *309*, 1387–1390. [CrossRef] [PubMed]
15. Janssen, P.H. Identifying the dominant soil bacterial taxa in libraries of 16S rRNA and 16S rRNA genes. *Appl. Environ. Microbiol.* **2006**, *72*, 1719–1728. [CrossRef] [PubMed]
16. Tanaka, Y.; Omura, S. Metabolism and products of actinomycetes: An introduction. *Actinomycetologica* **1990**, *4*, 13–14. [CrossRef]
17. Fenteany, G.; Standaert, R.F.; Lane, W.S.; Choi, S.; Corey, E.J.; Schreiber, S.L. Inhibition of proteasome activities and subunit-specific amino-terminal threonine modification by lactacystin. *Science* **1995**, *268*, 726–731. [CrossRef] [PubMed]
18. McNaught, K.S.P.; Perl, D.P.; Brownell, A.L.; Olanow, C.W. Systemic exposure to proteosome inhibitors causes a progressive model of Parkinson's disease. *Ann. Neurol.* **2004**, *56*, 149–162. [CrossRef] [PubMed]
19. Bove, J.; Zhou, C.; Jackson-Lewis, V.; Taylor, J.; Chu, Y.; Rideout, H.J.; Wu, D.-C.; Kordower, J.H.; Petrucelli, L.; Przedborski, S. Proteasome inhibition and Parkinson's disease modeling. *Ann. Neurol.* **2006**, *60*, 260–264. [CrossRef] [PubMed]
20. Kordower, J.H.; Kanann, N.M.; Chu, Y.; Babu, R.S.; Stansell, R.; Terpstra, B.T.; Sortwell, C.E.; Steece-Colllier, K.; Collier, T.J. Failure of proteasome inhibitor administration to provide a model of Parkinson's disease in rats and monkeys. *Ann. Neurol.* **2006**, *60*, 264–268. [CrossRef] [PubMed]
21. Landau, A.M.; Kouassi, E.; Siegrist-Johnstone, R.; Desbarats, J. Proteasome inhibitor model of Parkinson's disease in mice is confounded by neurotoxicity of the ethanol vehicle. *Mov. Disord.* **2007**, *22*, 403–407. [CrossRef] [PubMed]
22. Li, X.; Du, Y.; Fan, X.; Yang, D.; Luo, G.; Le, W. c-Jun N-terminal kinase mediates lactacystin-induced dopamine neuron degeneration. *J. Neuropathol. Exp. Neurol.* **2008**, *67*, 933–944. [CrossRef] [PubMed]
23. Kohbata, S.; Beaman, B.L. L-Dopa-responsive movement disorder caused by *Nocardia asteroides* localized in the brains of mice. *Infect. Immun.* **1991**, *59*, 181–191. [PubMed]
24. Tam, S.; Barry, D.P.; Beaman, L.; Beaman, B.L. Neuroinvasive *Nocardia asteroides* GUH-2 induces apoptosis in the substantia nigra in vivo and dopaminergic cells in vitro. *Exp. Neurol.* **2002**, *107*, 453–460. [CrossRef]
25. Ogata, S.A.; Beaman, B.L. Adherence of *Nocardia asteroides* within the murine brain. *Infect. Immun.* **1992**, *60*, 1800–1805. [PubMed]
26. Ogata, S.A.; Beaman, B.L. Site-specific growth of *Nocardia asteroides* in the murine brain. *Infect. Immun.* **1992**, *60*, 3262–3267. [PubMed]
27. Chapman, G.; Beaman, B.L.; Loeffler, D.A.; Camp, D.M.; Domino, E.F.; Dickson, D.W.; Ellis, W.G.; Chen, I.; Bachus, S.E.; LeWitt, P.A. In situ hybridization for detection of nocardial 16S rRNA: Reactivity within intracellular inclusions in experimentally infected cynomolgus monkeys–and in Lewy body-containing human brain specimens. *Exp. Neurol.* **2003**, *184*, 715–725. [CrossRef]
28. Barry, D.P.; Beaman, B.L. Modulation of eukaryotic cells apoptosis by members of the bacterial order Actinomycetales. *Apoptosis* **2006**, *11*, 1695–1707. [CrossRef] [PubMed]
29. Martinez, B.A.; Caldwell, K.A.; Caldwell, G.A. C. *C. elegans* as a model system to accelerate discovery for Parkinson disease. *Curr. Opin. Genet. Dev.* **2017**, *44*, 102–109. [CrossRef] [PubMed]
30. Cooper, A.A.; Gitler, A.D.; Cashikar, A.; Haynes, C.M.; Hill, K.J.; Bhullar, B.; Liu, K.; Xu, K.; Strathearn, K.E.; Liu, F.; et al. Alpha-synuclein blocks ER-golgi traffic and Rab1 rescues neuron loss in Parkinson's models. *Science* **2006**, *313*, 324–328. [CrossRef] [PubMed]
31. Gitler, A.D.; Bevis, B.J.; Shorter, J.; Strathearn, K.E.; Hamamichi, S.; Su, L.J.; Caldwell, K.A.; Caldwell, G.A.; Rochet, J.C.; McCaffery, J.M.; et al. The Parkinson's disease protein alpha-synuclein disrupts cellular Rab homeostasis. *Proc. Natl. Acad. Sci. USA* **2008**, *2008 105*, 145–150. [CrossRef]
32. Gitler, A.D.; Chesi, A.; Geddie, M.L.; Strathearn, K.E.; Hamamichi, S.; Hill, K.J.; Caldwell, K.A.; Caldwell, G.A.; Cooper, A.A.; Rochet, J.C.; et al. Alpha-synuclein is part of a diverse and highly conserved interaction network that includes PARK9 and manganese toxicity. *Nat. Genet.* **2009**, *41*, 308–315. [CrossRef] [PubMed]

33. Qiao, L.; Hamamichi, S.; Caldwell, K.A.; Caldwell, G.A.; Wilson, S.; Yacoubian, T.A.; Xie, Z.-L.; Speake, L.D.; Parks, R.; Crabtree, D.; et al. A neuroprotective role of lysosomal enzyme cathepsin D against alpha-synuclein pathogenesis. *Mol. Brain* **2008**, *1*, 17. [CrossRef] [PubMed]
34. Yacoubian, T.A.; Slone, S.R.; Harrington, A.J.; Hamamichi, S.; Schieltz, J.M.; Caldwell, K.A.; Caldwell, G.A.; Standaert, D.G. Differential neuroprotective effects of 14-3-3 proteins in models of Parkinson's disease. *Cell Death Dis.* **2010**, *1*, e2. [CrossRef] [PubMed]
35. Ruan, Q.; Harrington, A.J.; Caldwell, K.A.; Caldwell, G.A.; Standaert, D.G. VPS41, a protein involved in lysosomal trafficking, is protective in *Caenorhabditis elegans* and mammalian cellular models of Parkinson's disease. *Neurobiol. Dis.* **2010**, *37*, 330–338. [CrossRef] [PubMed]
36. Chalfie, M.; White, J. The nervous system. In *The Nematode Caenorhabditis Elegans*; Wood, W.B., Ed.; Cold Spring Harbor Laboratory Press: Cold Spring Harbor, NY, USA, 1988; pp. 337–391. ISBN 978-087969433-3.
37. Jonker, M.J.; Piskiewicz, A.M.; Ivorra i Castella, N.; Kammenga, J.E. Toxicity of binary mixtures of cadmium-copper and carbendazim-copper to the nematode *Caenorhabditis elegans*. *Environ. Toxicol. Chem.* **2004**, *23*, 1529–1537. [CrossRef] [PubMed]
38. Peres, T.V.; Schettinger, M.R.; Chen, P.; Carvalho, F.; Avila, D.S.; Bowman, A.B.; Aschner, M. Manganese-induced neurotoxicity: A review of its behavioral consequences and neuroprotective strategies. *BMC Pharmacol. Toxicol.* **2016**, *17*, 57. [CrossRef] [PubMed]
39. Locke, C.J.; Fox, S.A.; Caldwell, G.A.; Caldwell, K.A. Acetaminophen attenuates dopamine neuron degeneration in animal models of Parkinson's disease. *Neurosci. Lett.* **2008**, *439*, 129–133. [CrossRef] [PubMed]
40. Cao, S.; Hewett, J.W.; Yokoi, F.; Lu, J.; Buckley, A.C.; Burdette, A.J.; Chen, P.; Nery, F.C.; Li, Y.; Breakefield, X.O.; et al. Chemical enhancement of torsinA function in cell and animal models of torsion dystonia. *Dis. Model. Mech.* **2010**, *3*, 386–396. [CrossRef] [PubMed]
41. Tardiff, D.F.; Jui, N.T.; Khurana, V.; Tambe, M.A.; Thompson, M.L.; Chung, C.Y.; Kamadurai, H.B.; Kim, H.T.; Lancaster, A.K.; Caldwell, K.A.; et al. Yeast reveal a "druggable" Rsp5/Nedd4 network that ameliorates α-synuclein toxicity in neurons. *Science* **2013**, *342*, 979–983. [CrossRef] [PubMed]
42. Nass, R.; Hall, D.H.; Miller, D.M., 3rd; Blakely, R.D. Neurotoxin-induced degeneration of dopamine neurons in *Caenorhabditis elegans*. *Proc. Natl. Acad. Sci. USA* **2002**, *99*, 3264–3269. [CrossRef] [PubMed]
43. Ray, A.; Martinez, B.A.; Berkowitz, L.A.; Caldwell, G.A.; Caldwell, K.A. Mitochondrial dysfunction, oxidative stress, and neurodegeneration elicited by a bacterial metabolite in a *C. elegans* Parkinson's model. *Cell Death Dis.* **2014**, *5*, e984. [CrossRef] [PubMed]
44. Zhang, Y.; Lu, H.; Bargmann, C.I. Pathogenic bacteria induce aversive olfactory learning in *Caenorhabditis elegans*. *Nature* **2005**, *438*, 179–184. [CrossRef] [PubMed]
45. Caldwell, K.A.; Tucci, M.L.; Armagost, J.; Hodges, T.W.; Chen, J.; Memon, S.B.; Blalock, J.E.; DeLeon, S.M.; Findlay, R.H.; Ruan, Q.; et al. Investigating Bacterial Sources of Toxicity as an Environmental Contributor to Dopaminergic Neurodegeneration. *PLoS ONE* **2009**, *4*, e7227. [CrossRef] [PubMed]
46. Sulston, J.; Dew, M.; Brenner, S. Dopaminergic neurons in the nematode *Caenorhabditis elegans*. *J. Comp. Neurol.* **1975**, *163*, 215–226. [CrossRef] [PubMed]
47. Sanyal, S.; Wintle, R.F.; Kindt, K.S.; Nuttley, W.M.; Arvan, R.; Fitzmaurice, P.; Bigras, E.; Merz, D.C.; Hébert, T.E.; van der Kooy, D.; et al. Dopamine modulates the plasticity of mechanosensory responses in *Caenorhabditis elegans*. *EMBO J.* **2003**, *23*, 473–482. [CrossRef] [PubMed]
48. Singleton, A.B.; Farrer, M.; Johnson, J.; Singleton, A.; Hague, S.; Kachergus, J.; Hulihan, M.; Peuralinna, T.; Dutra, A.; Nussbaum, R.; et al. Alpha-synuclein locus triplication causes Parkinson's disease. *Science* **2003**, *302*, 841. [CrossRef] [PubMed]
49. Manning-Bog, A.B.; McCormack, A.L.; Li, J.; Uversky, V.N.; Fink, A.L.; Di Monte, D.A. The herbicide paraquat causes up-regulation and aggregation of alpha-synuclein in mice: Paraquat and alpha-synuclein. *J. Biol. Chem.* **2002**, *277*, 1641–1644. [CrossRef] [PubMed]
50. Uversky, V.B.; Li, J.; Fink, A.L. Pesticides directly accelerate the rate of alpha-synclein fibril formation: A possible factor in Parkinson's Disease. *FEBS Lett.* **2001**, *500*, 105–108. [CrossRef]
51. Martinez, B.A.; Kim, H.; Ray, A.; Caldwell, G.A.; Caldwell, K.A. A bacterial metabolite induces glutathione-tractable proteostatic damage, proteasomal disturbances, and PINK1-dependent autophagy in *C. elegans*. *Cell Death Dis.* **2015**, *6*, e1908–e1913. [CrossRef] [PubMed]
52. Hardy, J. Genetic analysis of pathways to Parkinson's disease. *Neuron* **2010**, *68*, 201–206. [CrossRef] [PubMed]

53. Saxena, S.; Caroni, P. Selective neuronal vulnerability in neurodegenerative diseases: From stressor thresholds to degeneration. *Neuron* **2011**, *71*, 35–48. [CrossRef] [PubMed]
54. Saha, S.; Guillily, M.D.; Ferree, A.; Lanceta, J.; Chan, D.; Ghosh, J.; Hsu, C.H.; Segal, L.; Raghavan, K.; Matsumoto, K.; et al. LRRK2 modulates vulnerability to mitochondrial dysfunction in *C. elegans*. *J. Neurosci.* **2009**, *29*, 9210–9218. [CrossRef] [PubMed]
55. Mortiboys, H.; Johansen, K.K.; Aasly, J.O.; Bandmann, O. Mitochondrial impairment in patients with Parkinson disease with the G2019S mutation in LRRK2. *Neurology* **2010**, *75*, 2017–2020. [CrossRef] [PubMed]
56. Ng, C.H.; Mok, S.Z.; Koh, C.; Ouyang, X.; Fivaz, M.L.; Tan, E.K.; Dawson, V.L.; Dawson, T.M.; Yu, F.; Lim, K.L. Parkin Protects against LRRK2 G2019S Mutant-Induced Dopaminergic Neurodegeneration in Drosophila. *J. Neurosci.* **2009**, *29*, 11257–11262. [CrossRef] [PubMed]
57. Jha, N.; Kumar, J.; Bonplueang, R.; Andersen, J.K. Glutathione decreases in dopaminergic PC12 cells interfere with the ubiquitin protein degradation pathway: Relevance for Parkinson's disease? *J. Neurochem.* **2002**, *80*, 555–561. [CrossRef] [PubMed]
58. Romero-Aristizabal, C.; Marks, D.S.; Fontana, W.; Apfeld, J. Regulated spatial organization and sensitivity of cytosolic protein oxidation in *Caenorhabditis elegans*. *Nat. Commun.* **2014**, *5*, 5020. [CrossRef] [PubMed]
59. Back, P.; Braeckma, B.P.; Matthijssens, F. ROS in Aging *Caenorhabditis elegans*: Damage or signaling. *Oxidative Med. Cell. Longev.* **2012**, *2012*, 608478. [CrossRef] [PubMed]
60. Springer, W.; Hoppe, T.; Schmidt, E.; Baumeister, R. A *Caenorhabditis elegans* Parkin mutant with altered solubility couples α-synuclein aggregation to proteotoxic stress. *Hum. Mol. Genet.* **2005**, *14*, 3407–3423. [CrossRef] [PubMed]
61. Yang, Y.; Ouyang, Y.; Yang, L.; Beal, M.F.; McQulbban, A.; Vogel, H.; Lu, B. Pink1 regulates mitochondrial dynamics through interaction with the fission/fusion machinery. *Proc. Natl. Acad. Sci. USA* **2008**, *105*, 7070–7075. [CrossRef] [PubMed]
62. Vives-Bauza, C.; Zhou, C.; Huang, Y.; Cui, M.; de Vries, R.L.A.; Kim, J.; May, J.; Tocilescu, M.A.; Liu, W.; Ko, H.S.; et al. PINK1-dependent recruitment of Parkin to mitochondria in mitophagy. *Proc. Natl. Acad. Sci. USA* **2010**, *107*, 378–383. [CrossRef] [PubMed]
63. Pargalija, D.; Klinkenberg, M.; Dominiguez-Bautista, J.; Hetzel, M.; Gispert, S.; Chimi, M.A.; Dröse, S.; Mai, S.; Brandt, U.; Auburger, G.; et al. Loss of PINK1 impairs stress-induced autophagy and cell survival. *PLoS ONE* **2014**, *9*, e95288.
64. Palikaras, K.; Lionaki, E.; Tavernarakis, N. Coordination of mitophagy and mitochondrial biogenesis during ageing in *C. elegans*. *Nature* **2015**, *521*, 525–528. [CrossRef] [PubMed]
65. Gosai, S.J.; Kwak, J.H.; Luke, C.J.; Long, O.S.; King, D.E.; Kovatch, K.J.; Johnston, P.A.; Shun, T.Y.; Lazo, J.S.; Perlmutter, D.H.; et al. Automated high-content live animal drug screening using *C. elegans* expressing the aggregation prone serpin α1-antitrypsin Z. *PLoS ONE* **2010**, *5*, e15460. [CrossRef] [PubMed]
66. Spires, T.L.; Hannan, A.J. Nature, nurture and neurology: Gene-environment interactions in neurodegenerative disease. *FASEB J.* **2005**, *272*, 2347–2361. [CrossRef] [PubMed]
67. Kim, I.; Rodriguez-Enriquez, S.; Lemasters, J.J. Selective degradation of mitochondria by mitophagy. *Arch. Biochem. Biophys.* **2007**, *462*, 245–253. [CrossRef] [PubMed]
68. Twig, G.; Hyde, B.; Shirihai, O.S. Mitochondrial fusion, fission and autophagy as a quality control axis: The bioenergetic view. *Biochim. Biophys. Acta Bioenerg.* **2008**, *1777*, 1092–1097. [CrossRef] [PubMed]
69. Lutz, A.K.; Exner, N.; Fett, M.E.; Schlehe, J.S.; Kloos, K.; Lämmermann, K.; Brunner, B.; Kurz-Drexler, A.; Vogel, F.; Reichert, A.S.; et al. Loss of Parkin or PINK1 function increases Drp1-dependent mitochondrial fragmentation. *J. Biol. Chem.* **2009**, *284*, 22938–22951. [CrossRef] [PubMed]
70. Haynes, C.M.; Ron, D. The mitochondrial UPR–protecting organelle protein homeostasis. *J. Cell Sci.* **2010**, *123*, 3849–3855. [CrossRef] [PubMed]
71. Murfitt, R.R.; Vogel, K.; Sanadi, D.R. Characterization of the mitochondria of the free-living nematode, *Caenorhabditis elegans*. *Comp. Biochem. Physiol.* **1976**, *53*, 423–430. [CrossRef]
72. Ved, R.; Saha, S.; Westlund, B.; Perier, C.; Burnam, L.; Sluder, A.; Hoener, M.; Rodrigues, C.M.; Alfonso, A.; Steer, C.; et al. Similar patterns of mitochondrial vulnerability and rescue induced by genetic modification of alpha-synuclein, parkin, and DJ-1 in *Caenorhabditis elegans*. *J. Biol. Chem.* **2005**, *280*, 42655–42668. [CrossRef] [PubMed]

73. Tieu, K.; Perier, C.; Caspersen, C.; Teismann, P.; Wu, D.C.; Yan, S.D.; Naini, A.; Vila, M.; Jackson-Lewis, V.; Ramasamy, R.; et al. D-beta-hydroxybutyrate rescues mitochondrial respiration and mitigates features of Parkinson disease. *J. Clin. Investig.* **2003**, *112*, 892–901. [CrossRef] [PubMed]
74. Grad, L.I.; Lemire, B.D. Mitochondrial complex I mutations in *Caenorhabditis elegans* produce cytochrome c oxidase stress and vitamin-responsive lactic acidosis. *Hum. Mol. Genet.* **2004**, *13*, 303–314. [CrossRef] [PubMed]
75. Ishihara, N.; Jofuku, A.; Eura, Y.; Mihara, K. Regulation of mitochondrial morphology by membrane potential, and DRP1-dependent division and FZO1-dependent fusion reaction in mammalian cells. *Biochim Biophys. Acta* **2003**, *301*, 891–898. [CrossRef]
76. Chan, D.C. Fusion and fission: Interlinked processes critical for mitochondrial health. *Annu. Rev. Genet.* **2012**, *46*, 265–287. [CrossRef] [PubMed]
77. Koopman, W.J.H.; Verkaart, S.; Visch, H.-J.; van der Westhuizen, F.H.; Murphy, M.P.; van den Heuvel, L.W.P.J.; Smeitink, J.A.M.; Willems, P.H.G.M. Inhibition of complex I of the electron transport chain causes O_2^--mediated mitochondrial outgrowth. *Am. J. Physiol. Cell Physiol.* **2005**, *288*, C1440–C1450. [CrossRef] [PubMed]
78. Arnold, B.; Cassady, S.J.; VanLaar, V.S.; Berman, S.B. Integrating multiple aspects of mitochondrial dynamics in neurons: Age-related differences and dynamic changes in a chronic rotenone model. *Neurobiol. Dis.* **2011**, *41*, 189–200. [CrossRef] [PubMed]
79. Wang, X.; Su, B.; Liu, W.; He, X.; Gao, Y.; Castellani, R.J.; Perry, G.; Smith, M.A.; Zhu, X. DLP1-dependent mitochondrial fragmentation mediates 1-methyl-4-phenylpyridinium toxicity in neurons: Implications for Parkinson's disease. *Aging Cell* **2011**, *10*, 807–823. [CrossRef] [PubMed]
80. Bajpai, P.; Sangar, M.C.; Singh, S.; Tang, W.; Bansal, S.; Chowdhury, G.; Cheng, Q.; Fang, J.-K.; Martin, M.V.; Guengerich, F.P.; et al. Metabolism of 1-methyl-4-phenyl-1,2,3,6-tetrahydropyridine by mitochondrion-targeted cytochrome P450 2D6: Implications in Parkinson disease. *J. Biol. Chem.* **2013**, *288*, 4436–4451. [CrossRef] [PubMed]
81. Peng, K.; Yang, L.; Wang, J.; Ye, F.; Dan, G.; Zhao, Y.; Cai, Y.; Cui, Z.; Ao, L.; Liu, J.; et al. The Interaction of Mitochondrial Biogenesis and Fission/Fusion Mediated by PGC-1α Regulates Rotenone-Induced Dopaminergic Neurotoxicity. *Mol. Neurobiol.* **2017**, *54*, 3783–3797. [CrossRef] [PubMed]
82. Wu, S.; Zhou, F.; Zhang, Z.; Xing, D. Mitochondrial oxidative stress causes mitochondrial fragmentation via differential modulation of mitochondrial fission–fusion proteins. *FEBS J.* **2011**, *278*, 941–954. [CrossRef] [PubMed]
83. Kim, H.; Perentis, R.J.; Caldwell, G.A.; Caldwell, K.A. Gene-by-environment interactions that disrupt mitochondrial homeostasis cause neurodegeneration in *C. elegans* Parkinson's models. *Cell Death Dis.* **2018**, *9*, 555. [CrossRef] [PubMed]
84. Harrington, A.J.; Yacoubian, T.A.; Slone, S.R.; Caldwell, K.A.; Caldwell, G.A. Functional analysis of VPS41-mediated neuroprotection in *Caenorhabditis elegans* and mammalian models of Parkinson's disease. *J. Neurosci.* **2012**, *32*, 2142–2153. [CrossRef] [PubMed]
85. Kanazawa, T.; Zappaterra, M.D.; Hasegawa, A.; Wright, A.P.; Newman-Smith, E.D.; Buttle, K.F.; McDonald, K.; Mannella, C.A.; van der Bliek, A.M. The *C. elegans* Opa1 Homologue EAT-3 Is Essential for Resistance to Free Radicals. *PLoS Genet.* **2008**, *4*, e1000022. [CrossRef] [PubMed]
86. Sandebring, A.; Thomas, K.J.; Beilina, A.; van der Brug, M.; Cleland, M.M.; Ahmad, R.; Miller, D.W.; Zambrano, I.; Cowburn, R.F.; Behbahani, H.; et al. Mitochondrial alterations in PINK1 deficient cells are influenced by calcineurin-dependent dephosphorylation of dynamin-related protein 1. *PLoS ONE* **2009**, *4*, e5701. [CrossRef] [PubMed]
87. Buhlman, L.; Damiano, M.; Bertolin, G.; Ferrando-Miguel, R.; Lombès, A.; Brice, A.; Corti, O. Functional interplay between Parkin and Drp1 in mitochondrial fission and clearance. *BBA Mol. Cell Res.* **2014**, *1843*, 2012–2026. [CrossRef] [PubMed]
88. Toyama, E.Q.; Herzig, S.; Courchet, J.; Lewis, L.; Losón, O.C.; Hellberg, K.; Young, N.P.; Chen, H.; Polleux, F.; Chan, D.C.; et al. AMP-activated protein kinase mediates mitochondrial fission in response to energy stress. *Science* **2016**, *351*, 275–281. [CrossRef] [PubMed]
89. Wang, C.; Youle, R. Cell biology: Form follows function for mitochondria. *Lett. Nat.* **2016**, *530*, 288–289. [CrossRef] [PubMed]
90. Watkins, A.L.; Ray, A.R.; Roberts, L.; Caldwell, K.A.; Olson, J.B. The Prevalence and Distribution of Neurodegenerative Compound-Producing Soil *Streptomyces* spp. *Sci. Rep.* **2016**, *6*, 22566. [CrossRef] [PubMed]

91. Challis, G.L.; Hopwood, D.A. Synergy and contingency as driving forces for the evolution of multiple secondary metabolite production by *Streptomyces* species. *Proc. Natl. Acad. Sci. USA* **2003**, *100*, 14555–14561. [CrossRef] [PubMed]
92. Van Lanen, S.G.; Shen, B. Microbial genomics for the improvement of natural product discovery. *Curr. Opin. Microbiol.* **2006**, *9*, 252–260. [CrossRef] [PubMed]
93. Dauer, W.; Przedborski, S. Parkinson's disease: Mechanisms and models. *Neuron* **2003**, *39*, 889–909. [CrossRef]
94. Lee, V.M.; Trojanowski, J.Q. Mechanisms of Parkinson's disease linked to pathological alpha-synuclein: New targets for drug discovery. *Neuron* **2006**, *52*, 33–38. [CrossRef] [PubMed]
95. Narendra, D.P.; Youle, R.J. Targeting mitochondrial dysfunction: Role for PINK1 and Parkin in mitochondrial quality control. *Antioxid. Redox Signal.* **2011**, *14*, 1929–1938. [CrossRef] [PubMed]
96. Vidal, M.; Cusick, M.E.; Barabasi, A.L. Interactome networks and human disease. *Cell* **2011**, *144*, 986–998. [CrossRef] [PubMed]
97. Towlson, E.K.; Vértes, P.E.; Yan, G.; Chew, Y.L.; Walker, D.S.; Schafer, W.R.; Barabási, A.L. *Caenorhabditis elegans* and the network control framework-FAQs. *Philos. Trans. R. Soc. Lond. B Biol. Sci.* **2018**, *373*, 20170372. [CrossRef] [PubMed]
98. Gao, A.W.; Smith, R.L.; van Weeghel, M.; Kamble, R.; Janssens, G. E.; Houtkooper, R.H. Identification of key pathways and metabolic fingerprints of longevity in *C. elegans*. *Exp. Gerontol.* **2018**, *113*, 128–140. [CrossRef] [PubMed]
99. Van Assche, R.; Temmerman, L.; Dias, D.A.; Boughton, B.; Boonen, K.; Braeckman, B.P.; Schoofs, L.; Roessner, U. Metabolic profiling of a transgenic *Caenorhabditis elegans* Alzheimer model. *Metabolomics* **2015**, *11*, 477–486. [CrossRef] [PubMed]
100. Gao, A.W.; Sterken, M.G.; Uit de Bos, J.; van Creij, J.; Kamble, R.; Snoek, B.L.; Kammenga, J.E.; Houtkooper, R.H. Natural genetic variation in *C. elegans* identified genomic loci controlling metabolite levels. *Genome Res.* **2018**, *28*, 1296–1308. [CrossRef] [PubMed]
101. Whitman, W.B.; Coleman, D.C.; Wiebe, W.J. Prokaryotes: The unseen majority. *Proc. Natl. Acad. Sci. USA* **1998**, *95*, 6578–6583. [CrossRef] [PubMed]

metabolites

MDPI

Article

Metabolomic Profiling of Bile Acids in an Experimental Model of Prodromal Parkinson's Disease

Stewart F. Graham [1,2,*], Nolwen L. Rey [3], Zafer Ugur [1], Ali Yilmaz [1], Eric Sherman [4], Michael Maddens [1,2], Ray O. Bahado-Singh [1,2], Katelyn Becker [3], Emily Schulz [3], Lindsay K. Meyerdirk [3], Jennifer A. Steiner [3], Jiyan Ma [3] and Patrik Brundin [3]

1 Beaumont Health, 3811 W. 13 Mile Road, Royal Oak, MI 48073, USA; Zafer.Ugur@beaumont.org (Z.U.); Ali.yilmaz@beaumont.org (A.Y.); mmaddens@beaumont.edu (M.M.); Ray.Bahado-Singh@beaumont.org (R.O.B.-S.)
2 Oakland University-William Beaumont School of Medicine, Rochester, MI 48309, USA
3 Center for Neurodegenerative Science, Van Andel Research Institute, Grand Rapids, MI 49503, USA; Nolwen.REY@cnrs.fr (N.L.R.); Katelyn.Becker@vai.org (K.B.); emily.schulz@vai.org (E.S.); Lindsay.Meyerdirk@vai.org (L.K.M.); Jennifer.Steiner@vai.org (J.A.S.); Jiyan.Ma@vai.org (J.M.); Patrik.Brundin@vai.org (P.B.)
4 University of Michigan, Ann Arbor, MI 48109, USA; ebsherm@umich.edu
* Correspondence: stewart.graham@beaumont.edu; Tel.: +1-248-551-2038

Received: 2 October 2018; Accepted: 26 October 2018; Published: 31 October 2018

Abstract: For people with Parkinson's disease (PD), considered the most common neurodegenerative disease behind Alzheimer's disease, accurate diagnosis is dependent on many factors; however, misdiagnosis is extremely common in the prodromal phases of the disease, when treatment is thought to be most effective. Currently, there are no robust biomarkers that aid in the early diagnosis of PD. Following previously reported work by our group, we accurately measured the concentrations of 18 bile acids in the serum of a prodromal mouse model of PD. We identified three bile acids at significantly different concentrations ($p < 0.05$) when mice representing a prodromal PD model were compared with controls. These include ω-murichoclic acid (MCAo), tauroursodeoxycholic acid (TUDCA) and ursodeoxycholic acid (UDCA). All were down-regulated in prodromal PD mice with TUDCA and UDCA at significantly lower levels (17-fold and 14-fold decrease, respectively). Using the concentration of three bile acids combined with logistic regression, we can discriminate between prodromal PD mice from control mice with high accuracy (AUC (95% CI) = 0.906 (0.777–1.000)) following cross validation. Our study highlights the need to investigate bile acids as potential biomarkers that predict PD and possibly reflect the progression of manifest PD.

Keywords: prodromal Parkinson's disease; bile acids; mass spectrometry; biomarkers; α-synuclein aggregates

1. Introduction

Parkinson's Disease (PD) is a common, long-term neurodegenerative disease. Adjusting for age and gender, the incidence of PD has been estimated to affect 1 in every 100 people over the age of 60 [1]. PD motor symptoms are believed to originate from striatal dopamine loss which occurs due to the death of dopaminergic neurons in the substantia nigra pars compacta (SNpc). The loss of dopaminergic neurons in the SNpc is the hallmark indicator for the post-mortem diagnosis of PD [2]. Lewy bodies and Lewy neurites, composed mainly of misfolded α-synuclein (α-syn) protein also feature in PD brains. Clinical diagnosis of PD is based on several criteria including bradykinesia in combination with rigidity, resting tremor, or both and response to dopaminergic drugs [3]. In addition to the classical motor symptoms, a wide range of non-motor symptoms and signs are apparent in PD patients [4], some of which are already present long before the onset of motor symptoms, in the PD prodrome [5]. However, misdiagnosis is common in the prodromal phase, when a potential disease-modifying treatment is thought to be most effective [6,7]. Currently, no robust biomarkers for early and more precise diagnosis of PD exist [8] and as several new potentially disease-modifying treatments emerge this is becoming a major unmet medical need [6,9].

In a previous study by our group, we identified Bile Acid metabolism as one of the major biochemical pathways to be perturbed in the brain of a mouse model of prodromal PD [10]. Bile acids are molecules derived from cholesterol in hepatocytes and are used to emulsify fats in the small intestine and promote fat digestion and absorption [11,12]. In addition to their role in lipid digestion and absorption, bile acids function as signaling molecules, participating as ligands in both membrane-bound receptors and nuclear hormone receptors [13,14]. It has been reported that certain bile acids, including ursodeoxycholic acid (UDCA) and tauroursodeoxycholic acid (TUDCA) can pass the blood–brain barrier [14] with their presence also being noted in cerebrospinal fluid (CSF), plasma, urine, and serum [15–18]. To date, several reports implicate bile acids in neurodegenerative diseases and suggest a possible role in modulating neuronal proliferation. One such study links statistically significant increases in levels of deoxycholic acid (DCA), glycodeoxycholic acid (GDCA), and lithocholic acid (LCA) in plasma, to Alzheimer's disease and mild cognitive impairment [19]. Abdelkader et al. observed a neuroprotective effect from administration of UDCA on a murine rotenone model of PD [20]. Further, it has been reported that cholic acid is a ligand for liver X receptors which promote ventral midbrain neurogenesis and cell survival [21]. Bile acids have also been reported to be potential biomarkers of other neurodegenerative diseases including Alzheimer's disease (AD) [22–24].

In the current study, we accurately measured the concentrations of 18 bile acids in the serum of a prodromal mouse model of PD. Following on from our previous metabolomics work using this model, we believe that bile acids may prove to be essential for the development of a robust biomarker panel capable of accurately diagnosing PD.

2. Results

2.1. Univariate Analysis

To investigate bile acids in a model of prodromal PD, we used a mouse model previously developed by our group which consists of WT mice injected with α-syn fibrils into the olfactory bulb [7,10]. The injection of α-syn fibrils leads to the propagation of α-syn aggregates throughout several interconnected regions in the brain. The progressive spreading of α-synucleinopathy shows many similarities with that which has been suggested to occur in PD [23,25–27]. Using mass spectrometry, we analyzed the serum of the α-syn fibrils-injected mice (PFF mice) and of α-syn monomers-injected mice (HuMonomers mice; controls), collected 3 months post injection.

Of the 18 bile acids profiled, all were within the limits of detection and quantification. Of these, we found three to be significantly perturbed in PFF mice compared to HuMonomers mice: Omega-murichoclic acid (MCAo), tauroursodeoxycholic acid (TUDCA) and ursodeoxycholic acid (UDCA) (Table 1). Of the three bile acids, we found UDCA and its taurine conjugated form TUDCA to be extremely decreased (17- and 14-fold, respectively) in the mice injected with PFFs.

Table 1. Results of the univariate analyses for bile acids measured in serum from mice injected with HuMonomers and PFFs. p-Values were calculated using the Wilcoxon–Mann–Whitney test. LOD-Limit of detection; LLOQ-Lower limit of quantification. Those bile acids highlighted in bold are considered statistically significantly different ($p < 0.05$; $q < 0.05$).

HMDB#	Name	Mean (SD) of HuMonomer	Mean (SD) of PFF	p-Value	q-Value (FDR)	Fold Change	LOD	LLOQ
HMDB0000619	Cholic Acid	11.09 (20.89)	10.12 (18.99)	0.24	0.39	1.10	0.004	0.03
HMDB0000518	Chenodeoxycholic acid	0.89 (1.22)	0.77 (1.53)	0.06	0.19	1.15	0.005	0.02
HMDB0000626	Deoxycholic acid	1.63 (2.07)	1.52 (2.61)	0.20	0.39	1.08	0.005	0.02
HMDB0000138	Glycocholic acid	0.07 (0.07)	0.06 (0.06)	0.67	0.85	1.14	0.003	0.03
HMDB0000637	Glycochenodeoxycholic acid	0.06 (0.14)	0.07 (0.14)	0.19	0.39	−1.07	0.01	0.02
HMDB0000631	Glycodeoxycholic acid	0.66 (0.77)	0.35 (0.46)	0.37	0.55	1.90	0.01	0.01
HMDB0000733	Hyodeoxycholic acid	0.65 (0.51)	0.44 (0.52)	0.04	0.16	1.47	0.005	0.02
HMDB0000761	Lithocholic acid	0.10 (0.13)	0.10 (0.15)	0.76	0.85	−1.04	0.002	0.01
HMDB0000506	Alpha-Muricholic acid	0.83 (1.42)	0.65 (1.23)	0.06	0.19	1.28	0.007	0.01
HMDB0000415	Beta-Muricholic acid	7.49 (10.54)	5.72 (8.760)	0.09	0.23	1.31	0.008	0.02
HMDB0000364	**Omega-Murichoclic acid**	**4.58 (2.04)**	**2.00 (2.03)**	**<0.0001**	**0.01**	**2.28**	0.007	0.01
HMDB0000036	Taurocholic acid	11.02 (17.81)	9.20 (20.59)	0.93	0.98	1.20	0.008	0.02
HMDB0000951	Taurochenodeoxycholic acid	0.75 (1.22)	0.79 (1.56)	0.99	0.99	−1.05	0.005	0.01
HMDB0000896	Taurodeoxycholic acid	0.29 (0.23)	0.35 (0.42)	0.74	0.85	−1.22	0.001	0.01
HMDB0000722	Taurolithocholic acid	0.01 (0.02)	0.02 (0.03)	0.40	0.55	−1.41	0.001	0.01
HMDB0000932	Tauromuricholic acid (sum of α and β)	1.07 (1.85)	0.42 (0.96)	0.22	0.39	2.52	0.001	0.01
HMDB0000874	**Tauroursodeoxycholic acid**	**1.67 (2.71)**	**0.12 (0.12)**	**<0.0001**	**<0.001**	**14.14**	0.001	0.01
HMDB0000946	**Ursodeoxycholic acid**	**0.55 (0.58)**	**0.03 (0.05)**	**<0.0001**	**<0.0001**	**17.55**	0.001	0.02

Figure 1 displays the Box and Whisker plots for the top three significantly different ($p < 0.05$; FDR < 0.05) metabolites in both the HuMonomer- and PFF-injected mice. As is evident from the plots, all are at significantly lower concentrations in PFF-injected mice.

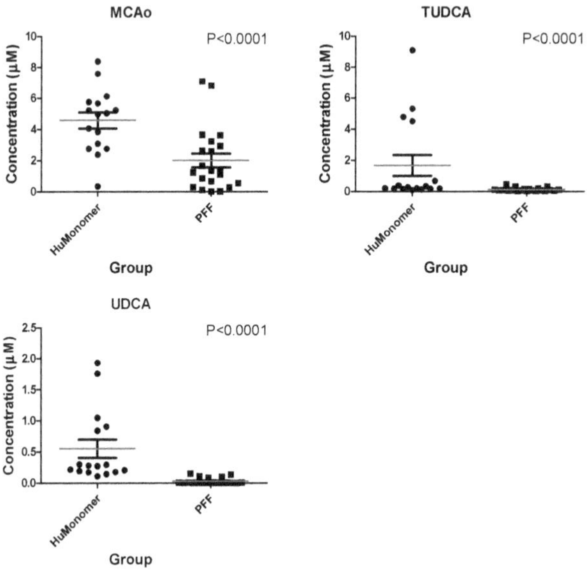

Figure 1. The mean distribution (±SEM) for each of the three significantly different bile acids between mice injected with HuMonomers and PFFs.

2.2. *Logistic Regression Analysis*

Using the concentrations of taurolithocholic acid (TLCA), glycochenodeoxycholic acid (GCDCA) and TUDCA, we developed a diagnostic algorithm capable of accurately differentiating between HuMonomer- and PFF-injected mice with 91.4% accuracy following 100-fold cross validations.

$$\text{logit}(P) = \log(P/(1 - P)) = -0.893 + 11.152\ \text{TLCA} + 8.917\ \text{GCDCA} - 18.221\ \text{TUDCA}$$

where P is Pr(y = 1 | x). The best threshold (or Cutoff) for the predicted P is 0.52. Original Label: 0/1 -> Labels in Logistic Regression: 0/1 Note) The class/response value is recommended as (Case: 1 and Control: 0).

Table 2 lists the summary of each feature used to produce the diagnostic algorithm. Table 3 details the performance values of the logistic regression model following 10-fold cross validation with Figure 2 displaying the ROC plot for said model. The model was significant following 1000-permutation tests with $p = 0.003$. Figure 2 displays the ROC curve for the logistic regression analysis following 10-fold cross validation.

Table 2. Logistic Regression Model—Summary of Each Feature.

	Estimate	Std. Error	*z* Value	Pr (>|z|)	Odds
(Intercept)	−0.893	2.857	−0.313	0.755	-
TLCA	11.152	7.264	1.535	0.125	69,675.46
GCDCA	8.917	9.571	0.932	0.352	7455.77
TUDCA	−18.221	7.762	−2.347	0.019	0

Table 3. The performance values for the logistic regression model.

	AUC	Sensitivity	Specificity
Training/Discovery	0.992 (0.985~0.998)	0.958 (0.929~0.986)	0.944 (0.907~0.982)
10-fold Cross-Validation	0.906 (0.777~1.000)	0.952 (0.952~1.000)	0.938 (0.819~1.000)

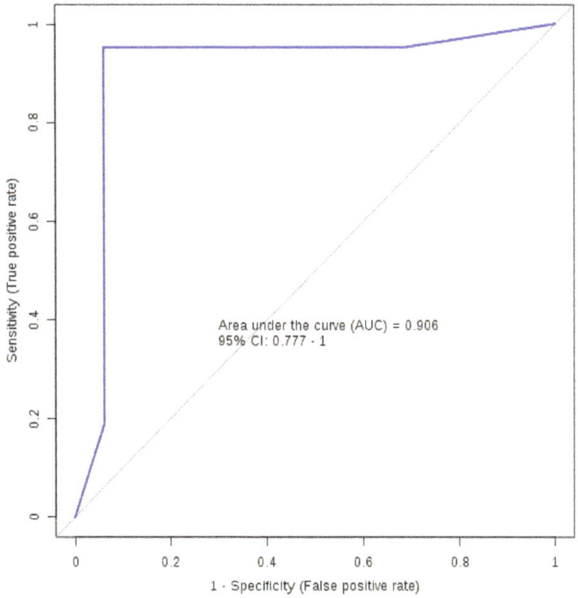

Figure 2. The ROC plot for the logistic regression diagnostic algorithm.

3. Discussion

This is the first study to accurately quantify bile acids from the serum of a validated mouse model of prodromal PD. Our study was primarily driven by the results from a previous study by our group [10]. In total, we profiled 18 bile acids of which only three were found to be statistically significantly different in PFF mice when compared with HuMonomer controls ($p < 0.05$). All three were found to be significantly decreased in PFF mouse serum, with TUDCA and UDCA at 14- and 17-fold lower concentrations, respectively.

Using the concertation of three bile acids (TLCA, GCDCA and TUDCA), we developed a predictive model capable of differentiating between PFF mice and HuMonomer controls with an AUC (95 % CI) = 0.906 (0.777–1.00) with high sensitivity and specificity values (0.952 (0.952–1.000) and 0.938 (0.819–1.000), respectively) following cross validation. This eclipses work previously reported by our group in which we report a predictive logistic regression model developed using the concentration of three phosphocholines and trans-4-hrdroxyproline [10]. This previous model achieved an AUC (95% CI) = 0.836 (0.696−0.9777) high sensitivity and specificity values (0.800 (0.800−0.975) and 0.889 (0.744−1.00), respectively); however, following cross validation, those results are less precise than what we report herein.

Bile acids play pivotal roles in many physiological and pathological activities which include acting as signaling molecules that regulate lipid, glucose and energy metabolism [28]; however, very little is known about the molecular mechanisms of bile acids in the central nervous system [29]. It has, however, been shown that following primary bile acid synthesis in the liver, bile acids are subsequently secreted into the gut where they are modified by the intestinal bacteria to produce secondary bile acids. These can be further modified in the liver or gut and may be conjugated with glycine or taurine [30].

Figure 3 displays a simplified depiction of the biochemistry. In Figure 3, we show which bile acids have been reported as being cytotoxic and neuroprotective [31,32]. Of the neuroprotective bile acids measured in this study, UDCA and TUDCA were found to be at markedly lower concentrations in the serum of PFF mice as compared to controls (17-fold and 14-fold, respectively). UDCA and TUDCA are secondary bile acids, produced in the gut and not in the liver. They have been reported to have neuroprotective effects in the brain, functioning partly as chaperones, decreasing the formation of toxic aggregates in protein folding disorders [33,34]. Further, they have also been reported to reduce reactive oxygen species formation [35], inhibit apoptosis [36] and prevent mitochondrial dysfunction [37].

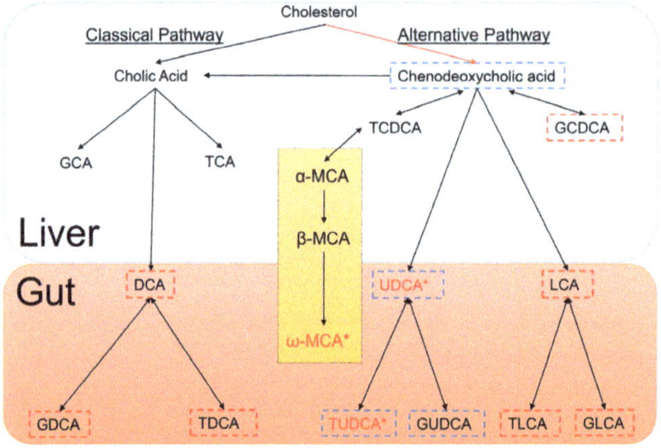

Figure 3. Depiction of Bile Acid Metabolism in the liver and gut of mice. Bile acids outlined in blue are neuroprotective, bile acids outlined in red are cytotoxic and those bile acids in red with an accompanying asterisk are statistically significantly different between HuMonomer- and PFF-injected mice. The section detailing Muricholic acid (MCA) only occurs in mice.

A recent emerging and exciting concept in health and disease is the ability of the guts microbiota to communicate with the brain and subsequently modulate behavior [38]. This bidirectional signaling axis between the gut and the brain is believed to be essential for conserving homeostasis which is regulated at the hormonal, immunological and neuronal levels (central and enteric nervous systems) [38]. While a lot of attention has been placed on the gut microbiome and neurodegenerative diseases, most of the reported studies have focused on the gut as being the driver. In this study, we show that by inducing α-synucleinopathy in the brain with PFFs to mirror what is observed in prodromal PD, we see a significant decrease in the concentrations of secondary bile acids which have neuroprotective properties. As depicted in Figure 3, the production of these secondary, neuroprotective bile acids only occurs in the gut by intestinal bacteria. So, is the formation of the α-syn aggregates in the brain directly affecting the PFF mouse gut bacteria and the formation of secondary bile acids deemed neuroprotective? Or is it possible that these neuroprotective bile acids are being degraded faster in the prodromal PD brain due to the developing α-synucleinopathy which subsequently leads to lower blood concentrations? Both hypotheses need further exploration in the future.

We report, for the first time, a bile acid biomarker panel capable of identifying mice with developing α-synucleinopathy. Using bile acids as biomarkers is a marked improvement on our previous metabolomics work and highlights the potential of bile acids for the prediction of those patients at greatest risk of developing PD, particularly in the prodromal phase when a treatment aiming at slowing disease progression is potentially most effective and might even delay the onset of motor symptoms [8,9]. Further, our results demonstrate a potential novel therapeutic area for prodromal PD and developing α-synucleinopathy which needs future exploration. More work is required to verify these initial hypotheses, using mouse models and, most importantly, large clinical cohorts of people who exhibit several signs of prodromal PD.

4. Materials and Methods

4.1. Animals

Under 12 h light/12 h dark cycles, C57Bl/6J mice (Jackson Laboratory) were housed four to five per cage with *ad libitum* access to food and water. As previously described by our group, all procedures relating to the animals followed The Guide for Care and Use of Laboratory Animals (National Research Council) and were validated by the Van Andel Research Institute's Institutional Animal Care and Use Committee (Animal Use Protocols 14-01-001 and 16-12-033).

4.2. Purification of Recombinant α-syn, Assembly of Preformed Fibrils and Stereotactic Injections

Recombinant α-syn purification, assembly of the fibrils and stereotactic injections were previously described by our group [7,10,39]. In brief, we cultured BL21 *E. coli* and induced them to express human α-syn. The bacteria were then pelleted, and lysed by sonication. We boiled the lysate for 10 min and collected the supernatant after centrifugation. The supernatant was then dialyzed overnight in 10 mM Tris, pH 7.5, 50 mM NaCl, and 1 mM EDTA. The lysate was then purified by chromatographic separation using a Superdex 200 Column (GE Healthcare Life Sciences, Marlborough, MA, USA) and a Hi-trap Q HP anion exchange column (GE Healthcare Life Sciences, Marlborough, MA, USA). Extracts from the different fractions were then migrated by SDS-PAGE and we identified the fractions containing α-syn after Coomassie staining. The selected fractions were then collected and dialyzed against PBS buffer (GE Healthcare Life Sciences, Marlborough, MA, USA). We then measured the final concentration of purified recombinant α-syn using a NanoDrop 2000 (Thermofisher Scientific, Waltham, MA, USA) and concentrated if needed. Aliquots were stored at −80 °C until use. For fibril assembly, purified recombinant α-syn was thawed and diluted to 5 mg/mL in PBS and under continuous shaking at 1000 rpm at 37 °C in a Thermomixer (Eppendorf, Hamburg, Germany) for 7 days. Fibrils were aliquoted and frozen at −80 °C until use.

Before injection, human α-syn fibrils (PFFs, 5 µg/µL) were thawed at RT and sonicated at RT as previously described in Graham et al., 2018 [10]. Human α-syn monomers (huMonomers) were thawed and we collected the supernatant after ultracentrifugation at 100,000 *g* for 30 min. We injected mice stereotactically with PFFs ($n = 20$) or huMonomers ($n = 20$) (0.8 µL, 5 µg/µL) in the OB (unilateral) of 2 months-old wild type mice as previously described [7,40]. Two mice injected with huMonomers were euthanized after developing severe dermatitis, unrelated to the surgical procedure.

We imaged the fibrils post-sonication by transmission electron microscopy to check the morphology of the fibrils. Human fibrils (after sonication) were diluted to 0.1 µg/µL into sterile PBS and negatively stained with 2% uranyl formate (Electron Microscopy Science, Hatfield, PA, USA, ref #22400). Grids were imaged using a FEI Tecnai G2 Spirit TWIN transmission electron microscope (FEI Company, Hillsboro, OR, USA) at 120 kV (Figure S1).

4.3. Serum Collection

Serum samples were acquired as previously described by our group [10]. Three months post-injection, mice were deeply anesthetized with sodium pentobarbital and we collected blood at final bleed by cardiac puncture in BD red top–vacutainer tubes. We kept the tubes at RT for 20–30 min to allow blood clot formation and then centrifuged them at 4500 *g* for 10 min at 15 °C. The serum was collected and transferred to pre-cooled vials, vortexed, aliquoted and frozen on crushed dry ice. Samples were then stored at −80 °C.

4.4. Bile Acid Quantification

Bile acids were analyzed using the Biocrates® Bile Acids Kit (Biocrates Life Science AG, Innsbruck, Austria) as described by our group previously [22]. In brief, data were acquired on a Waters TQ-S spectrometer coupled with an Acquity I-Class ultra-pressure liquid chromatography (UPLC) system. All serum specimens were acquired in accordance with the protocol as described in the Bile Acids kit manual. All data analysis was completed using the Biocrates MetIDQ software and TargetLynx (Waters, Milford, MA, USA).

4.5. Statistical Analysis

All data were analyzed using MetaboAnalyst (v4.0) [41]. A Wilcoxon–Mann–Whitney U-test was performed on all data acquired to determine whether there were any significantly different metabolites between prodromal PD model mice and age-matched controls injected with HuMonomers ($p < 0.05$; q-value < 0.05). Bonferroni-corrected p-values were used to correct for multiple comparisons.

Prior to logistic regression analyses, all data were normalized to the sum and autoscaled. To select the predictor variables used in the logistic regression analyses, Least Absolute Shrinkage and Selection Operator (LASSO) and stepwise variable selection were utilized for optimizing all the model components [42]. A k-fold cross-validation (CV) technique was used to show that the models were not over fit and to assess potential predictive accuracy in an independent sample [43]. Area under the curve (AUC (95% confidence interval)), sensitivity and specificity values were calculated to estimate the performance of the logistic regression and ROC analyses.

Supplementary Materials: The following are available online at http://www.mdpi.com/2218-1989/8/4/71/s1, Figure S1: Sonicated PFFs stained by uranyl formate, imaged by transmission electron microscopy to confirm their fibrillary nature.

Author Contributions: Designing Research Studies, S.F.G., N.L.R., J.A.S. and P.B.; Conducting Experiments, S.F.G., N.L.R., Z.U., A.Y., E.S. (Eric Sherman), K.B., E.S. (Emily Schulz), L.K.M. and J.A.S.; Analyzing Data, S.F.G.; Drafting the Manuscript, S.F.G., N.L.R. and P.B.; Writing the Manuscript, all authors contributed to the editing of the manuscript.

Acknowledgments: We would like to thank Anne Whitlaw for her charitable donation which helped to fund the metabolomics section of this work. In addition, this work was partly funded by the generous contribution made by the Fred A. & Barbara M. Erb Foundation. We acknowledge the Van Andel Research Institute and the many individuals and corporations that financially support research into neurodegenerative disease at the Institute. N.L.R. is supported by the Peter C. and Emajean Cook Foundation. P.B. is supported by grants from the National Institutes of Health (1R01DC016519-01 and 5R21NS093993-02). P.B. reports additional grants from Office of the Assistant Secretary of Defense for Health Affairs (Parkinson's Research Program, Award No. W81XWH-17-1-0534), The Michael J Fox Foundation, National Institutes of Health, Cure Parkinson's Trust, which are outside but relevant to the submitted work.

Conflicts of Interest: P.B. has received commercial support as a consultant from Renovo Neural, Inc., Fujifilm-Cellular Dynamics, Axial Biotherapeutics, Roche, Teva Inc., Lundbeck A/S, NeuroDerm, AbbVie, ClearView Healthcare, FCB Health, IOS Press Partners and Capital Technologies, Inc. He is conducting sponsored research on behalf of Roche and Lundbeck A/S. He has ownership interests in Acousort AB, Lund, Sweden. The other authors declare no conflict of interest.

References

1. De Lau, L.M.; Breteler, M.M. Epidemiology of Parkinson's disease. *Lancet Neurol.* **2006**, *5*, 525–535. [CrossRef]
2. Kalia, L.V.; Lang, A.E. Parkinson's disease. *Lancet* **2015**, *386*, 896–912. [CrossRef]
3. Tysnes, O.-B.; Storstein, A. Epidemiology of Parkinson's disease. *J. Neural Transm.* **2017**, *124*, 901–905. [CrossRef] [PubMed]
4. Schapira, A.H.V.; Chaudhuri, K.R.; Jenner, P. Non-motor features of Parkinson disease. *Nat. Rev. Neurosci.* **2017**, *18*, 435–450. [CrossRef] [PubMed]
5. Postuma, R.B.; Berg, D. Advances in markers of prodromal Parkinson disease. *Nat. Rev. Neurol.* **2016**, *12*, 622–634. [CrossRef] [PubMed]
6. Havelund, J.F.; Heegaard, N.H.H.; Faergeman, N.J.K.; Gramsbergen, J.B. Biomarker Research in Parkinson's Disease Using Metabolite Profiling. *Metabolites* **2017**, *7*, 42. [CrossRef] [PubMed]

7. Rey, N.L.; Steiner, J.A.; Maroof, N.; Luk, K.C.; Madaj, Z.; Trojanowski, J.Q.; Lee, V.M.-Y.; Brundin, P. Widespread transneuronal propagation of α-synucleinopathy triggered in olfactory bulb mimics prodromal Parkinson's disease. *J. Exp. Med.* **2016**, *213*, 1759–1778. [CrossRef] [PubMed]
8. Espay, A.J.; Schwarzschild, M.A.; Tanner, C.M.; Fernandez, H.H.; Simon, D.K.; Leverenz, J.B.; Merola, A.; Chen-Plotkin, A.; Brundin, P.; Erro, R.; et al. Biomarker-driven phenotyping in Parkinson's disease: A translational missing link in disease-modifying clinical trials. *Mov. Disord. Off. J. Mov. Dis. Soc.* **2017**, *32*, 319–324. [CrossRef] [PubMed]
9. Espay, A.J.; Brundin, P.; Lang, A.E. Precision medicine for disease modification in Parkinson disease. *Nat. Rev. Neurol.* **2017**, *13*, 119–126. [CrossRef] [PubMed]
10. Graham, S.F.; Rey, N.L.; Yilmaz, A.; Kumar, P.; Madaj, Z.; Maddens, M.; Bahado-Singh, R.O.; Becker, K.; Schulz, E.; Meyerdirk, L.K.; et al. Biochemical Profiling of the Brain and Blood Metabolome in a Mouse Model of Prodromal Parkinson's Disease Reveals Distinct Metabolic Profiles. *J. Proteom Res.* **2018**, *17*, 2460–2469. [CrossRef] [PubMed]
11. Camilleri, M.; Gores, G.J. Therapeutic targeting of bile acids. *Am. J. Phys. Gastrointest. Liver Phys.* **2015**, *309*, G209–G215. [CrossRef] [PubMed]
12. Thomas, C.; Pellicciari, R.; Pruzanski, M.; Auwerx, J.; Schoonjans, K. Targeting bile-acid signalling for metabolic diseases. *Nat. Rev. Drug Dis.* **2008**, *7*, 678–693. [CrossRef] [PubMed]
13. Perino, A.; Schoonjans, K. TGR5 and Immunometabolism: Insights from Physiology and Pharmacology. *Trends Pharmacol. Sci.* **2015**, *36*, 847–857. [CrossRef] [PubMed]
14. Parry, G.J.; Rodrigues, C.M.; Aranha, M.M.; Hilbert, S.J.; Davey, C.; Kelkar, P.; Low, W.C.; Steer, C.J. Safety, tolerability, and cerebrospinal fluid penetration of ursodeoxycholic Acid in patients with amyotrophic lateral sclerosis. *Clin. Neuropharmacol.* **2010**, *33*, 17–21. [CrossRef] [PubMed]
15. Mano, N.; Goto, T.; Uchida, M.; Nishimura, K.; Ando, M.; Kobayashi, N.; Goto, J. Presence of protein-bound unconjugated bile acids in the cytoplasmic fraction of rat brain. *J. Lipid Res.* **2004**, *45*, 295–300. [CrossRef] [PubMed]
16. Bron, B.; Waldram, R.; Silk, D.B.; Williams, R. Serum, cerebrospinal fluid, and brain levels of bile acids in patients with fulminant hepatic failure. *Gut* **1977**, *18*, 692–696. [CrossRef] [PubMed]
17. Olazaran, J.; Gil-de-Gomez, L.; Rodriguez-Martin, A.; Valenti-Soler, M.; Frades-Payo, B.; Marin-Munoz, J.; Antunez, C.; Frank-Garcia, A.; Acedo-Jimenez, C.; Morlan-Gracia, L.; et al. A blood-based, 7-metabolite signature for the early diagnosis of Alzheimer's disease. *J. Alzheimer's Dis.* **2015**, *45*, 1157–1173. [CrossRef] [PubMed]
18. Bathena, S.P.; Mukherjee, S.; Olivera, M.; Alnouti, Y. The profile of bile acids and their sulfate metabolites in human urine and serum. *J. Chromatogr. B Anal. Technol. Biomed. Life Sci.* **2013**, *942–943*, 53–62. [CrossRef] [PubMed]
19. Abdelkader, N.F.; Safar, M.M.; Salem, H.A. Ursodeoxycholic Acid Ameliorates Apoptotic Cascade in the Rotenone Model of Parkinson's Disease: Modulation of Mitochondrial Perturbations. *Mol. Neurobiol.* **2016**, *53*, 810–817. [CrossRef] [PubMed]
20. Theofilopoulos, S.; Wang, Y.; Kitambi, S.S.; Sacchetti, P.; Sousa, K.M.; Bodin, K.; Kirk, J.; Salto, C.; Gustafsson, M.; Toledo, E.M.; et al. Brain endogenous liver X receptor ligands selectively promote midbrain neurogenesis. *Nat. Chem. Biol.* **2013**, *9*, 126–133. [CrossRef] [PubMed]
21. Marksteiner, J.; Blasko, I.; Kemmler, G.; Koal, T.; Humpel, C. Bile acid quantification of 20 plasma metabolites identifies lithocholic acid as a putative biomarker in Alzheimer's disease. *Metabolomics* **2018**, *14*, 1. [CrossRef] [PubMed]
22. Pan, X.; Elliott, C.T.; McGuinness, B.; Passmore, P.; Kehoe, P.G.; Holscher, C.; McClean, P.L.; Graham, S.F.; Green, B.D. Metabolomic Profiling of Bile Acids in Clinical and Experimental Samples of Alzheimer's Disease. *Metabolites* **2017**, *7*, 28. [CrossRef] [PubMed]
23. Braak, H.; Ghebremedhin, E.; Rub, U.; Bratzke, H.; Del Tredici, K. Stages in the development of Parkinson's disease-related pathology. *Cell Tissue Res.* **2004**, *318*, 121–134. [CrossRef] [PubMed]
24. MahmoudianDehkordi, S.; Arnold, M.; Nho, K.; Ahmad, S.; Jia, W.; Xie, G.; Louie, G.; Kueider-Paisley, A.; Moseley, M.A.; Thompson, J.W.; et al. Altered bile acid profile associates with cognitive impairment in Alzheimer's disease-An emerging role for gut microbiome. *Alzheimer's Dement. J. Alzheimer's Assoc.* **2018**. [CrossRef] [PubMed]

25. Braak, H.; Del Tredici, K. Neuropathological Staging of Brain Pathology in Sporadic Parkinson's disease: Separating the Wheat from the Chaff. *J. Parkinson's Dis.* **2017**, *7*, S71–S85. [CrossRef] [PubMed]
26. Beach, T.G.; White, C.L., 3rd; Hladik, C.L.; Sabbagh, M.N.; Connor, D.J.; Shill, H.A.; Sue, L.I.; Sasse, J.; Bachalakuri, J.; Henry-Watson, J.; et al. Olfactory bulb alpha-synucleinopathy has high specificity and sensitivity for Lewy body disorders. *Acta Neuropathol.* **2009**, *117*, 169–174. [CrossRef] [PubMed]
27. Beach, T.G.; Adler, C.H.; Lue, L.; Sue, L.I.; Bachalakuri, J.; Henry-Watson, J.; Sasse, J.; Boyer, S.; Shirohi, S.; Brooks, R.; et al. Unified staging system for Lewy body disorders: Correlation with nigrostriatal degeneration, cognitive impairment and motor dysfunction. *Acta Neuropathol.* **2009**, *117*, 613–634. [CrossRef] [PubMed]
28. Liu, Y.; Rong, Z.; Xiang, D.; Zhang, C.; Liu, D. Detection technologies and metabolic profiling of bile acids: A comprehensive review. *Lipid Health Dis.* **2018**, *17*, 121. [CrossRef] [PubMed]
29. Lieu, T.; Jayaweera, G.; Bunnett, N.W. GPBA: A GPCR for bile acids and an emerging therapeutic target for disorders of digestion and sensation. *Br. J. Pharmacol.* **2014**, *171*, 1156–1166. [CrossRef] [PubMed]
30. Hofmann, A.F. The continuing importance of bile acids in liver and intestinal disease. *Arch. Int. Med.* **1999**, *159*, 2647–2658. [CrossRef]
31. Benedetti, A.; Alvaro, D.; Bassotti, C.; Gigliozzi, A.; Ferretti, G.; La Rosa, T.; Di Sario, A.; Baiocchi, L.; Jezequel, A.M. Cytotoxicity of bile salts against biliary epithelium: A study in isolated bile ductule fragments and isolated perfused rat liver. *Hepatology* **1997**, *26*, 9–21. [CrossRef] [PubMed]
32. Mello-Vieira, J.; Sousa, T.; Coutinho, A.; Fedorov, A.; Lucas, S.D.; Moreira, R.; Castro, R.E.; Rodrigues, C.M.; Prieto, M.; Fernandes, F. Cytotoxic bile acids, but not cytoprotective species, inhibit the ordering effect of cholesterol in model membranes at physiologically active concentrations. *Biochim. Biophys. Acta* **2013**, *1828*, 2152–2163. [CrossRef] [PubMed]
33. Geier, A.; Wagner, M.; Dietrich, C.G.; Trauner, M. Principles of hepatic organic anion transporter regulation during cholestasis, inflammation and liver regeneration. *Biochim. Biophys. Acta* **2007**, *1773*, 283–308. [CrossRef] [PubMed]
34. Cortez, L.M.; Campeau, J.; Norman, G.; Kalayil, M.; Van der Merwe, J.; McKenzie, D.; Sim, V.L. Bile Acids Reduce Prion Conversion, Reduce Neuronal Loss, and Prolong Male Survival in Models of Prion Disease. *J. Virol.* **2015**, *89*, 7660–7672. [CrossRef] [PubMed]
35. Rodrigues, C.M.; Fan, G.; Wong, P.Y.; Kren, B.T.; Steer, C.J. Ursodeoxycholic acid may inhibit deoxycholic acid-induced apoptosis by modulating mitochondrial transmembrane potential and reactive oxygen species production. *Mol. Med.* **1998**, *4*, 165–178. [CrossRef] [PubMed]
36. Rodrigues, C.M.; Sola, S.; Sharpe, J.C.; Moura, J.J.; Steer, C.J. Tauroursodeoxycholic acid prevents Bax-induced membrane perturbation and cytochrome C release in isolated mitochondria. *Biochemistry* **2003**, *42*, 3070–3080. [CrossRef] [PubMed]
37. Rodrigues, C.M.; Fan, G.; Ma, X.; Kren, B.T.; Steer, C.J. A novel role for ursodeoxycholic acid in inhibiting apoptosis by modulating mitochondrial membrane perturbation. *J. Clin. Investig.* **1998**, *101*, 2790–2799. [CrossRef] [PubMed]
38. Cryan, J.F.; O'Mahony, S.M. The microbiome-gut-brain axis: From bowel to behavior. *Neurogastroenterol. Motil.* **2011**, *23*, 187–192. [CrossRef] [PubMed]
39. Rey, N.L.; George, S.; Steiner, J.A.; Madaj, Z.; Luk, K.C.; Trojanowski, J.Q.; Lee, V.M.-Y.; Brundin, P. Spread of aggregates after olfactory bulb injection of α-synuclein fibrils is associated with early neuronal loss and is reduced long term. *Acta Neuropathol.* **2018**, *135*, 65–83. [CrossRef] [PubMed]
40. Rey, N.L.; Petit, G.H.; Bousset, L.; Melki, R.; Brundin, P. Transfer of human alpha-synuclein from the olfactory bulb to interconnected brain regions in mice. *Acta Neuropathol.* **2013**, *126*, 555–573. [CrossRef] [PubMed]
41. Xia, J.; Sinelnikov, I.V.; Han, B.; Wishart, D.S. MetaboAnalyst 3.0—Making metabolomics more meaningful. *Nucleic Acid Res.* **2015**, *43*, W251–W257. [CrossRef] [PubMed]
42. Tibshirani, R. Regression Shrinkage and Selection via the Lasso. *J. R. Stat. Soc. Ser. B* **1996**, *58*, 267–288.
43. Xia, J.; Psychogios, N.; Young, N.; Wishart, D.S. MetaboAnalyst: A web server for metabolomic data analysis and interpretation. *Nucleic Acid Res.* **2009**, *37*, W652–W660. [CrossRef] [PubMed]

metabolites

MDPI

Review

Metabolomics and Age-Related Macular Degeneration

Connor N. Brown [1], Brian D. Green [2], Richard B. Thompson [3], Anneke I. den Hollander [4], Imre Lengyel [1,*] and on behalf of the EYE-RISK consortium †

1 Wellcome-Wolfson Institute for Experimental Medicine (WWIEM), Queen's University Belfast, Belfast BT9 7BL, UK; cbrown88@qub.ac.uk
2 Institute for Global Food Security (IGFS), Queen's University Belfast, Belfast BT9 6AG, UK; b.green@qub.ac.uk
3 Department of Biochemistry and Molecular Biology, School of Medicine, University of Maryland, Baltimore, MD 21201, USA; rthompson@som.umaryland.edu
4 Department of Ophthalmology, Radboud University Nijmegen Medical Centre, Nijmegen 6525 EX, The Netherlands; a.denhollander@antrg.umcn.nl
* Correspondence: i.lengyel@qub.ac.uk; Tel.: +44-289-097-6027
† Membership of the Eye-Risk Consortium is provided in the Acknowledgments.

Received: 21 November 2018; Accepted: 20 December 2018; Published: 27 December 2018

Abstract: Age-related macular degeneration (AMD) leads to irreversible visual loss, therefore, early intervention is desirable, but due to its multifactorial nature, diagnosis of early disease might be challenging. Identification of early markers for disease development and progression is key for disease diagnosis. Suitable biomarkers can potentially provide opportunities for clinical intervention at a stage of the disease when irreversible changes are yet to take place. One of the most metabolically active tissues in the human body is the retina, making the use of hypothesis-free techniques, like metabolomics, to measure molecular changes in AMD appealing. Indeed, there is increasing evidence that metabolic dysfunction has an important role in the development and progression of AMD. Therefore, metabolomics appears to be an appropriate platform to investigate disease-associated biomarkers. In this review, we explored what is known about metabolic changes in the retina, in conjunction with the emerging literature in AMD metabolomics research. Methods for metabolic biomarker identification in the eye have also been discussed, including the use of tears, vitreous, and aqueous humor, as well as imaging methods, like fluorescence lifetime imaging, that could be translated into a clinical diagnostic tool with molecular level resolution.

Keywords: age-related macular degeneration; metabolomics; metabolism; biomarkers; drusen; retinal pigment epithelium

1. Introduction

Age-related macular degeneration (AMD) accounts for 8.7% of the world's total blindness and is the leading cause of irreversible visual impairment in the Western world of people aged 65 and older [1,2]. The total number of individuals with this condition is expected to rise to 196 million by 2020 and 288 million by 2040 [2]. Choroidal neovascularization (CNV) is the most aggressive form of advanced AMD and results in the rapid loss of central vision. The other form of advanced AMD, geographic atrophy (GA), is characterized by the progressive loss of central vision due to the death of retinal pigment epithelium (RPE) and photoreceptor cells. At early stages of AMD, accumulation of intracellular lipofuscin in the RPE and the build-up of extracellular deposits under the RPE occurs [3]. Treatment strategies are available for CNV, but not for GA. Antibodies against vascular endothelial growth factor (VEGF) can halt progression of CNV, but the effect is usually not permanent and the disease usually progresses to macular atrophy after the anti-VEGF treatment [4,5].

Age is a predominant risk factor contributing to the development of any AMD, with the prevalence reaching nearly 30% over the age of 85 years in a European population [6]. Environmental factors, such as diet (fat intake and antioxidants) and lifestyle, particularly smoking [7,8], also contribute to the risk of developing the disease. There are several genetic variants that are associated with an increased prevalence of AMD [9]. The two most significant of these are polymorphisms in the *CFH* and *ARMS2* (age-related maculopathy susceptibility 2) genes. The *CFH* gene encodes for complement factor H, a glycoprotein which has an integral role in the regulation of the alternative complement pathway [10]. The *ARMS2* gene [11,12] encodes the ARMS2 protein, which helps to initiate complement activation from the surface of retinal monocytes and microglia by binding to the surface of apoptotic and necrotic cells [13]. More than 50 genetic variants at 34 loci have been associated with AMD development [9]. A large proportion of these genetic variants are located in or near genes of the complement system, lipid metabolism, and extracellular remodeling [14]. These genetic associations, along with risk factors related to diet, serum cholesterol, and triglyceride levels for late-stage AMD [15], highlight the potential importance of studying metabolomic changes to identify systemic molecular biomarkers associated with these risks. In addition, the cellular interactions and exchange of metabolites between the retina, RPE, and choroid complex also provide an important basis of studying local metabolomic alterations in AMD. Considering that the metabolome is closer to the molecular phenotype than either the genome, the transcriptome, or the proteome [16], the study of metabolites could lead to better prediction of the resulting phenotype than other -omics approaches. With AMD progression increasingly associated with metabolic dysfunction, there is the potential for systemic and local metabolic biomarkers to provide opportunities to detect and follow the progression of the disease at an early stage. Systemic biomarkers and the role of lipid metabolism in AMD has been comprehensively highlighted [17,18], whilst general reviews have briefly covered the metabolomics of AMD, alongside other ocular diseases, and the limitations associated with such studies [19–21].

The aim of this review is to specifically highlight metabolic processes occurring within the retina, including those related to the development of AMD. It will also provide an in-depth overview of metabolomics studies conducted in AMD. The possible biofluids and metabolomics methods will also be discussed, as well as the potential utility of metabolomics for discovering biomarkers and identifying new therapeutic approaches for AMD and related diseases.

2. Metabolic Processes in the Posterior Eye

2.1. Energy Sources in the Retina

The complex cellular interactions in the retina give rise to a unique metabolic environment, which is influenced by a range of external factors, including the differential blood supply to the layers of the retina and the detection of light or darkness by the photoreceptors [22]. Due to the number of cell types, there are various metabolic processes which occur throughout the vertebrate retina [23]. Glucose is the primary fuel source for the photoreceptors in the retina, supplied by the choriocapillaris through the Bruch's membrane (BrM) and the RPE. The metabolic environment of the retina is diverse and increases in complexity with the laminated morphology present between the cells. For instance, the glucose concentration decreases from the RPE surface to the retina surface in vitro [24,25]. Aerobic glycolysis is the primary form of energy metabolism in the retina [26], where glucose is converted to lactate at a comparable rate to cancer cells [27], even when there is plenty of oxygen. In avascular retinas [28], an alternative energy source to adenosine triphosphate (ATP) is required as ATP is exposed to highly active ATP degrading ion pumps between the centrally located photoreceptor mitochondria and the photoreceptor synaptic terminal [29,30]. The phosphocreatine shuttle is instead used [30] and although this is not essential in vascularized mouse retinas, an isoform of the creatine kinase is still localized at the photoreceptor synaptic terminal. The energy metabolism of the retina is unique in that it is dependent on the light or dark state of the tissue and respiration is more uncoupled from ATP synthesis than in other tissues [31].

RPE cells appear to be different from photoreceptors in that they are specialized to utilize reductive carboxylation as a source of energy [32]. This minimizes RPE glucose consumption for the reduced form of nicotinamide adenine dinucleotide phosphate (NADP(H)) generation, so efficient glucose transport occurs between the choroid and retina. This process is disrupted by excessive oxidative stress and mitochondrial dysfunction, which can be a result of an inability of mitochondria to access necessary substrates, as well as defective electron transport and ATP-synthesis systems [33]. Glycolysis and reductive carboxylation become hindered [32] as pyridine nucleotides, such as reduced nicotinamide adenine dinucleotide (NAD(H)), are depleted, which leads to RPE and retinal degeneration [34,35]. When glycolysis is promoted in these cells, photoreceptors die [36,37], leading to the conclusion that the retina and RPE contribute specific metabolic functions, which maintain their own functional ecosystem [38], both in vivo and in vitro.

2.2. *Lipofuscin Accumulation in the RPE*

Lipofuscin is a lipid-containing, pigmented granule, which accumulates in various tissues throughout the body as a result of aging, and can be found as an accumulation in the RPE [39]. Lipofuscin accumulation is a potential risk factor contributing to the development of AMD [40]. RPE cells convert the condensation product of all-*trans*-retinal and phosphatidylethanolamine to *N*-retinyl-*N*-retinylidene ethanolamine (A2E), which is a major component, and the main chromophore, of lipofuscin [41,42]. There is direct evidence that A2E alters cholesterol metabolism in the RPE, contributing to AMD [43]. In addition, A2E directly causes RPE cytotoxicity by inducing apoptosis through specific inhibition of cytochrome *c* oxidase (COX) [44,45], which leads to the inhibition of oxygen consumption and light homeostasis. This increase in oxidative stress leads to mitochondrial dysfunction, which releases two apoptosis-promoting molecules, cytochrome *c* and apoptosis inducing factor (AIF) [46,47], from the mitochondria of RPE cells [45]. Lipofuscin could, therefore, contribute to the disruption of mitochondrial function and oxidative stress in AMD. This is further highlighted through links that lipofuscin has with the essential trace element, zinc, which is highly concentrated in the RPE.

In the retina, zinc is required for the metabolism of ingested photoreceptor outer segments (POS) by the RPE [48] and provides protection against oxidative stress [49]. For this reason, it has been suggested that zinc deficiency is linked with AMD, with oral supplementation providing a protective effect [50–52]. This decreases the risk of progression from intermediate stages of the disease to the neovascular form in clinical trials [53], but the direct effects of zinc deficiency in the retina remain to be explored extensively. Julien et al. [54] identified an accumulation of lipofuscin and lipofuscin-like products in the RPE of zinc-deficient rats, which may contribute to AMD progression [40]. The authors also proposed that this zinc deficiency contributed to lipofuscin accumulation due to the oxidative stress and functional deficiency of RPE lysosomes as a result of lipid membrane damage caused by lipid peroxidation from zinc deficiency [54]. This ultimately led to incomplete degradation of POS in the dysfunctional RPE lysosomes [55]. This has been further demonstrated in more recent studies exploring the mechanisms of zinc deficiency and supplementation [52,56].

There is also evidence which suggests that the link between lipofuscin accumulation in the RPE and AMD progression is more tenuous. Although A2E is known to be produced in the RPE, it has been shown that, in humans, A2E is preferentially located in the RPE cells of the peripheral retina, rather than the central area of the retina containing the macula [57]. Therefore, in fundus autofluorescence (FAF) imaging, the higher levels of lipofuscin fluorescence associated with the central area of the RPE cannot be exclusively attributed to the concentration of A2E. Other studies have also contradicted the aforementioned results of an increased signal on FAF by quantitatively determining that the fluorescence signal associated with lipofuscin and A2E decreases from subgroups of early to late AMD patients compared to controls [58,59]. As metabolic waste products, including lipofuscin, have previously been associated with AMD progression, clinical trials began to investigate the treatment of GA by limiting the metabolic waste accumulation within the RPE [60,61]. These studies showed no significant reduction in the rate of progression of GA, further highlighting the inconsistent evidence regarding lipofuscin accumulation and its role in the development of AMD.

2.3. *Sub-RPE Accumulations*

2.3.1. Lipid Accumulation

The trafficking and accumulation of lipids and lipid metabolites have long been associated with BrM aging [62–64]. As identification of these lipids has progressed, studies identified esterified and unesterified cholesterol (EC and UC, respectively), which are also known as cholesteryl ester (CE) and free cholesterol (FC), respectively [65,66]. Oil red O staining demonstrated that the EC accumulated in the macula seven-fold more than in the periphery of the retina [64,67]. Although EC accumulates exclusively at the BrM, UC and other phospholipids are also found within cellular and intracellular membranes [68]. The association between cholesterol and lipids in the retina, their link to AMD, and potential therapeutic target strategies have previously been reviewed [18,69,70] (Figure 1).

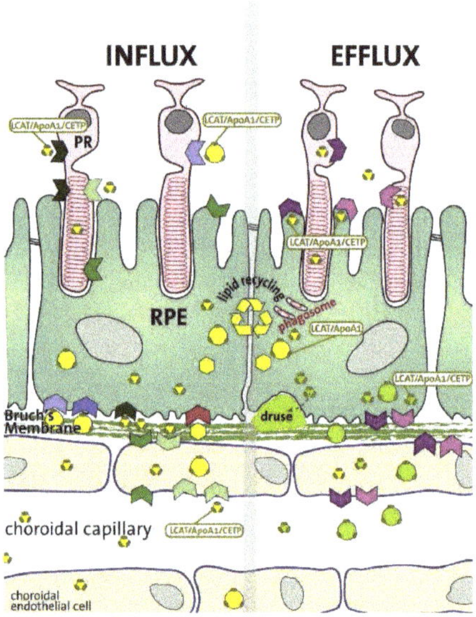

Figure 1. The metabolic flux of various lipids and their associated proteins in the retinal pigment epithelium (RPE). In addition to anabolic and catabolic lipid metabolism, the RPE also functions as a transfer site for lipids and proteins (yellow and green spheres) between the circulation and the photoreceptors. The influx of lipids from the RPE to the photoreceptors is represented on the left side, whilst the efflux of lipids is represented on the right side of the image. There is substantial recycling of lipids by the RPE, which are continuously provided through the phagocytosed membrane discs of photoreceptor outer segments (POS). The oxidized lipid species either enter the circulation as lipoprotein particles (green spheres) or are basally deposited into the sub-RPE space leading to the formation of drusen. LCAT, lecithin-cholesterol acyltransferase; APOA1, apolipoprotein A1; CETP, cholesteryl ester transfer protein. Colored arrows represent lipid receptors and their direction of transport. Reproduced with permission from van Leeuwen et al. [18].

2.3.2. Advanced Glycation End Product Accumulation

Advanced glycation end products (AGEs), oxidized products of non-enzymatic, extracellular protein and lipid glycosylation, have been associated with and implicated in AMD progression for several decades. In one early investigation, the accumulation of these molecules was associated with soft, macular drusen and RPE cells [71]. This accumulation was proposed to contribute to the neovascularization associated with AMD, and more recently, they have been found to accumulate in the BrM [72]. In this localized region, they inhibit protein function and are associated with age-related damage. ARPE-19 cells have also been grown in the presence of AGEs, which leads to an additional increase in the accumulation of lipofuscin, which has its own implication in the pathogenesis of AMD previously discussed.

As these chemical modifications are promoted by smoking, a major risk factor for AMD development [73], it is not surprising that AGEs and their receptors (RAGEs) are suggested to promote the development of AMD. A more recent study investigated the activation of the AGE receptor (RAGE) when the threshold for AGE accumulation is reached [74]. The activation of RAGE is associated with transitioning an acute inflammatory response to a more chronic disease, as indicated in this study, which demonstrated that RAGE was significantly associated with CNV in mice.

2.3.3. Drusen Accumulation and Development

One of the main issues surrounding the prevention of AMD is that the early stages of the disease are often asymptomatic. This means that AMD is often identified when a patient has already progressed to the intermediate stage of the disease and might be suffering from partial sight loss. Early stage AMD can still be characterized by the presence of sub-RPE deposits, which accumulate with age and are associated with the thickening of Bruch's membrane. Sub-RPE deposits are focal, termed drusen, or diffuse, termed basal laminar (BLamD) or linear (BLinD) deposits. Their formation can contribute to RPE detachment and photoreceptor death [75–77]. Lipids, proteins, minerals, and cellular debris are constituents of all sub-RPE deposits, but the composition of the various types can differ [78]. Hard drusen are associated with the normal aging process while intermediate and large soft drusen, as well as BLinD and BLamD, are implicated as contributing to increased disease susceptibility, especially when they are present in the macula [79]. Using new clinical imaging modalities, sub-RPE deposits are becoming better phenotyped in clinical settings [80] and, in combination with laboratory imaging, can help to develop better diagnosis [81].

Although drusen form in the natural aging process, when they increase in number and size, they are critical for the development of late-stage AMD. Indeed, an increase in drusen volume, as measured with spectral domain optical coherence tomography (SD-OCT) in vivo, has been shown to increase the risk of progression from the intermediate stages to advanced AMD [82,83]. Various research groups have begun assessing the molecular constituents contributing to this pathological accumulation. The molecular composition of drusen has been studied in the past, with a major focus on the proteins present in drusen. Common components repeatedly found include apolipoprotein E [84]; amyloid components, including amyloid β [85,86]; complement components; and vitronectin [87,88]. Although the protein composition of sub-RPE deposits is not limited to these proteins, it is worth noting that they were also found to be coating hydroxyapatite (HAP or $Ca_5(PO_4)_3OH$) spherules [89] and large HAP nodules [81]. These studies suggest that metabolic changes associated with mineral formation are involved. It has been shown that HAP can be deposited onto cholesterol-containing extracellular lipid droplets, after which proteins can bind and oligomerize to start forming the sub-RPE deposits (Figure 2). This HAP deposition and subsequent loss of permeability of the BrM can be modelled in primary cell culture [90], therefore, the metabolic changes associated with mineralization can now be explored.

2.3.4. Trace Metal Homeostasis

Altered metal ion homeostasis has also been implicated in AMD. Apart from zinc and calcium, which were discussed in Section 2.2. and Section 2.3.3., respectively, another metal ion commonly implicated in AMD is iron.

Iron is well known to contribute to retinal degeneration [91], where it causes damage through the induction of oxidative stress. Within the RPE, the accumulation of iron contributes to the buildup of lipofuscin [92]. In the sub-RPE space, iron accumulation can interfere with molecular pathways, such as the complement system [93].

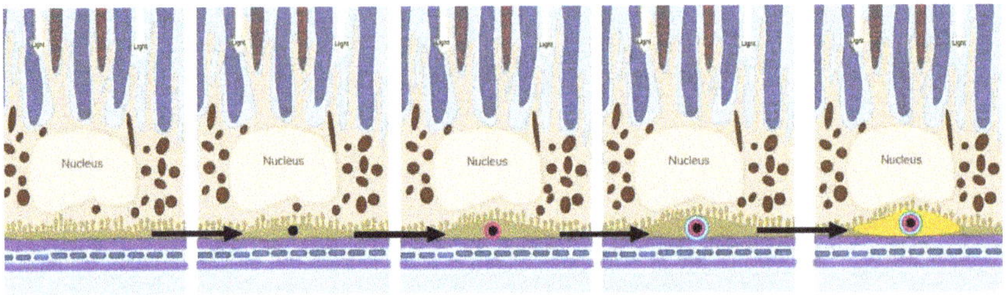

Figure 2. Model of sub-RPE deposit formation. Graphical overview of the proposed mechanism for the growth of sub-RPE deposits containing hydroxyapatite (reproduced with permission from Thompson et al. [89]). Micrometer-sized, cholesterol-containing extracellular lipid droplets (black) provide a site of hydroxyapatite (HAP) (magenta) precipitation. Deposit growth follows, with the binding of various proteins (blue) to the surface of HAP, facilitating a self-driven oligomerization process forming the macroscopic sub-RPE deposits (yellow). The brown particles within the RPE represent melanocytes.

When the metal ion content of sub-RPE deposits were determined in the macula, equator, and far periphery of the retina [94], iron appeared to have the lowest concentration when compared to zinc or calcium, regardless of the geographical locations.

While the underlying mechanisms for the positive effects of zinc are not yet fully understood, recent evidence suggests that there are multiple effects on the RPE of externally added zinc [52]. However, uncontrolled regulation of zinc levels can have negative effects at both ends of the spectrum. Zinc deficiency appears to be linked to lipofuscin accumulation in RPE cells [54] and it has been shown to impair the phagocytic and lysosomal activity of RPE cells through lipid peroxidation [56]. Alternatively, an excess of zinc contributes to RPE cytotoxicity [95] and can lead to zinc deposition, notably in the sub-RPE space [96]. There is also evidence suggesting that zinc is involved in the oligomerization of CFH and the modulation of the complement cascade and contributes to the protein content of sub-RPE deposits [93].

2.4. Choroid-BrM-RPE Interaction

The choroid is the vascularized layer of the eye, which is located between the retina and the sclera, with the innermost layer (choriocapillaris) located on the basal side of the BrM. The retina of humans and other non-human primates is supported by the underlying choroidal vasculature and the retinal vasculature, which supports the inner retina. These dense capillaries are fenestrated on the retinal surface to supply oxygen and nutrients to the RPE and photoreceptors, as well as removing waste products generated in these highly metabolic cells. Studies on the blood flow of the choroid have been carried out [97], where it has previously been determined that a small portion of the choroidal blood flow reaches the retinal photoreceptors. This is primarily due to the high blood flow relative to the small tissue mass [98,99], which exists to account for the poor (< 1 volume %) oxygen extraction from the blood to supply the outer retinal structures [100,101]. The high metabolic demands of the photoreceptor inner segments means that, under normal conditions, there is a large oxygen supply from the choroid to help overcome the issue of distance [102]. This causes an issue when there is any disruption to the distance between the choroid and photoreceptors (such as in the presence of sub-RPE drusen), as the lack of oxygen will lead to progressive photoreceptor degeneration. Chirco et al. [103] have recently reviewed the changes that occur to the choroid with aging, at both a structural and molecular level, and how this relates to AMD disease progression.

The blood-retinal barrier formed by the tight junctions of the RPE is important for the supply of nutrients to the photoreceptors. The BrM was first thought to form part of this barrier, but it became apparent in early studies that it was relatively permeable compared to the RPE and, instead, was more important in the removal of waste products from the retina. Unfortunately, as the BrM thickens with age, it becomes increasingly impermeable to the low concentrations of waste products that begin to accumulate in the sub-RPE space. The opposite has been found to be the case for the choroid, which appears to thin with age, particularly at the fovea [104]. This effect happens alongside a decrease in von Willebrand factor and human leukocyte antigen (HLA) class I proteins [105], both vascular specific proteins. This suggests that there is a dedifferentiation of the endothelial cells in eyes with early AMD and, along with the loss of the RPE cell layer, provides an insight into the steady metabolic dysregulation associated with the development of AMD.

3. Metabolomics in AMD

3.1. Introduction into Metabolomics

Metabolomics can be defined as the measurement of all small molecule metabolites within a biological system [106], including those that are environmental in origin. There is an increasing understanding of the components associated with the transcriptome and proteome, but the metabolome offers an integrated perspective of cellular processes and the effect environmental factors may have on the biological state (Figure 3). Compared with the other '-omics' approaches for investigating changes in the physiology of individuals and populations [107], metabolomics remains in its infancy. However, its popularity has rapidly increased over the last two decades. Mass spectrometry (MS) and nuclear magnetic resonance (NMR) spectroscopy are the predominant platforms that are used to obtain metabolomic profiles from a diverse range of sample types. To increase its resolution and sensitivity, MS is combined with different separation techniques, such as gas chromatography (GC) and high-performance liquid chromatography (HPLC) [108]. NMR spectroscopy is useful for metabolomic investigations because of its reliability and for the structural information it provides. NMR is not as sensitive as MS, making it more useful for quantifying high abundance metabolites. Furthermore, NMR does not always offer the high-throughput data acquisition that is common with mass spectrometers. However, there is evidence that this barrier is now being overcome [109].

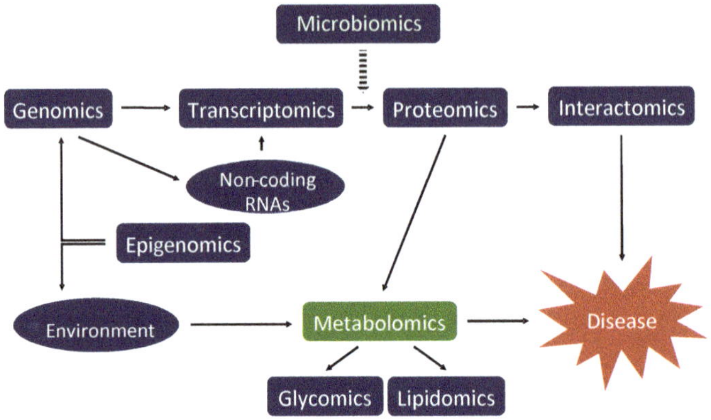

Figure 3. Overview of the various omics approaches that can be applied to assess the biological components that contribute to multifactorial diseases, such as age-related macular degeneration (AMD). Metabolomics is highlighted in green as an approach to test the effect of the environment and processes in the body on the development of disease. Microbiomics is growing in importance in understanding disease, but how it influences the whole system in a human, or in fact in animals, is yet to be determined. Adapted from Lauwen et al. [110].

Sample preparation is critically important in metabolomic investigations because metabolites could be introduced, altered, or removed during processing, which may not reflect the actual molecular state of an organism. For this reason, there is a preference for the use of biofluid samples over tissues or cells for some applications because they require less processing [111–113]. As the biofluids are in contact with various organs throughout the body, they also provide a more representative global metabolic profile. Even with the best sample preparation, there are still limitations as the current techniques of subsequent separation and detection methods are unable to identify all in vivo metabolites, primarily due to the heterogeneous and diverse chemistry of the currently known metabolites [114,115]. As these separation and detection platforms improve, there will be increasing numbers of studies investigating greater depths of metabolomics knowledge and understanding, as already evidenced by the rapid expansion of identified human metabolites from just over 6,800 to over 110,000 metabolite entities [116,117].

3.2. Retinal Tissues

In AMD, metabolomics studies have been conducted using samples obtained from animal models and human patients. A summary of untargeted cellular, human, and animal AMD metabolomics studies and the specific techniques used can be found in Table 1. One such study explored the metabolic changes associated with photoreceptor degeneration, a consequence of late stage AMD and GA, as well as whether induced pluripotent stem cell (iPSC)-generated RPE cell grafts altered this metabolomic profile [118]. The metabolites from rat eyes at different time points post-graft were processed by HPLC-QTOF-MS (quadrupole time-of-flight mass spectrometry) and were analyzed against the METLIN database. When comparing 52-week old dystrophic rats against age-matched control rats, significant changes were reportedly observed at 3 weeks, with most changes occurring at 52 weeks. Of the 203 metabolites which significantly changed in the diseased samples, more than half were phospholipids and various oxidized species. Glycerophosphocholines were the most abundant subclass that changed, followed by glycerophosphoethanolamines, long-chain acylcarnitines, monoglycerols, fatty acid amides, and long-chain polyunsaturated fatty acids (LC-PUFAs). There was a general increase in the fold-change of phospholipids in the dystrophic tissue, suggesting a higher level of lipid metabolism in the diseased eyes. Alternatively, there appeared to be a downregulation of the acylcarnitines, which are important for energy homeostasis and RPE function, and of the docosahexaenoic acid ω-3 fatty acid (DHA), which is necessary for retinal homeostasis [119,120]. DHA is the precursor of various phospholipids found enriched in the outer segment membranes of the retina, including lysophosphatidylcholine (lysoPC), lysophosphatidylethanolamine (lysoPE), and lysophosphatidylserine (lysoPS), all of which were downregulated in a similar manner as DHA.

A similar decrease was observed over time for all-*trans*-retinal (atRAL) in Royal College of Surgeons (RCS) rats, which is indicative of photoreceptor degeneration [121], along with an increase in the toxic A2E fluorophore. This resulted from inefficient clearance of atRAL in the dysregulated visual cycle [122]. The final part of their study identified that stem cell-derived RPE rescued the loss of DHA-lipids associated with the diseased phenotype, suggesting RPE transplantation could be a metabolic mediator with therapeutic benefits [123].

Table 1. Summary of untargeted metabolomic studies investigating AMD, including details of the tissue/biofluid profiled, and methods and instrumentation used.

Subjects and Biofluid Used	Number of Identified Metabolites	Metabolite Separation Method (Chromatography)	Detection Instrument Used	Reference
Mouse eye lysates	Not reported	Imtakt Scherzo SM-18 150 × 2 mm column Agilent 1200 capillary LC	Agilent 6538 UHD-QTOF MS (ESI+/ESI-)	[36]
Mouse eye lysates	203	XBridge C18 column (3.5 μm, 135 Å, 150 mm × 1.0 mm) Agilent 1260 HPLC	Agilent 6538 UHD Accurate Mass Q-TOF (ESI+)	[118]
Human serum (NVAMD); Mouse eye lysates	Not reported	Aeris Peptide XB-C18 column (3.6 μm, 100 × 2.10 mm) Phenomenex HPLC	Untargeted: Thermo Scientific LTQ Velos Orbitrap (ESI+/ESI-) Targeted: Thermo Scientific LXQ (ESI-)	[124]
hfRPE cells; Mouse retina; Apical/basal secretomes	202 (101 in medium; 53 changed substantially)	Ethylene bridged hybrid Amid column (1.7 μm, 2.1 mm × 150 mm) Agilent 1260 HPLC	AB Sciex QTrap 5500 (MRM)	[125]
Human plasma (NVAMD)	1168 (94 differed significantly)	Hamilton PRPX-110S (2.1 cm × 10 cm) anion exchange column	Thermo LTQ-FT spectrometer (ESI+)	[126]
Human plasma (NVAMD)	864 (10 differed significantly)	Acquity HSS T3 UPLC column (1.8 μm, 2.1 × 100 mm) Agilent 1290 Infinity UHPLC	AB SCIEX Triple 6600 TOF (ESI+/ESI-)	[127]
Human plasma	1188 spectra (30 low-M_w metabolites)	N/A	NMR: Bruker Avance DRX 500 spectrometer (300 K) operating at 500.13 MHz for protein, with a 5 mm TXI probe	[128]
Human plasma (nonexudative)	698 endogenous (87 associated with AMD)	C18 (acidic positive and basic negative ionization); HILIC (negative ionization) Waters ACQUITY ultra-UPLC (Metabolon, Inc.)	Thermo Scientific Q-Exactive (HESI-II), Orbitrap mass analyzer	[129]
WT mouse plasma (MS) and urine (NMR)	MS: 309 NMR: 47	Polar and nonpolar lipids: ACQUITY BEH C8 column (1.7 μm, 100 × 2.1 mm) Simadzu Nexera X2 UHPLC Hydrophilic metabolites: Atlantis HILIC column (3 μm, 150 × 2 mm) Shimadzu Nexera X2 UHPLC Additional polar metabolites: Phenomenex Luna NH2 column (150 × 2.0 mm) ACQUITY UHPLC	Polar and nonpolar lipids: Thermo Scientific Exactive Plus Orbitrap MS (ESI+) Hydrophilic metabolites: Thermo Fisher Scientific Q Exactive hybrid quadrupole Orbitrap MS (ESI+) Additional polar metabolites: AB SCIEX 5500 QTRAP MS (ESI- and MRM) NMR: Bruker Avance 600 spectrometer	[130]
Human plasma (NVAMD)	159 features differed (39 with medium to high confidence)	Hamilton PRP-X110S, 2.1 × 10 cm (anion exhchange) [a] Higgins Analytical C18 column, 2.1 × 10 cm (reverse phase) [a]	Thermo Scientific LTQ Velos Orbitrap MS (ESI+)	[131]

[a] Chromatography methods not stated, therefore, the above description was obtained from references provided in the study text.

Comparable results have been obtained when using mouse models of photoreceptor degeneration to compare healthy mice and AMD patients. Orban et al. [124] used tandem MS to analyze the *Abca4*$^{-/-}$ *Rdh8*$^{-/-}$ mice, which exhibit retinal degeneration under light-induced photoreceptor damage. This revealed that 11-*cis*-retinal and DHA levels were both decreased, whilst intense light produced increased levels of prostaglandin G2. The animal model metabolite differences also appeared in serum from AMD patients suffering from the nonexudative form of the disease in the dysregulation of DHA. Intense light also reduced the levels of DHA in wild-type mice, but this did not lead to photoreceptor degeneration. One of the differences between the disease seen in mice and the AMD patients was the statistically significant increase in arachidonic acid (AA) in the human serum, which was not present in the mice eyes. This study also employed a statistical model as a predictive tool for the identification of AMD, which gave a 74% chance of identifying AMD patients when compared to controls using only DHA and AA. The inclusion of the AMD-associated Ala^{69}-Ser variant in the *ARMS2* gene did not significantly improve this model. This was primarily because only a small correlation was found between the genetic mutation and AMD, meaning that, in these samples, the genetic screen for this amino acid transition was not a good predictor of AMD development. This variant has previously been shown to be associated with AMD progression and could be used as a predictive tool [11], which indicates using multiple models with independent biomarkers are a better diagnostic strategy to employ.

As it is well documented that the environment can impact the metabolome, it may also be worth noting how different tissue collection methods are employed for metabolomics studies. One study has highlighted the potential impact of anesthesia and euthanasia on metabolomics studies carried out on tissues [132]. Here, it is highlighted that there are tissue-specific changes dependent on the method of tissue sampling. For example, after euthanasia, skeletal muscles showed higher levels of glucose-6-phosphate, whereas nucleotide and purine derived metabolites accumulated in the heart and liver, when compared to anesthetized animal tissues. Although the authors recommend utilizing anesthesia for tissue collection, this may not be feasible for every study and, instead, highlights a point of consideration in tissue metabolomics studies.

Another consideration for metabolomics studies is post mortem time, especially in humans. There is evidence from GC-MS and ultra-high performance LC-MS (UHPLC-MS) studies in the eye that statistically significant post-mortem changes can be observed in a number of metabolites, although most metabolites were stable for up to eight hours post-mortem [133].

3.3. RPE Cells

Although several cellular models have been developed to replicate the RPE layer in vivo, the complex cellular environment of the retina means the exact environment is difficult to mimic. The human fetal RPE (hfRPE) cellular model is possibly the most reliable model of RPE cell function as it appears to replicate the physiological RPE morphology that other cell models and cell lines do not [134–136]. The use of this cell model as an accurate indicator of cellular metabolism may be helpful for metabolomics investigations [137]. The retina contains many different cell types, which have preferential metabolic pathways for energy metabolism, such as aerobic glycolysis in photoreceptors and reductive carboxylation in the RPE. In addition to the metabolism studies mentioned previously [23,125,138], these primary cell culture models also provide a platform for studying the development of drusen deposits in vivo [139], which are important factors in the development of GA. This has been expanded upon in recent years, with the identification of individual components of the drusen already mentioned [78,89,90]. Of interest is the accumulation of HAP, which is not commonly found in healthy soft tissues. The accumulation of high concentrations of phosphate and calcium, together with other divalent ions, in the sub-RPE space provided a new insight into the development and progression of sub-RPE deposits. It is perhaps unsurprising that there is an accumulation of phosphate and calcium, given that the retina is one of the most metabolically active tissues in the body [140–142], and the RPE has a very high calcium content [143]. Sphingolipids are one metabolite class that have been demonstrated to both contribute to, and protect against, photoreceptor apoptosis [144–146], as well as potentially having a role in choroidal and retinal neovascularization [147,148]. Another role that could be related to the deposition of calcium-phosphate at the RPE basement membrane is the role sphingolipids and their potential kinases have in calcium mobilization, particularly in response to the influx of calcium via transient receptor potential (TRP) channels [146,149–151].

Each of the cell types within the retina must communicate with one another to maintain a regular homeostatic environment and function properly to maintain viability. Recently, Chao et al. [125] investigated the consumption of nutrients and the subsequent transport of metabolites through the RPE cell layer. Through LC-MS/MS, 120 metabolites were identified in the culture medium of hfRPE cells and three were identified as being the most heavily consumed nutrients. Glucose and taurine showed the highest consumption from the apical medium, with proline consumed from both the apical and basal medium after 24 hours. Through isotopic labelling, they were also able to identify that metabolic intermediates from both glucose and proline metabolism were preferentially exported to the apical side of the culture, which is then imported into the neural retina when co-cultured. Proline is an energy source, which is utilized by both the citric acid cycle and the reductive carboxylation pathway. This provides a valuable insight into the unique metabolism that may occur in vivo, demonstrating the use of alternative metabolic pathways in RPE cells and indicating metabolite secretions, which may present as biomarkers that could change when the RPE is in a diseased state. This publication further proved human primary RPE cells are suitable model systems to study the dynamic changes in and around the RPE [125]. Further manipulation of culture conditions will now be able to study differences in the metabolism, which could mimic those found at the photoreceptor/RPE/choroid interface.

3.4. RPE Cells and the Retina

The metabolic connection between RPE cells and photoreceptors means that the loss of the RPE cell layer is detrimental, not only to the choriocapillaris, but also as a contribution to photoreceptor degeneration in the late stages of AMD [152–155]. Knowing that the RPE cells provide a homeostatic platform for the underlying photoreceptors, and that their degeneration contributes to the pathogenesis of AMD, Kurihara et al. [36] investigated the effects of oxidative stress on this cellular interaction as an early indicator of the disease. Using a murine model and primary RPE cells cultured from mice, the authors found that a hypoxic environment, induced from choriocapillaris vasodilation, led to the accumulation of lipid molecules, BrM thickening, RPE hypertrophy, and significant photoreceptor degeneration [36]. In addition to the pathological changes associated in the different cell types, glucose metabolism also became impaired within the hypoxic RPE. This was evidenced by the shift from oxidative phosphorylation to glycolytic metabolic pathways, under hypoxic environments, in vivo. This was supported by increased apical RPE uptake of glucose in vitro, which subsequently reduced available glucose for photoreceptor metabolism. These results are similar to those suggested in other studies looking directly at the effects of RPE and photoreceptor metabolism [125,138]. Through untargeted MS analysis, it was revealed that various forms of acylcarnitines were also significantly different in the von Hippel Lindau (*Vhl*) knock-out mice [36]. When combined with the gene knockout for hypoxia-inducible transcription factor 2 (HIF2α), the levels of acylcarnitines were of a similar level to that in control mice. This is similar to results obtained in other animal models using the same two genetic mutations [156,157], which suggests that HIF2α could play a crucial role in retinal lipid regulation as the combined genetic knockout of *Vhl*/*Hif2α* partially restores the wild-type lipid homeostasis. This is just one example that investigated the effects of oxidative stress on the RPE, but the overall literature in oxidative stress and the RPE were recently reviewed in the context of neovascular AMD (NVAMD) [158].

3.5. Biofluids

One of the first investigations into the metabolic changes associated with oxidative stress in AMD was performed in blood, but later studies on different biofluid metabolites were also investigated. These are summarized in Table 2. In the first study, the authors targeted metabolites, which were products of thiol redox reactions and lipid peroxidation [159]. Blood plasma samples from AMD patients and control individuals were assessed for cysteine (Cys), cystine (CySS), glutathione (GSH), isofurans (IsoFs), and F_2-isoprostanes (F_2-IsoPs), chosen based on previous results demonstrating that increased oxidation of these metabolites were associated with AMD risk factors [160–162]. Using HPLC and gas chromatography MS (GC-MS), they identified that only the levels of CySS were significantly higher (9.1%) when comparing AMD patients and controls. The levels of CySS were also significantly greater between neovascular and advanced AMD patients compared to controls. However, when the analysis was adjusted for age, gender, and smoking, the results were no longer significant and so the results are only suggestive of possible systemic metabolite changes.

An additional group used a targeted approach to investigate the association of long-chain ω-3 PUFAs, triglycerides, and high (HDL) and low (LDL) density lipoprotein-cholesterols in patients with NVAMD [163]. Serum and red blood cell membrane (RBCM) fatty acids were determined by GC, whereas enzymatic colorimetric and electrophoretic methods were employed to measure triglycerides and the serum lipoprotein-cholesterols. Compared to control samples from individuals with no history of ocular diseases, the NVAMD patients had significantly lower plasma triglycerides, serum eicosapentaenoic acid (EPA), RBCM EPA, and DHA after adjustment for age and sex. Similar levels of plasma total, HDL-, and LDL-cholesterol, as well as serum DHA and EPA+DHA (omega-3 index), were observed in both NVAMD patients and the controls. These results build on a previous population-based study by the same group [164], which showed a trend of a decreased risk to progress to NVAMD in patients with higher plasma ω-3 LC-PUFAs. A decrease was also found between the plasma EPA in the Alienor study [164], which was not associated with AMD, as opposed to the significant association of AMD with serum EPA in the more recent study [163]. Decreased levels of DHA in association with AMD has previously been highlighted in Section 3.2. [118,124].

The same group that investigated oxidative stress in AMD patients [159] progressed to an untargeted metabolomic investigation of plasma in NVAMD patients and age-matched controls with no clinical signs of AMD [126]. Using LC-FTMS (Fourier-transform mass spectrometry), there appeared to be significant changes in the intensity of 94 metabolites between the two cohorts. A more detailed analysis revealed a total of 40 metabolites that overlapped between the log2 transformed and non-transformed data. Cluster analysis of these 40 metabolites showed significant increases in certain peptides, modified amino acids, and natural products, defined as either a metabolite synthesized by a living organism or a metabolite specifically involved in secondary or specialized metabolism [165]. Significant decreases were found in bile acids, vitamin D-related metabolites, and dipeptides (histidine-arginine and tryptophan-phenylalanine) in NVAMD patients compared to controls. Pathway analysis of these metabolites revealed that phenylalanine and dopaquinone in the tyrosine metabolism pathway, which aids in the synthesis of melanin [166], or aspartate and glutamine in the urea cycle pathway, a natural process to remove excess nitrogen from the body, are involved.

Table 2. Metabolites in human biofluids differing between AMD patients and control individuals. The number of individuals assessed in each study is indicated in parentheses.

Reference	Biofluid	Comparison	Metabolomic Technique Employed	Definitively Identified Metabolites	Level in AMD Cohort Compared to Controls
[124]	Blood serum	NVAMD patients (*n* = 22) and age-matched control patients (*n* = 22)	LC-MS	Docosahexaenoic acid	Lower
				Amino acids	Higher
				Prostaglandin G2	Higher [NS]
[159]	Blood plasma	Intermediate AMD (drusen), late AMD (GA and CNV) (*n* = 77) and non-AMD control patients (*n* = 75)	HPLC and GC-negative-ion chemical ionization (NICI)-MS	Cystine Isofurans	Higher [NS]
[126]	Blood plasma	NVAMD patients (*n* = 26) and age-matched control patients (*n* = 19)	LC- Fourier transform MS (FTMS)	Acetylphenylalanine Dipeptide; Tripeptides (modified cysteine and alanine [a]) Sethoxydim Tripeptides [b] Tripeptides (acetyltryptophan[a]) Flavones; halofenozide	Higher
				Glycocholic acid Vitamin D-related metabolites; phytochemicals [b] Glycodeoxycholic acid+H^+; Glycoursodeoxycholic acid+H^+ Glycodeoxycholic acid+Na^+; Glycoursodeoxycholic acid+Na^+ Sencrassidol Didemethylsimmondsin Dipeptides [b]	Lower
[127]	Blood plasma	NVAMD patients (*n* = 20) and age-matched control patients (*n* = 20)	UPLC-TOF-MS	*N*-Acetyl-L-alanine L-Tyrosine L-Phenylalanine L-Methionine L-Arginine Isomaltose	Higher
				*N*1-Methyl-2-pyridone-5-carboxamide L-Palmitoylcarnitine Hydrocortisone Biliverdin	Lower

Table 2. *Cont.*

Reference	Biofluid	Comparison	Metabolomic Technique Employed	Definitively Identified Metabolites	Level in AMD Cohort Compared to Controls
[167]	Blood serum	NVAMD patients (n = 20), PCV patients (n = 20), and age-matched controls (n = 20)	UPLC-QTOF-MS	Glycerophospholipids [c] Phosphatidylcholine Covalently modified amino acids [c] Di/tri-peptides [c] Tripeptides [c] ω-3 and ω-6 PUFAs [c] Pinolenic acid Docoxahexaenoic acid Eicosatetraenoic acid Carnitine sp. [c]	Higher
[129]	Blood plasma	AMD patients (n = 314) and age-matched controls (n = 82), both across two locations	HILIC- and UPLC-MS	Creatine [d]	Higher
				Oleic acid [d] $N(CH_3)_3$ choline HDL [d] Acetate [d] Dimethylsulfone [d]	Higher [f]
				Pyruvate [d] Glutamine [e] Unsaturated F.A. [e] Unsaturated F.A. LDL + VLDL [e]	Higher [g]
				Unsaturated F.A. [d] Unsaturated F.A. LDL + VLDL [d]	Lower [f]
				Histidine [d] Acetoacetate [d] β-hydroxybutyrate [d]	Lower [h]
				Unsaturated F.A. LDL + VLDL [e]	Lower [f]
				Glutamine [e] Histidine [e] CH_2CH_2COOR F.A. [e] $CH_2CH_2C{=}C$ F.A. [e] Albumin lysil [e]	Lower [g]
				Alanine [e] Histidine [e] Glyceryl C1,3H′ [e]	Lower [h]

Table 2. *Cont.*

Reference	Biofluid	Comparison	Metabolomic Technique Employed	Definitively Identified Metabolites	Level in AMD Cohort Compared to Controls
[128]	Blood plasma	AMD patients (n = 89) and age-matched control patients (n = 30)	NMR	*N*2-methylguanosine	Higher
				1-Stearoyl-2-oleoyl GPC	Lower [NS]
				1-Linoleoyl-2-arachidonoyl GPC Stearoyl-arachidonoyl glycerol Oleoyl-olyeol-glycerol Dihomo-linolenoyl carnitine 1-Stearoyl-2-arachidonoyl GPC Linoleoyl-linolenoyl glycerol 1-Stearoyl-2-linoleoyl-GPI [f] Oleoyl-linoleoyl-glycerol Oleoylcarnitine Ximenoylcarnitine 1-Stearoyl-2-arachidonoyl GPI [i]	Lower
[131]	Blood plasma	NVAMD patients (n = 100) and control patients (n = 192)	LC-MS and LC-MS/MS	L-Oxalylalbizziine [j] Isopentyl beta-D-glucoside [j] LysoPC(P-18:0) [j] LysoPC(P-18:1(9Z)) [j] LysoPC(16:1(9Z)) [j] Darunavir [j] Bepridil [j] 912-Hexadecadienoylcarnitine [j] 456-Trimethylscutellarein 7-glucoside [j] 1-Lyso-2-arachidonoyl-phosphidate [j] Americanin B [j] Corchoroside A [j] *N*-Ornithyl-L-taurine [j]	Higher
				Lyciumoside III [j] Phosphatidylethanolamine [f,j] Phytosphingosine [j] Lenticin [j]	Lower
				9-Hexadecenoylcarnitine [k] Heptadecanoyl carnitine [k] 11Z-Octadecenylcarnitine [k] L-Palmitoylcarnitine [k] Stearoylcarnitine [k]	Higher

Table 2. *Cont.*

Reference	Biofluid	Comparison	Metabolomic Technique Employed	Definitively Identified Metabolites	Level in AMD Cohort Compared to Controls
[168,169]	Blood serum and urine	Neovascular and nonexudative AMD patients (n = 104) [l]	NMR	Arginine	Higher [m,NS]
				Glucose Lactate Glutamine Reduced glutathione	Lower [m,NS]

[a] When no specific metabolites were given (e.g., dipeptides), correlated Metlin matches with the same *m/z* are given; [b] No correlated Metlin matches; [c] Full list of 197 differing metabolites can be found in Supplementary Table 1 of [167]; [d] Metabolites obtained from the Coimbra cohort studied; [e] Metabolites obtained from the Boston cohort studied; [f] Effect size difference between early AMD patients vs controls; [g] Effect size difference between intermediate and early AMD patients; [h] Effect size difference between late and intermediate AMD patients; [i] No metabolite identity given; [j] Metabolites identified through high-resolution LC-MS; [k] Metabolite identity confirmed through high-resolution LC-MS/MS; [l] Specific patient cohort information not available; [m] When comparing NVAMD patients to nonexudative AMD patients. Abbreviations: AMD = age-related macular degeneration; F.A. = fatty acids; GA = geographic atrophy; GPC = glycerol-3-phosphocholine; GPI = glycosylphosphatidylinositol (assumed, no definition given); HDL = high-density lipoproteins; LC-MS = liquid chromatography-mass spectrometry; LDL = low-density lipoproteins; NVAMD = neovascular AMD; VLDL = very low-density lipoproteins; NS = not significant.

Similarly, Luo et al. [127] assessed the plasma metabolomics profile of a Chinese cohort of patients suffering from NVAMD and healthy controls using UHPLC-QTOF MS. In this study, there were 10 metabolites that differed significantly between the two groups, the majority of which were amino acids, and the most significant finding was an increase in L-phenylalanine in AMD patients. A further analysis of the associated metabolic pathways revealed that most metabolites belonged to the amino acid biosynthesis pathway. The most common metabolites found across the metabolic pathways had also been previously shown as having a significant change by Osborn and colleagues [126].

Untargeted metabolomic studies assessing metabolic differences between healthy individuals and AMD patients are increasing in frequency. One such study used LC-MS/MS analysis to assess the blood serum of 60 individuals characterized as either healthy controls, patients suffering with CNV, or patients suffering with polypoidal choroidal vasculopathy (PCV) [167], a subtype of AMD commonly found in Asian populations [170]. It was shown that glycerophospholipids, amino acids, di/tripeptides, ω-3 and -6 PUFAs, and various carnitine species were all elevated in both CNV and PCV patient samples. In total, there were 197 significantly altered metabolites across both conditions compared to serum metabolites from controls. Only one metabolite (pinolenic acid) was shown to be different between CNV and PCV patients, which suggests there could be significant metabolic overlap between these two diseases [170]. The results from this study provided additional knowledge to previous research, which compared the serum lipid profiles of PCV patients and compared these to controls [171]. A total of 41 metabolites were significantly altered in PCV patients, which included increases in 18 phosphatidylcholines (PCs), eight sphingomyelins (SMs), three lysoPCs, three platelet-activating factors (PAFs), one lysophosphatidic acid (LPAs), and one phytosphingosine. Significant decreases were found in one PC, three LPAs, two sphingosines, and one phosphatidylethanolamine (PE) [171].

The results from the PCV and AMD comparison study [170] were expanded in 2017, where investigations into the plasma metabolomics profiles of AMD patients and healthy age-matched controls were conducted using NMR [128] and UPLC-MS analysis [129]. For the plasma samples analyzed by NMR, two large cohorts of AMD patients were recruited at two separate study locations (Coimbra and Boston) and were characterized by the severity of their AMD progression. These were compared against each other, as well as being compared against control samples from both study cohorts. There were noticeable differences in the metabolic profiles between AMD severity stages, including between samples from early-stage AMD and the control groups. There were higher levels of circulating creatine, acetate, dimethyl sulfone, C18 cholesterol, and high-density lipoprotein (HDL)-choline, whilst there were lower levels of unsaturated fatty acids between controls and the early AMD Coimbra cohort [128]. Across the AMD severity stages, there were minor differences in low-molecular weight (M_W) metabolites, such as higher pyruvate for intermediate AMD and lower levels of histidine, acetoacetate, and β-hydroxybutyrate for late AMD. Samples from the Boston cohort showed slight differences in the low-M_{W-} metabolites, with higher and lower levels of glutamine for early and intermediate AMD, respectively. Lower histidine levels were found in intermediate and late AMD, with lower levels of alanine in late AMD. As evidenced by the metabolites found between the two different cohorts, there appears to be a geographic influence on the metabolites that present as significantly different in AMD patients, again highlighting the influence of the environment on metabolomics studies and disease profile.

A subsequent study by the same group analyzed plasma metabolites from patients with different stages of AMD progression and age-matched controls, using UHPLC-MS [129]. Here, 87 metabolites were identified as differing between AMD and controls. Most of the metabolites were members of the lipid super-pathway (82.8%), followed by amino acids (5.7%). Similarly, six out of seven of the most significantly different metabolites were lipids, the exception being adenosine. In terms of increasing AMD severity groups, 48 metabolites significantly differed and all but one of the most significant metabolites were involved in lipid pathways. Pathway analysis demonstrated that most were involved in glycerophospholipid metabolism. It is important to note that Osborn et al. [126] were unable to distinguish the lipids that were present in their samples and so they were not analyzed and cannot be compared to the study by Laíns et al [129].

Recently, the limitations of the original paper by Osborn et al. [126] have been improved on in a study, which investigated the metabolites and associated metabolic pathways of a larger cohort of NVAMD patients [131]. Plasma samples were collected from NVAMD patients, who exhibited extensive CNV, subretinal hemorrhaging or fibrosis, or photocoagulation scarring in one or both eyes. Control individuals were identified as having fewer than 10 small drusen and no macular pigment changes in both eyes. Untargeted metabolomics was carried out using LC-MS, which identified 10,917 unique metabolite features. Analysis of these features highlighted 159 metabolites that were distinguishable between NVAMD patients and controls. There was an increase in 110 of these metabolites in NVAMD patients, with 49 showing decreased levels, compared to controls. Further analysis identified 39 metabolites with medium to high confidence. Only metabolites that have been exclusively identified have been listed in Table 2. Acylcarnities, amino acids, bile acids, lysophospholipids, and phospholipids were amongst those annotated and, following Bonferroni-corrected pathway analysis, the carnitine shuttle pathway was revealed to be significantly altered in NVAMD patients. Further LC-MS/MS analysis confirmed the identity of five of the six carnitine shuttle pathway metabolites, which all showed a significant increase in NVAMD patients. This larger cohort study builds on previous NVAMD metabolomics research, which have also identified acylcarnitine and bile acid alterations in the patient population [126,127].

Preliminary studies published as abstracts on metabolite profiles in AMD have also been compared using NMR metabolite profiles of laser-induced mouse models of NVAMD and human AMD patients, identifying lactate as an important metabolite in both cases [172,173]. Preliminary studies in urine have also been carried out [168,169]. Here, the NMR profiles for NVAMD cluster well, whereas the clustering of the nonexudative AMD patients was more diffuse. This may suggest a more heterogeneous population and highlights differences between AMD subgroups. In NVAMD patients, there were notable increases in arginine and decreases in glucose, lactate, glutamine, and glutathione. Results obtained in urine showed an overlap between neovascular and nonexudative AMD, suggesting that the two forms of the disease could be linked [168,169]. The metabolomics profile of subretinal fluid has also recently been investigated [174], in the context of other ocular diseases, with 651 metabolites identified in a small sample size of three patients.

It is perhaps not surprising that lipids are the most consistent metabolites that show changes in the different stages of AMD. The role that lipids play in the pathogenesis of AMD is becoming clearer though it is still to be fully elucidated [175]. The most significantly associated metabolites in these studies belonged to the glycerophospholipid family, which provide structural stability and fluidity to neural membranes. The most likely source of these are the degrading cell membranes and photoreceptor outer segments' discs. Although these studies provide a basis for the identity of potential metabolic biomarkers present in the circulation and tissues of AMD patients and models of AMD disease states, they may not be directly relatable to the process occurring in the retina.

4. Alternative Approaches for AMD Metabolomics Studies

4.1. Tears

Tears are a biofluid originating from the anterior of the eye and are a potential source of metabolite biomarkers directly linked to ophthalmology. Although relatively small in volume [176], technological advancements now make it possible to characterize the proteomic [177,178], lipidomic [179], and metabolomic [180] composition of the human tear. As a source of biomarkers, the tear has been used to assess the disease profiles of various ocular diseases, including, but not limited to, dry eye disease, keratoconus, trachoma, and diabetic retinopathy, which have been reviewed elsewhere [181]. The majority of tear analysis studies have focused on the proteome as the relative amount of protein is greater than metabolites. Along with this, a single technique is unable detect all tear metabolites due to no standardized collection, analysis, or identification methods. Despite this issue and not being in direct contact with the retina, tears provide a non-invasive source of metabolomic biomarkers [182].

One of the first exclusive characterizations of the human tear metabolome aimed to use a standard clinical method of tear collection and to develop an analytical platform, which could be applied to characterize the global repertoire of human tear metabolites [180]. In this study, tears of healthy individuals were collected using the clinically utilized Schirmer strips, separated by UFLC (ultra-fast LC) and analyzed by Q-TOF MS/MS. This untargeted method of metabolite analysis identified 60 metabolites from a range of 16 compound classes. Of these metabolites, 44 were 'novel' as they did not correspond with metabolites identified in previous targeted studies (see Table 1 in [180]). This set the precedent for what could be achieved in tear metabolome studies, but it was clear that this method did not measure some well-known metabolites (e.g., measurement of glucose and ascorbic acid was affected by background interference).

Very few lipid species were identified, but within the literature, others have identified several classes of lipids using targeted analysis. These include free cholesterol [183], phosphatidylcholines [183,184], SMs [183,184], wax esters [183,185], lysoPC [186,187], triacylglycerides, ceramides, and phosphatidylethanolamines [186]. A further study investigated lipid composition during collection with Schirmer strips using untargeted analysis [179]. Tears were collected either by capillary tube or Schirmer strip and extracted lipids were analyzed using HPLC-MS. Over 600 lipid species across 17 lipid classes were detected, the majority of which were categorized as either wax esters or cholesteryl esters.

Contact of parts of the strip with the eye [188], particularly the meibum [189], clearly contributes a large proportion of free cholesterols, sphingolipids, and phospholipids, and therefore the tear lipidome is less complex than the meibum lipidome. These results were consistent with previous studies investigating lipid classes in human tears, but this was the first extensive characterization of the tear lipidome, where novel metabolites were also identified [179]. This includes cholesteryl sulfates, which, as a stabilizing agent [190], could contribute to the amphiphilic sublayer of the tear film. This study indicates the limitations of tear collection methodologies within the context of metabolomic investigations. Schirmer strip collections yield the highest absolute amounts of lipids and are routinely employed in the clinic. Although the level of background noise was high in blank Schirmer strips, no endogenous tear lipids were detected [179]. Interestingly, the strips act as a chromatographic system for lipid metabolites [179,186]. The aqueous fraction of the tear travels further along the strip than the non-polar lipids. The strips can capture an accurate representation of the lipidomic profile of tears and their relative concentrations when compared to spiking with artificial tear solutions [179]. It should be noted that tears used in this study were obtained from patients with dry eye syndrome [179] and so their metabolomic profiles and relative metabolite concentrations may differ from other patient groups.

Tear glucose levels have been considered as a non-invasive method of detecting the early stages of diabetes [191] and they are increasingly being investigated for their use as a sensor for diabetes mellitus [192]. It is well known that tear glucose levels are variable diurnally [193] and from which eye the sample is taken [192]. However, there is evidence suggesting that the tear glucose levels are reflective of blood glucose levels using enzyme-based and amperometric biosensors [194,195], suggesting future clinical investigations of tear metabolites are worthwhile. However, there have been questions raised as to the reliability of the results obtained and whether they are comparable to blood glucose levels [196,197], and thus needs further investigation.

Tear samples appear to be a reasonable reservoir of metabolites, their collection is non-invasive, and the samples are easy to handle and cheap to transport, indicating that tear sampling could become a reliable source for metabolic markers for eye as well as other diseases [198]. Saliva has also been studied as a source of potential biofluid metabolites [199–201], but there have been no studies investigating the salivary metabolome in eye diseases.

4.2. Vitreous and Aqueous Humor

In a similar manner to tears, the vitreous and aqueous humors might be a representative ocular biofluid to use for metabolomics studies in AMD. Although they have to be obtained through invasive procedures, often requiring collection during surgery, they are also being used as surrogate sources of ocular disease biomarkers. Vitreous humor has been studied for metabolite changes in diabetic retinopathy [202–207], proliferative vitreoretinopathy [207,208], rhegmatogenous retinal detachment both associated and not associated with choroidal detachment [202,207–209], and uveitis [207]. Young et al. [207] investigated the metabolomic profiles of vitreous humor obtained from a variety of inflammatory eye diseases. They were able to demonstrate clear and specific differences in the metabolites obtained from each disease, with a high sensitivity in clinically relevant samples. In a non-clinical context, vitreous humor from sheep, pigs, and rabbits was profiled using targeted methods [210]. This revealed that acetylcholine esterase activity varied across species, but less so between breeds of rabbit. Untargeted LC-MS analysis also found differences in metabolite profiles that may simply reflect the diets of each animal, and this may have relevance to studies in humans. However, it should be noted that, based on a study on rats, only 1.6% of the total metabolic profile overlapped between the vitreous and the retina [133] although this might change in pathological states.

Aqueous humor has also been explored as a source of metabolite biomarkers in different diseases, but the number of studies is more limited. In an acute model, glaucoma changes in glucose and citrate levels were detected [211]. In a chronic glaucoma mice model, an increase of sphingolipid and ceramide species was found [212]. Furthermore, in a chronic rat model for glaucoma, increases in acetoacetate, citrate, and various amino acids, including alanine, lysine, and valine, as well as a decrease in glucose levels, were found using ^{1}H-NMR [213]. For acute and chronic glaucoma studies in human aqueous, humor phospholipids [214], cholesterol [215], sphingolipid, and ceramide species [215,216] profiles have been demonstrated to be dysregulated.

Aqueous humor has also been used in the identification of metabolic changes associated with myopia [217,218]. Using a dual platform of capillary electrophoresis–mass spectrometry (CE-MS) and LC-MS, one of the studies identified 40 metabolites [217]. Of these, 20 were deemed to be significantly different between varying stages of myopia. Increases in arginine, citrulline, and sphinganine were associated with high myopia while increases in aminoundecanoic acid and dihydro-retinoic acid were associated with low myopia [217]. The other study used GC/TOF-MS and compared the metabolites present in the aqueous humor of patients with high myopia and compared these to controls [218]. A total of 242 metabolites were identified, with significant increases observed in 27 and significant decreases in two metabolites [218].

Metabolomic profiles in human aqueous humor have been compared with serum metabolites from the same patients [219]. The most notable metabolite to differ between the two biofluids was ascorbate, attributed to the ascorbate-specific pumps at the blood-aqueous border. Other differences were attributed to the differential metabolic activity of the compartmentalized ocular tissues, but the potential for post-mortem artefacts has also been raised [219]. Due to the invasiveness of obtaining vitreous and/or aqueous humor, their use will probably be limited to those where surgical intervention is a necessity.

4.3. In vivo Imaging

A different approach to obtaining metabolomic information, and of interest for retinal studies, is the use of in vivo fluorescence imaging. Imaging approaches offer the advantages of data collection with high spatial resolution, modest intervention, high safety, and low cost of data. Due to their accessibility, the eye, and retina in particular, are well suited for optical studies compared to most organs. This has fueled the development of numerous approaches for diagnosis and therapy in the eye. Many metabolites highlighted in Table 1, and in previous sections, have little or no fluorescence emission at wavelengths longer than 300 nm, nor unique IR nor Raman spectra, limiting the scope of fluorescence imaging as a broad metabolomic approach. However, the nicotinamide adenine dinucleotide coenzyme NADP(H), and to a lesser extent flavin mononucleotide (FMN), flavin adenine dinucleotide (FAD), and pyridoxal/pyridoxamine, exhibit useful visible fluorescence under certain conditions [220]. These molecules are also intimately linked to numerous metabolic pathways and offer information about metabolite fluxes through those pathways. Of interest are the nicotinamide adenine dinucleotides, NAD(H) and NADP(H), which are the principal carriers of reducing equivalents within cells. NAD(H) is crucial for energy production in the cell as it is the primary transporter of electrons for oxidative phosphorylation, whereas NADP(H) provides the reducing equivalents for the neutralization of cellular reactive oxygen species (ROS). The importance of NAD(H) to the energetic state of cells was recognized decades ago, where the proportions of NAD(H) and NADH were measured spectrophotometrically and by fluorescence [221]. Naturally, as a cell acquires energy through different routes (oxidative phosphorylation, glycolysis, fermentation, reductive carboxylation) and from different fuels (glucose, fatty acids, ketone bodies), different pathways are activated or deactivated, and the broader metabolome will reflect this.

Recently, more powerful techniques have been developed to study NADP(H) and other fluorophores. For example, the fluorescence emission spectra of NAD(H) and NADP(H), free in solution, are very similar (a broad peak around 460 nm after excitation near 340 nm) and so the proportions of each cannot be distinguished based on spectra alone. However, the reduced forms of both exhibit different fluorescence lifetimes when free in solution and bound to proteins [222,223]. The fluorescence lifetime is the average amount of time the fluorophore spends in the excited state between excitation and emission, and is typically in the range of nanoseconds [224]. This can either be measured in the time domain with a technique known as time correlated single photon counting (TCSPC) [225], or in the frequency domain by phase fluorometry [224]. While the lifetime is a general property of each fluorophore, for some fluorophores, the lifetime is very sensitive to environmental conditions. When free in solution, NAD(H) has a low quantum yield and a mixture of fluorescence lifetimes between 0.3 and 0.8 nsec. When protein-bound, the lifetime ranges from 1.5 to 6 nsec. The development of fluorescence lifetime-based imaging microscopy (FLIM) [224,225] made it possible to collect specimen images where the contrast comes from differences in lifetime, not intensity. Studies have appeared demonstrating that FLIM can distinguish between free and protein-bound NAD(H) [226] and between intracellular NAD(H) and NADP(H) [227], a result not possible when only detecting their spectrally identical fluorescence [228]. The metabolic state and composition of a tissue can also be assessed by measuring the autofluorescence lifetime of endogenous fluorophores [229–232], including protein-bound NADP(H) [223].

In in vitro studies, FLIM has been utilized to assess different structures within RPE cells [233], as well as in retinal tissue from donors with AMD [234]. Miura et al. [233] identified that the fluorescence lifetime observed inside and surrounding cultured RPE cells increases significantly when oxidative stress is induced. This enabled the discrimination between the granules associated with increases in metabolic stress from the melanosomes seen under normal conditions. From retinal tissues [234], the RPE and BrM could be discriminated from one another using their fluorescence lifetimes. This was primarily because of the presence of lipofuscin in the RPE, which had a shorter lifetime than other emitters. Where it was present, drusen could also be discriminated from the RPE and the BrM. In addition, it was found that different drusen had different lifetime distributions, indicating that different fluorophores, or the same fluorophores in different environments, contribute to the FLIM image [234]. Schweitzer et al. [234] also found lipofuscin-like FLIM signatures in structures within drusen, which they interpreted as indicating a role of lipofuscin in the formation of sub-RPE deposits.

The analysis of the ocular fundus has been extended to include fluorescence lifetime imaging ophthalmoscopy (FLIO), a technique analogous to FLIM [235]. Fluorophores that are interesting for AMD are redox coenzymes, NAD(H) and FAD, as well as lipofuscin [236], AGEs [71], and collagen. The excitation and emission spectra of these endogenous fundus fluorophores have been investigated in vivo, which showed that it would not be possible to distinguish them based on autofluoresecnce [231]. However, changes in fluorescence lifetime, especially when the metabolic environment is altered, could identify novel signatures that are associated with AMD [231]. When performing FLIO, it is worth considering the strong fluorescence of the lens [237], which cannot be entirely suppressed, but which can be overcome using specific analysis software [238]. Multiphoton excitation may help minimize such background fluorescence, but the safety considerations should be made when utilizing focused, high peak power picosecond lasers for in vivo studies [239,240].

Following on from the in vitro observations described by Miura et al. [233], the in vivo FLIO has also been utilized as a tool to investigate the effects of induced oxidative stress on RPE degeneration and photoreceptor loss in a mouse model [241]. Following intravenous sodium iodate injection, which induces RPE degeneration, the retinal autofluorescence lifetimes increased over a 28-day period compared to control mice. In contrast, intraperitoneal injection of *N*-methyl-*N*-nitrosourea, which causes the specific degeneration of the photoreceptors, resulted in shorter lifetimes being observed [241]. From these results, the authors suggested that short lifetimes are present in the RPE, but are altered with the overlying retinal structures. FLIO imaging has also been used to distinguish AMD patients from controls [242]. It was found that images from AMD eyes had a significantly longer retinal fluorescence lifetime, and they found that the lifetimes may vary from drusen to drusen, although distinguishing drusen on FLIO images appears to be challenging [242]. Similar observations were made in another study [243]. However, lifetime changes are present in 36% of the healthy controls, which suggests these observations could be the result of aging [243]. Whether these fluorescence lifetime changes are associated with the metabolomics changes mentioned above will need to be investigated further.

The ability to distinguish between drusen subtypes and different structures in the retina is complementary to the multimodal imaging approaches described previously [81,244]. These studies indicate that fluorescence lifetime imaging approaches (FLIM and FLIO) could become tools for studying not only the biology and metabolomics of the retina, but also the pathologic processes that lead to diseases, such as AMD. Several comprehensive reviews detailing the principles of fluorescence lifetime imaging and its application as a clinical tool has recently been published [245,246].

5. Conclusions

In summary, the complex repertoire of metabolites found in the retina is only beginning to be revealed, and there is scope to identify those associated with the development of AMD. Clearly there is a dysregulation of lipids in patients suffering from AMD, and various models exist for investigating AMD biomarkers. Considering the close association between photoreceptor degeneration, RPE health, sub-RPE deposit formation, and choroidal changes, it is perhaps not surprising to find lipids and molecules associated with metabolic fluxes in the studies highlighted. Plasma metabolomics could be a convenient tool for analyzing a wide range of factors potentially contributing to AMD, however, the number of published studies compared to other '-omics' methodologies still remains low [110]. Further efforts are clearly needed here whilst due consideration of the blood-retinal barrier is required. It is still unclear whether the systemic biomarkers identified in biofluids can be representative of the changes in the eye. Alternative biofluids, such as tear, saliva, and vitreous, are also viable possibilities for detecting local and systemic changes in AMD patients.

Some of the current barriers associated with large-scale epidemiological metabolite studies are being overcome by advances in technology for untargeted metabolite analysis [247]. Comprehensive study design is crucial for high accuracy and good quality metabolomics data collection, which means other potential limitations include, but are not limited to, subject selection, sample selection, collection, handling, storage, and preparation [21,248]. As part of overcoming these limitations, it will become necessary to establish reference lists of metabolites obtained from various biofluids in larger cohorts of healthy, age-similar patients. Similar cell-based libraries are being developed for proteomic integration [249]. For metabolomics, this will require additional advances in the analytical technologies, as well as strong collaborations to systematically ensure a standardized benchmark can be obtained.

The introduction and application of lifetime based clinical imaging with FLIO now allows a non-invasive approach for the identification of metabolic changes in the retina. This is a new methodology that needs further validation, but the potential to use this approach for clinical practice is very appealing. Apart from the imaging of endogenous fluorophores, there is a possibility to deliver markers that change their fluorescent lifetime once they are bound to specific metabolites. While this approach is still in its infancy [250], once proven to be safe, it could provide new insight into the metabolic machinery in health and diseases.

Apart from the comparisons between disease and control, it will be interesting to study metabolic changes associated with dietary intervention in AMD [251,252]. A recent study on an animal model investigated this [130]. Plasma and urine samples were obtained from wild-type mice, which were fed diets with either a high- or low-glycemic index and subsequently analyzed by LC-MS and proton NMR. A total of 330 metabolites were found in plasma and urine (309 in plasma, 47 in urine, with 26 found in both). The mice fed the high-glycemic diet showed higher levels of lipids, including phosphatidylcholine, C3 carnitine, and lysoPE. Higher levels of the protein derivatives of 2-ω-carboxyethyl pyrrole (CEP) were also found [253,254]. These changes were associated with dysfunction of the RPE and degradation of the retinal cell structures. Therefore, dietary studies relevant for AMD can now be investigated both in animals and perhaps in humans [255].

Identifying biomarkers for such a multifactorial disease as AMD remains a significant challenge. There are noticeable overlaps between risk factors for AMD and many other diseases (as reviewed by Kersten et al. [17]). Therefore, metabolomics investigations can provide a further source of information that can be integrated into distinguishing AMD from other comorbidities. Further studies will shed light on potential metabolites, which may be used as early stage biomarkers, perhaps even being applied as precision medicine tools in the future treatment of AMD [256–258].

Author Contributions: Conceptualization, C.N.B., B.D.G. and I.L.; Original Draft Preparation, C.N.B.; Review & Editing, C.N.B., B.D.G., R.B.T., A.I.d.H. and I.L. and the Eye-Risk Consortium members; Supervision, B.D.G. and I.L.; Funding Acquisition, I.L.

Funding: This review article was supported by a Department for Education PhD studentship to C.N.B. and the Eye-Risk project funded by the European Union's Horizon 2020 research and innovation programme [634479].

Acknowledgments: The authors would like to thank Marc Biarnés, Eszter Emri, Tunde Peto and Marius Ueffing and the participants of the Eye-Risk consortium Ilhan E. Acar, Soufiane Ajana, Blanca Arango-Gonzalez, Angela Armento, Franz Badura, Karl U. Bartz-Schmidt, Vaibhav Bhatia, Shomi S. Bhattacharya, Marc Biarnés, Anna Borrell, Sofia M. Calado, Johanna M. Colijn, Audrey Cougnard-Grégoire, Sascha Dammeier, Anita de Breuk, Berta De la Cerda, Cécile Delcourt, Anneke I. den Hollander, Francisco J. Diaz-Corrales, Sigrid Diether, Eszter Emri, Tanja Endermann, Lucia L. Ferraro, Míriam Garcia, Thomas J. Heesterbeek, Sabina Honisch, Carel B. Hoyng, Ellen Kilger, Caroline C. W. Klaver, Elöd Körtvely, Hanno Langen, Claire Lastrucci, Imre Lengyel, Phil Luthert, Magda Meester-Smoor, Bénédicte M. J. Merle, Jordi Monés, Everson Nogoceke, Tunde Peto, Frances M. Pool, Eduardo Rodríguez-Bocanegra, Luis Serrano, Jose Sousa, Eric F. Thee, Marius Ueffing, Timo Verzijden, Markus Zumbansen for their valuable comments during the preparation of this review.

Conflicts of Interest: The authors declare no conflict of interest. The funders had no role in the design of the study; in the collection, analyses, or interpretation of data; in the writing of the manuscript, and in the decision to publish the results.

References

1. Friedman, D.S.; O'Colmain, B.J.; Muñoz, B.; Tomany, S.C.; McCarty, C.; DeJong, P.T.V.M.; Nemesure, B.; Mitchell, P.; Kempen, J.; Congdon, N. Prevalence of age-related macular degeneration in the United States. *Arch. Ophthalmol.* **2004**, *122*, 564–572. [CrossRef]
2. Wong, W.L.; Su, X.; Li, X.; Cheung, C.M.G.; Klein, R.; Cheng, C.-Y.; Wong, T.Y. Global prevalence of age-related macular degeneration and disease burden projection for 2020 and 2040: A systematic review and meta-analysis. *Lancet Glob. Health* **2014**, *2*, e106–e116. [CrossRef]

3. Spraul, C.W.; Lang, G.E.; Grossniklaus, H.E.; Lang, G.K. Histologic and morphometric analysis of the choroid, Bruch's membrane, and retinal pigment epithelium in postmortem eyes with age-related macular degeneration and histologic examination of surgically excised choroidal neovascular membranes. *Surv. Ophthalmol.* **1999**, *44*, S10–S32. [CrossRef]
4. Chakravarthy, U.; Harding, S.P.; Rogers, C.A.; Downes, S.M.; Lotery, A.J.; Culliford, L.A.; Reeves, B.C. Alternative treatments to inhibit VEGF in age-related choroidal neovascularisation: 2-year findings of the IVAN randomised controlled trial. *Lancet* **2013**, *382*, 1258–1267. [CrossRef]
5. Grunwald, J.E.; Daniel, E.; Huang, J.; Ying, G.-S.; Maguire, M.G.; Toth, C.A.; Jaffe, G.J.; Fine, S.L.; Blodi, B.; Klein, M.L.; et al. Risk of geographic atrophy in the comparison of age-related macular degeneration treatments trials. *Ophthalmology* **2014**, *121*, 150–161. [CrossRef]
6. Colijn, J.M.; Buitendijk, G.H.S.S.; Prokofyeva, E.; Alves, D.; Cachulo, M.L.; Khawaja, A.P.; Cougnard-Gregoire, A.; Merle, B.M.J.J.; Korb, C.; Erke, M.G.; et al. Prevalence of age-related macular degeneration in Europe: The past and the future. *Ophthalmology* **2017**, *124*, 1753–1763. [CrossRef]
7. Chakravarthy, U.; Augood, C.; Bentham, G.C.; de Jong, P.T.V.M.; Rahu, M.; Seland, J.; Soubrane, G.; Tomazzoli, L.; Topouzis, F.; Vingerling, J.R.; et al. Cigarette smoking and age-related macular degeneration in the EUREYE study. *Ophthalmology* **2007**, *114*, 1157–1163. [CrossRef]
8. Christen, W.G.; Glynn, R.J.; Manson, J.E.; Ajani, U.A.; Buring, J.E. A prospective study of cigarette smoking and risk of age-related macular degeneration in men. *JAMA* **1996**, *276*, 1147–1151. [CrossRef]
9. Fritsche, L.G.; Igl, W.; Bailey, J.N.C.; Grassmann, F.; Sengupta, S.; Bragg-Gresham, J.L.; Burdon, K.P.; Hebbring, S.J.; Wen, C.; Gorski, M.; et al. A large genome-wide association study of age-related macular degeneration highlights contributions of rare and common variants. *Nat. Genet.* **2016**, *48*, 134–143. [CrossRef]
10. Hageman, G.S.; Anderson, D.H.; Johnson, L.V.; Hancox, L.S.; Taiber, A.J.; Hardisty, L.I.; Hageman, J.L.; Stockman, H.A.; Borchardt, J.D.; Gehrs, K.M.; et al. A common haplotype in the complement regulatory gene factor H (HF1/CFH) predisposes individuals to age-related macular degeneration. *Proc. Natl. Acad. Sci. USA* **2005**, *102*, 7227–7232. [CrossRef]
11. Kanda, A.; Chen, W.; Othman, M.; Branham, K.E.H.; Brooks, M.; Khanna, R.; He, S.; Lyons, R.; Abecasis, G.R.; Swaroop, A. A variant of mitochondrial protein LOC387715/ARMS2, not HTRA1, is strongly associated with age-related macular degeneration. *Proc. Natl. Acad. Sci. USA* **2007**, *104*, 16227–16232. [CrossRef] [PubMed]
12. Fritsche, L.G.; Loenhardt, T.; Janssen, A.; Fisher, S.A.; Rivera, A.; Keilhauer, C.N.; Weber, B.H.F. Age-related macular degeneration is associated with an unstable *ARMS2* (*LOC387715*) mRNA. *Nat. Genet.* **2008**, *40*, 892–896. [CrossRef] [PubMed]
13. Micklisch, S.; Lin, Y.; Jacob, S.; Karlstetter, M.; Dannhausen, K.; Dasari, P.; von der Heide, M.; Dahse, H.-M.; Schmölz, L.; Grassmann, F.; et al. Age-related macular degeneration associated polymorphism rs10490924 in *ARMS2* results in deficiency of a complement activator. *J. Neuroinflamm.* **2017**, *14*. [CrossRef] [PubMed]
14. Miller, J.W. Age-Related Macular Degeneration Revisited–Piecing the Puzzle: The LXIX Edward Jackson Memorial Lecture. *Am. J. Ophthalmol.* **2013**, *155*, 1–35.e13. [CrossRef] [PubMed]
15. Chakravarthy, U.; Wong, T.Y.; Fletcher, A.; Piault, E.; Evans, C.; Zlateva, G.; Buggage, R.; Pleil, A.; Mitchell, P. Clinical risk factors for age-related macular degeneration: A systematic review and meta-analysis. *BMC Ophthalmol.* **2010**, *10*. [CrossRef] [PubMed]
16. Balashova, E.E.; Maslov, D.L.; Lokhov, P.G. A metabolomics approach to pharmacotherapy personalization. *J. Pers. Med.* **2018**, *8*, 28. [CrossRef] [PubMed]
17. Kersten, E.; Paun, C.C.; Schellevis, R.L.; Hoyng, C.B.; Delcourt, C.; Lengyel, I.; Peto, T.; Ueffing, M.; Klaver, C.C.W.; Dammeier, S.; et al. Systemic and ocular fluid compounds as potential biomarkers in age-related macular degeneration. *Surv. Ophthalmol.* **2018**, *63*, 9–39. [CrossRef]
18. Van Leeuwen, E.M.; Emri, E.; Merle, B.M.J.; Colijn, J.M.; Kersten, E.; Cougnard-Gregoire, A.; Dammeier, S.; Meester-Smoor, M.; Pool, F.M.; de Jong, E.K.; et al. A new perspective on lipid research in age-related macular degeneration. *Prog. Retin. Eye Res.* **2018**. [CrossRef]
19. Chen, L.; Gao, Y.; Wang, L.Z.; Cheung, N.; Tan, G.S.W.; Cheung, G.C.M.; Beuerman, R.W.; Wong, T.Y.; Chan, E.C.Y.; Zhou, L. Recent advances in the applications of metabolomics in eye research. *Anal. Chim. Acta* **2018**, *1037*, 28–40. [CrossRef]
20. Tan, S.Z.; Begley, P.; Mullard, G.; Hollywood, K.A.; Bishop, P.N. Introduction to metabolomics and its applications in ophthalmology. *Eye* **2016**, *30*, 773–783. [CrossRef]

21. Laíns, I.; Gantner, M.; Murinello, S.; Lasky-Su, J.A.; Miller, J.W.; Friedlander, M.; Husain, D. Metabolomics in the study of retinal health and disease. *Prog. Retin. Eye Res.* **2018**. [CrossRef] [PubMed]
22. Dowling, J.E. Visual Adaptation. In *The Retina: An Approachable Part of the Brain*; Belknap Press of Harvard University Press: Cambridge, MA, USA, 2012; ISBN 9780674061545.
23. Hurley, J.B.; Lindsay, K.J.; Du, J. Glucose, lactate, and shuttling of metabolites in vertebrate retinas. *J. Neurosci. Res.* **2015**, *93*, 1079–1092. [CrossRef] [PubMed]
24. Macgregor, L.C.; Rosecan, L.R.; Laties, A.M.; Matschinsky, F.M. Altered retinal metabolism in diabetes. I. Microanalysis of lipid, glucose, sorbitol, and *myo*-inositol in the choroid and in the individual layers of the rabbit retina. *J. Biol. Chem.* **1986**, *261*, 4046–4051. [PubMed]
25. Adler, A.J.; Southwick, R.E. Distribution of glucose and lactate in the interphotoreceptor matrix. *Ophthalmic Res.* **1992**, *24*, 243–252. [CrossRef] [PubMed]
26. Krebs, H.A. On the metabolism of the retina. *Biochem. Z.* **1927**, *189*, 57–59.
27. Winkler, B.S. Glycolytic and oxidative metabolism in relation to retinal function. *J. Gen. Physiol.* **1981**, *77*, 667–692. [CrossRef] [PubMed]
28. Stone, J.; Van Driel, D.; Valter, K.; Rees, S.; Provis, J. The locations of mitochondria in mammalian photoreceptors: Relation to retinal vasculature. *Brain Res.* **2008**, *1189*, 58–69. [CrossRef]
29. Molday, L.L.; Wu, W.W.H.; Molday, R.S. Retinoschisin (RS1), the protein encoded by the X-linked retinoschisis gene, is anchored to the surface of retinal photoreceptor and bipolar cells through its interactions with a Na/K ATPase-SARM1 complex. *J. Biol. Chem.* **2007**, *282*, 32792–32801. [CrossRef]
30. Linton, J.D.; Holzhausen, L.C.; Babai, N.; Song, H.; Miyagishima, K.J.; Stearns, G.W.; Lindsay, K.; Wei, J.; Chertov, A.O.; Peters, T.A.; et al. Flow of energy in the outer retina in darkness and in light. *Proc. Natl. Acad. Sci. USA* **2010**, *107*, 8599–8604. [CrossRef]
31. Du, J.; Rountree, A.; Cleghorn, W.M.; Contreras, L.; Lindsay, K.J.; Sadilek, M.; Gu, H.; Djukovic, D.; Raftery, D.; Satrústegui, J.; et al. Phototransduction influences metabolic flux and nucleotide metabolism in mouse retina. *J. Biol. Chem.* **2016**, *291*, 4698–4710. [CrossRef]
32. Du, J.; Yanagida, A.; Knight, K.; Engel, A.L.; Vo, A.H.; Jankowski, C.; Sadilek, M.; Tran, V.T.B.; Manson, M.A.; Ramakrishnan, A.; et al. Reductive carboxylation is a major metabolic pathway in the retinal pigment epithelium. *Proc. Natl. Acad. Sci. USA* **2016**, *113*, 14710–14715. [CrossRef] [PubMed]
33. Pieczenik, S.R.; Neustadt, J. Mitochondrial dysfunction and molecular pathways of disease. *Exp. Mol. Pathol.* **2007**, *83*, 84–92. [CrossRef] [PubMed]
34. Zabka, T.S.; Singh, J.; Dhawan, P.; Liederer, B.M.; Oeh, J.; Kauss, M.A.; Xiao, Y.; Zak, M.; Lin, T.; McCray, B.; et al. Retinal toxicity, *in vivo* and *in vitro*, associated with inhibition of nicotinamide phosphoribosyltransferase. *Toxicol. Sci.* **2015**, *144*, 163–172. [CrossRef] [PubMed]
35. Bai, S.; Sheline, C.T. NAD^+ maintenance attenuates light induced photoreceptor degeneration. *Exp. Eye Res.* **2013**, *108*, 76–83. [CrossRef] [PubMed]
36. Kurihara, T.; Westenskow, P.D.; Gantner, M.L.; Usui, Y.; Schultz, A.; Bravo, S.; Aguilar, E.; Wittgrove, C.; Friedlander, M.S.; Paris, L.P.; et al. Hypoxia-induced metabolic stress in retinal pigment epithelial cells is sufficient to induce photoreceptor degeneration. *Elife* **2016**, *5*, e14319. [CrossRef] [PubMed]
37. Zhao, C.; Yasumura, D.; Li, X.; Matthes, M.; Lloyd, M.; Nielsen, G.; Ahern, K.; Snyder, M.; Bok, D.; Dunaief, J.L.; et al. mTOR-mediated dedifferentiation of the retinal pigment epithelium initiates photoreceptor degeneration in mice. *J. Clin. Investig.* **2011**, *121*, 369–383. [CrossRef] [PubMed]
38. Kanow, M.A.; Giarmarco, M.M.; Jankowski, C.S.R.; Tsantilas, K.; Engel, A.L.; Du, J.; Linton, J.D.; Farnsworth, C.C.; Sloat, S.R.; Rountree, A.; et al. Biochemical adaptations of the retina and retinal pigment epithelium support a metabolic ecosystem in the vertebrate eye. *Elife* **2017**, *6*, e28899. [CrossRef]
39. Feeney, L. Lipofuscin and melanin of human retinal pigment epithelium. Fluorescence, enzyme cytochemical, and ultrastructural studies. *Investig. Ophthalmol. Vis. Sci.* **1978**, *17*, 583–600.
40. Dorey, C.K.; Wu, G.; Ebenstein, D.; Garsd, A.; Weiter, J.J. Cell loss in the aging retina. Relationship to lipofuscin accumulation and macular degeneration. *Investig. Ophthalmol. Vis. Sci.* **1989**, *30*, 1691–1699.
41. Wolf, G. Lipofuscin and macular degeneration. *Nutr. Rev.* **2003**, *61*, 342–346. [CrossRef]
42. Sparrow, J.R.; Fishkin, N.; Zhou, J.; Cai, B.; Jang, Y.P.; Krane, S.; Itagaki, Y.; Nakanishi, K. A2E, a byproduct of the visual cycle. *Vis. Res.* **2003**, *43*, 2983–2990. [CrossRef]

43. Lakkaraju, A.; Finnemann, S.C.; Rodriguez-Boulan, E. The lipofuscin fluorophore A2E perturbs cholesterol metabolism in retinal pigment epithelial cells. *Proc. Natl. Acad. Sci. USA* **2007**, *104*, 11026–11031. [CrossRef] [PubMed]
44. Shaban, H.; Gazzotti, P.; Richter, C.; Remé, C.E.; Grimm, C.; Wenzel, A.; Jä, M.; Esser, P.; Kociok, N.; Leist, M.; et al. Cytochrome *c* oxidase inhibition by *N*-retinyl-*N*-retinylidene ethanolamine, a compound suspected to cause age-related macula degeneration. *Arch. Biochem. Biophys.* **2001**, *394*, 111–116. [CrossRef] [PubMed]
45. Suter, M.; Remé, C.; Grimm, C.; Wenzel, A.; Jäättela, M.; Esser, P.; Kociok, N.; Leist, M.; Richter, C. Age-related macular degeneration. The lipofusion component *N*-retinyl-*N*-retinylidene ethanolamine detaches proapoptotic proteins from mitochondria and induces apoptosis in mammalian retinal pigment epithelial cells. *J. Biol. Chem.* **2000**, *275*, 39625–39630. [CrossRef] [PubMed]
46. Leist, M.; Single, B.; Naumann, H.; Fava, E.; Simon, B.; Kühnle, S.; Nicotera, P. Inhibition of mitochondrial ATP generation by nitric oxide switches apoptosis to necrosis. *Exp. Cell Res.* **1999**, *249*, 396–403. [CrossRef] [PubMed]
47. Susin, S.A.; Lorenzo, H.K.; Zamzami, N.; Marzo, I.; Snow, B.E.; Brothers, G.M.; Mangion, J.; Jacotot, E.; Costantini, P.; Loeffler, M.; et al. Molecular characterization of mitochondrial apoptosis-inducing factor. *Nature* **1999**, *397*, 441–446. [CrossRef] [PubMed]
48. Tate, D.J.; Miceli, M.V.; Newsome, D.A.; Alcock, N.W.; Oliver, P.D. Influence of zinc on selected cellular functions of cultured human retinal pigment epithelium. *Curr. Eye Res.* **1995**, *14*, 897–903. [CrossRef]
49. Tate, D.J.; Miceli, M.V.; Newsome, D.A. Zinc protects against oxidative damage in cultured human retinal pigment epithelial cells. *Free Radic. Biol. Med.* **1999**, *26*, 704–713. [CrossRef]
50. Nicolas, M.G.; Fujiki, K.; Murayama, K.; Suzuki, M.T.; Shindo, N.; Hotta, Y.; Iwata, F.; Fujimura, T.; Yoshikawa, Y.; Cho, F.; et al. Studies on the mechanism of early onset macular degeneration in cynomolgus monkeys. II. Suppression of metallothionein synthesis in the retina in oxidative stress. *Exp. Eye Res.* **1996**, *62*, 399–408. [CrossRef]
51. Newsome, D.A.; Swartz, M.; Leone, N.C.; Elston, R.C.; Miller, E. Oral zinc in macular degeneration. *Arch. Ophthalmol.* **1988**, *106*, 192–198. [CrossRef]
52. Rajapakse, D.; Curtis, T.; Chen, M.; Xu, H. Zinc protects oxidative stress-induced RPE death by reducing mitochondrial damage and preventing lysosome rupture. *Oxidative Med. Cell. Longev.* **2017**, *2017*, 1–12. [CrossRef] [PubMed]
53. Age-Related Eye Disease Study Research Group. A randomized, placebo-controlled, clinical trial of high-dose supplementation with vitamins C and E, beta carotene, and zinc for age-related macular degeneration and vision loss: AREDS report no. 8. *Arch. Ophthalmol.* **2001**, *119*, 1417–1436. [CrossRef]
54. Julien, S.; Biesemeier, A.; Kokkinou, D.; Eibl, O.; Schraermeyer, U. Zinc deficiency leads to lipofuscin accumulation in the retinal pigment epithelium of pigmented rats. *PLoS ONE* **2011**, *6*, e29245. [CrossRef] [PubMed]
55. Kennedy, C.J.; Rakoczy, P.E.; Robertson, T.A.; Papadimitriou, J.M.; Constable, I.J. Kinetic studies on phagocytosis and lysosomal digestion of rod outer segments by human retinal pigment epithelial cells *in vitro*. *Exp. Cell Res.* **1994**, *210*, 209–214. [CrossRef] [PubMed]
56. Tokarz, P.; Kaarniranta, K.; Blasiak, J. Role of antioxidant enzymes and small molecular weight antioxidants in the pathogenesis of age-related macular degeneration (AMD). *Biogerontology* **2013**, *14*, 461–482. [CrossRef]
57. Adler, L., IV; Boyer, N.P.; Anderson, D.M.; Spraggins, J.M.; Schey, K.L.; Hanneken, A.; Ablonczy, Z.; Crouch, R.K.; Koutalos, Y. Determination of *N*-retinylidene-*N*-retinylethanolamine (A2E) levels in central and peripheral areas of human retinal pigment epithelium. *Photochem. Photobiol. Sci.* **2015**, *14*, 1983–1990. [CrossRef] [PubMed]
58. Gliem, M.; Müller, P.L.; Finger, R.P.; McGuinness, M.B.; Holz, F.G.; Charbel Issa, P. Quantitative fundus autofluorescence in early and intermediate age-related macular degeneration. *JAMA Ophthalmol.* **2016**, *134*, 817–824. [CrossRef]
59. Orellana-Rios, J.; Yokoyama, S.; Agee, J.M.; Challa, N.; Freund, K.B.; Yannuzzi, L.A.; Smith, R.T. Quantitative fundus autofluorescence in non-neovascular age-related macular degeneration. *Ophthalmic Surg. Lasers Imaging Retin.* **2018**, *49*, S34–S42. [CrossRef]
60. Wong, W.T.; Kam, W.; Cunningham, D.; Harrington, M.; Hammel, K.; Meyerle, C.B.; Cukras, C.; Chew, E.Y.; Sadda, S.R.; Ferris, F.L. Treatment of geographic atrophy by the topical administration of OT-551: Results of a phase II clinical trial. *Investig. Ophthalmol. Vis. Sci.* **2010**, *51*, 6131–6139. [CrossRef]

61. Mata, N.L.; Lichter, J.B.; Vogel, R.; Han, Y.; Bui, T.V.; Singerman, L.J. Investigation of oral fenretinide for treatment of geographic atrophy in age-related macular degeneration. *Retina* **2013**, *33*, 498–507. [CrossRef]
62. Pauleikhoff, D.; Harper, C.A.; Marshall, J.; Bird, A.C. Aging changes in Bruch's membrane: A histochemical and morphologic study. *Ophthalmology* **1990**, *97*, 171–178. [CrossRef]
63. Sheraidah, G.; Steinmetz, R.; Maguire, J.; Pauleikhoff, D.; Marshall, J.; Bird, A.C. Correlation between lipids extracted from Bruch's membrane and age. *Ophthalmology* **1993**, *100*, 47–51. [CrossRef]
64. Curcio, C.A.; Millican, C.L.; Bailey, T.; Kruth, H.S. Accumulation of cholesterol with age in human Bruch's membrane. *Investig. Ophthalmol. Vis. Sci.* **2001**, *42*, 265–274.
65. Li, C.-M.; Clark, M.E.; Rudolf, M.; Curcio, C.A. Distribution and composition of esterified and unesterified cholesterol in extra-macular drusen. *Exp. Eye Res.* **2007**, *85*, 192–201. [CrossRef] [PubMed]
66. Curcio, C.A.; Presley, J.B.; Malek, G.; Medeiros, N.E.; Avery, D.V.; Kruth, H.S. Esterified and unesterified cholesterol in drusen and basal deposits of eyes with age-related maculopathy. *Exp. Eye Res.* **2005**, *81*, 731–741. [CrossRef] [PubMed]
67. Haimovici, R.; Gantz, D.L.; Rumelt, S.; Freddo, T.F.; Small, D.M. The lipid composition of drusen, Bruch's membrane, and sclera by hot stage polarizing light microscopy. *Investig. Ophthalmol. Vis. Sci.* **2001**, *42*, 1592–1599.
68. Rudolf, M.; Curcio, C.A. Esterified cholesterol is highly localized to Bruch's membrane, as revealed by lipid histochemistry in wholemounts of human choroid. *J. Histochem. Cytochem.* **2009**, *57*, 731–739. [CrossRef]
69. Pikuleva, I.A.; Curcio, C.A. Cholesterol in the retina: The best is yet to come. *Prog. Retin. Eye Res.* **2014**, *41*, 64–89. [CrossRef]
70. Shen, J.; He, J.; Wang, F. Association of lipids with age-related macular degeneration. *Discov. Med.* **2016**, *22*, 129–145.
71. Ishibashi, T.; Murata, T.; Hangai, M.; Nagai, R.; Horiuchi, S.; Lopez, P.F.; Hinton, D.R.; Ryan, S.J. Advanced glycation end products in age-related macular degeneration. *Arch. Ophthalmol.* **1998**, *116*, 1629–1632. [CrossRef]
72. Glenn, J.V.; Mahaffy, H.; Wu, K.; Smith, G.; Nagai, R.; Simpson, D.A.C.; Boulton, M.E.; Stitt, A.W. Advanced glycation end product (AGE) accumulation on Bruch's membrane: Links to age-related RPE dysfunction. *Investig. Ophthalmol. Vis. Sci.* **2009**, *50*, 441–451. [CrossRef] [PubMed]
73. Yamada, Y.; Ishibashi, K.; Ishibashi, K.; Bhutto, I.A.; Tian, J.; Lutty, G.A.; Handa, J.T. The expression of advanced glycation endproduct receptors in rpe cells associated with basal deposits in human maculas. *Exp. Eye Res.* **2006**, *82*, 840–848. [CrossRef] [PubMed]
74. Chen, M.; Glenn, J.V.; Dasari, S.; McVicar, C.; Ward, M.; Colhoun, L.; Quinn, M.; Bierhaus, A.; Xu, H.; Stitt, A.W. RAGE regulates immune cell infiltration and angiogenesis in choroidal neovascularization. *PLoS ONE* **2014**, *9*, e89548. [CrossRef] [PubMed]
75. Bird, A.C. Towards an understanding of age-related macular disease. *Eye* **2003**, *17*, 457–466. [CrossRef] [PubMed]
76. Balaratnasingam, C.; Yannuzzi, L.A.; Curcio, C.A.; Morgan, W.H.; Querques, G.; Capuano, V.; Souied, E.; Jung, J.; Freund, K.B. Associations between retinal pigment epithelium and drusen volume changes during the lifecycle of large drusenoid pigment epithelial detachments. *Investig. Ophthalmol. Vis. Sci.* **2016**, *57*, 5479–5489. [CrossRef] [PubMed]
77. Sarks, S.; Cherepanoff, S.; Killingsworth, M.; Sarks, J. Relationship of basal laminar deposit and membranous debris to the clinical presentation of early age-related macular degeneration. *Investig. Ophthalmol. Vis. Sci.* **2007**, *48*, 968–977. [CrossRef] [PubMed]
78. Wang, L.; Clark, M.E.; Crossman, D.K.; Kojima, K.; Messinger, J.D.; Mobley, J.A.; Curcio, C.A. Abundant lipid and protein components of drusen. *PLoS ONE* **2010**, *5*, e10329. [CrossRef] [PubMed]
79. Lotery, A.; Trump, D. Progress in defining the molecular biology of age related macular degeneration. *Hum. Genet.* **2007**, *122*, 219–236. [CrossRef]
80. Khan, K.N.; Mahroo, O.A.; Khan, R.S.; Mohamed, M.D.; McKibbin, M.; Bird, A.; Michaelides, M.; Tufail, A.; Moore, A.T. Differentiating drusen: Drusen and drusen-like appearances associated with ageing, age-related macular degeneration, inherited eye disease and other pathological processes. *Prog. Retin. Eye Res.* **2016**, *53*, 70–106. [CrossRef]

81. Tan, A.C.S.; Pilgrim, M.G.; Fearn, S.; Bertazzo, S.; Tsolaki, E.; Morrell, A.P.; Li, M.; Messinger, J.D.; Dolz-Marco, R.; Lei, J.; et al. Calcified nodules in retinal drusen are associated with disease progression in age-related macular degeneration. *Sci. Transl. Med.* **2018**, *10*, eaat4544. [CrossRef]
82. Schlanitz, F.G.; Baumann, B.; Kundi, M.; Sacu, S.; Baratsits, M.; Scheschy, U.; Shahlaee, A.; Mittermüller, T.J.; Montuoro, A.; Roberts, P.; et al. Drusen volume development over time and its relevance to the course of age-related macular degeneration. *Br. J. Ophthalmol.* **2017**, *101*, 198–203. [CrossRef] [PubMed]
83. Nathoo, N.A.; Or, C.; Young, M.; Chui, L.; Fallah, N.; Kirker, A.W.; Albiani, D.A.; Merkur, A.B.; Forooghian, F. Optical coherence tomography-based measurement of drusen load predicts development of advanced age-related macular degeneration. *Am. J. Ophthalmol.* **2014**, *158*, 757–761. [CrossRef] [PubMed]
84. Umeda, S.; Suzuki, M.T.; Okamoto, H.; Ono, F.; Mizota, A.; Terao, K.; Yoshikawa, Y.; Tanaka, Y.; Iwata, T. Molecular composition of drusen and possible involvement of anti-retinal autoimmunity in two different forms of macular degeneration in cynomolgus monkey (*Macaca fascicularis*). *FASEB J.* **2005**, *19*, 1683–1685. [CrossRef] [PubMed]
85. Anderson, D.H.; Talaga, K.C.; Rivest, A.J.; Barron, E.; Hageman, G.S.; Johnson, L.V. Characterization of β amyloid assemblies in drusen: The deposits associated with aging and age-related macular degeneration. *Exp. Eye Res.* **2004**, *78*, 243–256. [CrossRef] [PubMed]
86. Ratnayaka, J.A.; Serpell, L.C.; Lotery, A.J. Dementia of the eye: The role of amyloid beta in retinal degeneration. *Eye* **2015**, *29*, 1013–1026. [CrossRef] [PubMed]
87. Hageman, G.S.; Mullins, R.F.; Russell, S.R.; Johnson, L.V.; Anderson, D.H. Vitronectin is a constituent of ocular drusen and the vitronectin gene is expressed in human retinal pigmented epithelial cells. *FASEB J.* **1999**, *13*, 477–484. [CrossRef] [PubMed]
88. Anderson, D.H.; Hageman, G.S.; Mullins, R.F.; Neitz, M.; Neitz, J.; Ozaki, S.; Preissner, K.T.; Johnson, L.V. Vitronectin gene expression in the adult human retina. *Investig. Ophthalmol. Vis. Sci.* **1999**, *40*, 3305–3315.
89. Thompson, R.B.; Reffatto, V.; Bundy, J.G.; Kortvely, E.; Flinn, J.M.; Lanzirotti, A.; Jones, E.A.; McPhail, D.S.; Fearn, S.; Boldt, K.; et al. Identification of hydroxyapatite spherules provides new insight into subretinal pigment epithelial deposit formation in the aging eye. *Proc. Natl. Acad. Sci. USA* **2015**, *112*, 1565–1570. [CrossRef]
90. Pilgrim, M.G.; Lengyel, I.; Lanzirotti, A.; Newville, M.; Fearn, S.; Emri, E.; Knowles, J.C.; Messinger, J.D.; Read, R.W.; Guidry, C.; et al. Subretinal pigment epithelial deposition of drusen components including hydroxyapatite in a primary cell culture model. *Investig. Ophthalmol. Vis. Sci.* **2017**, *58*, 708–719. [CrossRef]
91. Brewer, G.J. Risks of copper and iron toxicity during aging in humans. *Chem. Res. Toxicol.* **2010**, *23*, 319–326. [CrossRef]
92. Kurz, T.; Karlsson, M.; Brunk, U.T.; Nilsson, S.E. ARPE-19 retinal pigment epithelial cells are highly resistant to oxidative stress and exercise strict control over their lysosomal redox-active iron. *Autophagy* **2009**, *5*, 494–501. [CrossRef]
93. Nan, R.; Gor, J.; Lengyel, I.; Perkins, S.J. Uncontrolled zinc- and copper-induced oligomerisation of the human complement regulator factor H and its possible implications for function and disease. *J. Mol. Biol.* **2008**, *384*, 1341–1352. [CrossRef] [PubMed]
94. Flinn, J.M.; Kakalec, P.; Tappero, R.; Jones, B.; Lengyel, I. Correlations in distribution and concentration of calcium, copper and iron with zinc in isolated extracellular deposits associated with age-related macular degeneration. *Metallomics* **2014**, *6*, 1223–1228. [CrossRef] [PubMed]
95. Wood, J.P.M.; Osborne, N.N. Zinc and energy requirements in induction of oxidative stress to retinal pigmented epithelial cells. *Neurochem. Res.* **2003**, *28*, 1525–1533. [CrossRef] [PubMed]
96. Lengyel, I.; Flinn, J.M.; Peto, T.; Linkous, D.H.; Cano, K.; Bird, A.C.; Lanzirotti, A.; Frederickson, C.J.; van Kuijk, F.J.G.M. High concentration of zinc in sub-retinal pigment epithelial deposits. *Exp. Eye Res.* **2007**, *84*, 772–780. [CrossRef] [PubMed]
97. Friedman, E.; Kopald, H.H.; Smith, T.R.; Mimura, S. Retinal and choroidal blood flow determined with krypton-85 anesthetized animals. *Investig. Ophthalmol. Vis. Sci.* **1964**, *3*, 539–547.
98. Alm, A.; Bill, A. Blood flow and oxygen extraction in the cat uvea at normal and high intraocular pressures. *Acta Physiol. Scand.* **1970**, *80*, 19–28. [CrossRef] [PubMed]
99. Roth, S.; Pietrzyk, Z. Blood flow after retinal ischemia in cats. *Investig. Ophthalmol. Vis. Sci.* **1994**, *35*, 3209–3217.

100. Alm, A.; Bill, A. The oxygen supply to the retina, II. Effects of high intraocular pressure and of increased arterial carbon dioxide tension on uveal and retinal blood flow in cats. *Acta Physiol. Scand.* **1972**, *84*, 306–319. [CrossRef]
101. Wang, L.; Kondo, M.; Bill, A. Glucose metabolism in cat outer retina. Effects of light and hyperoxia. *Investig. Ophthalmol. Vis. Sci.* **1997**, *38*, 48–55.
102. Linsenmeier, R.A.; Padnick–Silver, L. Metabolic dependence of photoreceptors on the choroid in the normal and detached retina. *Investig. Ophthalmol. Vis. Sci.* **2000**, *41*, 3117–3123.
103. Chirco, K.R.; Sohn, E.H.; Stone, E.M.; Tucker, B.A.; Mullins, R.F. Structural and molecular changes in the aging choroid: Implications for age-related macular degeneration. *Eye* **2017**, *31*, 10–25. [CrossRef] [PubMed]
104. Wakatsuki, Y.; Shinojima, A.; Kawamura, A.; Yuzawa, M. Correlation of aging and segmental choroidal thickness measurement using swept source optical coherence tomography in healthy eyes. *PLoS ONE* **2015**, *10*, e0144156. [CrossRef] [PubMed]
105. Yuan, X.; Gu, X.; Crabb, J.W.S.; Yue, X.; Shadrach, K.; Hollyfield, J.G.; Crabb, J.W.S. Quantitative proteomics: Comparison of the macular Bruch membrane/choroid complex from age-related macular degeneration and normal eyes. *Mol. Cell. Proteom.* **2010**, *9*, 1031–1046. [CrossRef] [PubMed]
106. Boccard, J.; Veuthey, J.-L.; Rudaz, S. Knowledge discovery in metabolomics: An overview of MS data handling. *J. Sep. Sci.* **2010**, *33*, 290–304. [CrossRef] [PubMed]
107. Kristal, B.S.; Vigneau-Callahan, K.E.; Matson, W.R. Simultaneous analysis of the majority of low-molecular-weight, redox-active compounds from mitochondria. *Anal. Biochem.* **1998**, *263*, 18–25. [CrossRef] [PubMed]
108. Goodacre, R.; Vaidyanathan, S.; Dunn, W.B.; Harrigan, G.G.; Kell, D.B. Metabolomics by numbers: Acquiring and understanding global metabolite data. *Trends Biotechnol.* **2004**, *22*, 245–252. [CrossRef]
109. Würtz, P.; Kangas, A.J.; Soininen, P.; Lawlor, D.A.; Davey Smith, G.; Ala-Korpela, M. Quantitative serum nuclear magnetic resonance metabolomics in large-scale epidemiology: A primer on -omic technologies. *Am. J. Epidemiol.* **2017**, *186*, 1084–1096. [CrossRef]
110. Lauwen, S.; de Jong, E.K.; Lefeber, D.J.; den Hollander, A.I. Omics biomarkers in ophthalmology. *Investig. Ophthalmol. Vis. Sci.* **2017**, *58*, BIO88–BIO98. [CrossRef]
111. Brindle, J.T.; Antti, H.; Holmes, E.; Tranter, G.; Nicholson, J.K.; Bethell, H.W.L.; Clarke, S.; Schofield, P.M.; McKilligin, E.; Mosedale, D.E.; et al. Rapid and noninvasive diagnosis of the presence and severity of coronary heart disease using ^{1}H-NMR-based metabonomics. *Nat. Med.* **2002**, *8*, 1439–1445. [CrossRef]
112. Lindon, J.C.; Holmes, E.; Nicholson, J.K. So what's the deal with metabonomics? *Anal. Chem.* **2003**, *75*, 384 A–391 A. [CrossRef]
113. Nicholson, J.K.; Wilson, I.D. Understanding "global" systems biology: Metabonomics and the continuum of metabolism. *Nat. Rev. Drug Discov.* **2003**, *2*, 668–676. [CrossRef] [PubMed]
114. Rustam, Y.H.; Reid, G.E. Analytical challenges and recent advances in mass spectrometry based lipidomics. *Anaytical Chem.* **2018**, *90*, 374–397. [CrossRef] [PubMed]
115. Cui, L.; Lu, H.; Lee, Y.H. Challenges and emergent solutions for LC-MS/MS based untargeted metabolomics in diseases. *Mass Spectrom. Rev.* **2018**, *37*, 772–792. [CrossRef] [PubMed]
116. Wishart, D.S.; Knox, C.; Guo, A.C.; Eisner, R.; Young, N.; Gautam, B.; Hau, D.D.; Psychogios, N.; Dong, E.; Bouatra, S.; et al. HMDB: A knowledgebase for the human metabolome. *Nucleic Acids Res.* **2009**, *37*, D603–D610. [CrossRef] [PubMed]
117. Wishart, D.S.; Feunang, Y.D.; Marcu, A.; Guo, A.C.; Liang, K.; Vázquez-Fresno, R.; Sajed, T.; Johnson, D.; Li, C.; Karu, N.; et al. HMDB 4.0: The human metabolome database for 2018. *Nucleic Acids Res.* **2018**, *46*, D608–D617. [CrossRef] [PubMed]
118. Wang, J.; Westenskow, P.D.; Fang, M.; Friedlander, M.; Siuzdak, G. Quantitative metabolomics of photoreceptor degeneration and the effects of stem cell-derived retinal pigment epithelium transplantation. *Philos. Trans. A Math. Phys. Eng. Sci.* **2016**, *374*, 20150376. [CrossRef]
119. Chen, Y.; Houghton, L.A.; Brenna, J.T.; Noy, N. Docosahexaenoic acid modulates the interactions of the interphotoreceptor retinoid-binding protein with 11-*cis*-retinal. *J. Biol. Chem.* **1996**, *271*, 20507–20515. [CrossRef]
120. Chen, Y.; Saari, J.C.; Noy, N. Interactions of all-trans-retinol and long-chain fatty acids with interphotoreceptor retinoid-binding protein1. *Biochemistry* **1993**, *32*, 11311–11318. [CrossRef]

121. Chen, Y.; Okano, K.; Maeda, T.; Chauhan, V.; Golczak, M.; Maeda, A.; Palczewski, K. Mechanism of all-*trans*-retinal toxicity with implications for stargardt disease and age-related macular degeneration. *J. Biol. Chem.* **2012**, *287*, 5059–5069. [CrossRef]
122. Sparrow, J.R.; Wu, Y.; Kim, C.Y.; Zhou, J. Phospholipid meets all-*trans*-retinal: The making of RPE bisretinoids. *J. Lipid Res.* **2010**, *51*, 247–261. [CrossRef] [PubMed]
123. Carr, A.-J.F.; Smart, M.J.K.; Ramsden, C.M.; Powner, M.B.; da Cruz, L.; Coffey, P.J. Development of human embryonic stem cell therapies for age-related macular degeneration. *Trends Neurosci.* **2013**, *36*, 385–395. [CrossRef] [PubMed]
124. Orban, T.; Johnson, W.M.; Dong, Z.; Maeda, T.; Maeda, A.; Sakai, T.; Tsuneoka, H.; Mieyal, J.J.; Palczewski, K. Serum levels of lipid metabolites in age-related macular degeneration. *FASEB J.* **2015**, *29*, 4579–4588. [CrossRef] [PubMed]
125. Chao, J.R.; Knight, K.; Engel, A.L.; Jankowski, C.; Wang, Y.; Manson, M.A.; Gu, H.; Djukovic, D.; Raftery, D.; Hurley, J.B.; et al. Human retinal pigment epithelial cells prefer proline as a nutrient and transport metabolic intermediates to the retinal side. *J. Biol. Chem.* **2017**, *292*, 12895–12905. [CrossRef]
126. Osborn, M.P.; Park, Y.; Parks, M.B.; Burgess, L.G.; Uppal, K.; Lee, K.; Jones, D.P.; Brantley, M.A. Metabolome-wide association study of neovascular age-related macular degeneration. *PLoS ONE* **2013**, *8*, e72737. [CrossRef]
127. Luo, D.; Deng, T.; Yuan, W.; Deng, H.; Jin, M. Plasma metabolomic study in Chinese patients with wet age-related macular degeneration. *BMC Ophthalmol.* **2017**, *17*, 165. [CrossRef]
128. Laíns, I.; Duarte, D.; Barros, A.S.; Martins, A.S.; Gil, J.; Miller, J.W.J.B.; Marques, M.; Mesquita, T.; Kim, I.K.; da Luz Cachulo, M.; et al. Human plasma metabolomics in age-related macular degeneration (AMD) using nuclear magnetic resonance spectroscopy. *PLoS ONE* **2017**, *12*, e0177749. [CrossRef]
129. Laíns, I.; Kelly, R.S.; Miller, J.W.J.B.; Silva, R.; Vavvas, D.G.; Kim, I.K.; Murta, J.N.; Lasky-Su, J.; Miller, J.W.J.B.; Husain, D.; et al. Human plasma metabolomics study across all stages of age-related macular degeneration identifies potential lipid biomarkers. *Ophthalmology* **2017**, *125*, 245–254. [CrossRef]
130. Rowan, S.; Jiang, S.; Korem, T.; Szymanski, J.; Chang, M.-L.; Szelog, J.; Cassalman, C.; Dasuri, K.; McGuire, C.; Nagai, R.; et al. Involvement of a gut-retina axis in protection against dietary glycemia-induced age-related macular degeneration. *Proc. Natl. Acad. Sci. USA* **2017**, *114*, E4472–E4481. [CrossRef]
131. Mitchell, S.L.; Uppal, K.; Williamson, S.M.; Liu, K.; Burgess, L.G.; Tran, V.; Umfress, A.C.; Jarrell, K.L.; Cooke Bailey, J.N.; Agarwal, A.; et al. The carnitine shuttle pathway is altered in patients with neovascular age-related macular degeneration. *Investig. Ophthalmol. Vis. Sci.* **2018**, *59*, 4978–4985. [CrossRef]
132. Overmyer, K.A.; Thonusin, C.; Qi, N.R.; Burant, C.F.; Evans, C.R. Impact of anesthesia and euthanasia on metabolomics of mammalian tissues: Studies in a C57BL/6J mouse model. *PLoS ONE* **2015**, *10*, e0117232. [CrossRef] [PubMed]
133. Tan, S.Z.; Mullard, G.; Hollywood, K.A.; Dunn, W.B.; Bishop, P.N. Characterisation of the metabolome of ocular tissues and post-mortem changes in the rat retina. *Exp. Eye Res.* **2016**, *149*, 8–15. [CrossRef] [PubMed]
134. Hu, J.; Bok, D. A cell culture medium that supports the differentiation of human retinal pigment epithelium into functionally polarized monolayers. *Mol. Vis.* **2001**, *7*, 14–19. [PubMed]
135. Maminishkis, A.; Chen, S.; Jalickee, S.; Banzon, T.; Shi, G.; Wang, F.E.; Ehalt, T.; Hammer, J.A.; Miller, S.S. Confluent monolayers of cultured human fetal retinal pigment epithelium exhibit morphology and physiology of native tissue. *Investig. Ophthalmol. Vis. Sci.* **2006**, *47*, 3612–3624. [CrossRef] [PubMed]
136. Maminishkis, A.; Miller, S.S. Experimental models for study of retinal pigment epithelial physiology and pathophysiology. *J. Vis. Exp.* **2010**, e2032. [CrossRef] [PubMed]
137. Adijanto, J.; Philp, N.J. Cultured primary human fetal retinal pigment epithelium (hfRPE) as a model for evaluating RPE metabolism. *Exp. Eye Res.* **2014**, *126*, 77–84. [CrossRef]
138. Du, J.; Linton, J.D.; Hurley, J.B. Probing metabolism in the intact retina using stable isotope tracers. *Methods Enzymol.* **2015**, *561*, 149–170. [CrossRef]
139. Johnson, L.V.; Forest, D.L.; Banna, C.D.; Radeke, C.M.; Maloney, M.A.; Hu, J.; Spencer, C.N.; Walker, A.M.; Tsie, M.S.; Bok, D.; et al. Cell culture model that mimics drusen formation and triggers complement activation associated with age-related macular degeneration. *Proc. Natl. Acad. Sci. USA* **2011**, *108*, 18277–18282. [CrossRef]
140. Anderson, B.; Saltzman, H.A. Retinal oxygen utilization measured by hyperbaric blackout. *Arch. Ophthalmol.* **1964**, *72*, 792–795. [CrossRef]

141. Ames, A.; Li, Y.Y.; Heher, E.C.; Kimble, C.R. Energy metabolism of rabbit retina as related to function: High cost of Na+ transport. *J. Neurosci.* **1992**, *12*, 840–853. [CrossRef]
142. Yu, D.-Y.; Cringle, S.J. Oxygen distribution and consumption within the retina in vascularised and avascular retinas and in animal models of retinal disease. *Prog. Retin. Eye Res.* **2001**, *20*, 175–208. [CrossRef]
143. Hess, H.H. The high calcium content of retinal pigmented epithelium. *Exp. Eye Res.* **1975**, *21*, 471–479. [CrossRef]
144. Chen, H.; Tran, J.-T.A.; Eckerd, A.; Huynh, T.-P.; Elliott, M.H.; Brush, R.S.; Mandal, N.A. Inhibition of de novo ceramide biosynthesis by FTY720 protects rat retina from light-induced degeneration. *J. Lipid Res.* **2013**, *54*, 1616–1629. [CrossRef] [PubMed]
145. Abrahan, C.E.; Miranda, G.E.; Agnolazza, D.L.; Politi, L.E.; Rotstein, N.P. Synthesis of sphingosine is essential for oxidative stress-induced apoptosis of photoreceptors. *Investig. Ophthalmol. Vis. Sci.* **2010**, *51*, 1171–1180. [CrossRef] [PubMed]
146. Rotstein, N.P.; Miranda, G.E.; Abrahan, C.E.; German, O.L. Regulating survival and development in the retina: Key roles for simple sphingolipids. *J. Lipid Res.* **2010**, *51*, 1247–1262. [CrossRef] [PubMed]
147. Xie, B.; Shen, J.; Dong, A.; Rashid, A.; Stoller, G.; Campochiaro, P.A. Blockade of sphingosine-1-phosphate reduces macrophage influx and retinal and choroidal neovascularization. *J. Cell. Physiol.* **2009**, *218*, 192–198. [CrossRef] [PubMed]
148. Caballero, S.; Swaney, J.; Moreno, K.; Afzal, A.; Kielczewski, J.; Stoller, G.; Cavalli, A.; Garland, W.; Hansen, G.; Sabbadini, R.; et al. Anti-sphingosine-1-phosphate monoclonal antibodies inhibit angiogenesis and sub-retinal fibrosis in a murine model of laser-induced choroidal neovascularization. *Exp. Eye Res.* **2009**, *88*, 367–377. [CrossRef] [PubMed]
149. Yonamine, I.; Bamba, T.; Nirala, N.K.; Jesmin, N.; Kosakowska-Cholody, T.; Nagashima, K.; Fukusaki, E.; Acharya, J.K.; Acharya, U. Sphingosine kinases and their metabolites modulate endolysosomal trafficking in photoreceptors. *J. Cell Biol.* **2011**, *192*, 557–567. [CrossRef]
150. Kennedy, B.G.; Torabi, A.J.; Kurzawa, R.; Echtenkamp, S.F.; Mangini, N.J. Expression of transient receptor potential vanilloid channels TRPV5 and TRPV6 in retinal pigment epithelium. *Mol. Vis.* **2010**, *16*, 665–675.
151. Wimmers, S.; Strauss, O. Basal calcium entry in retinal pigment epithelial cells is mediated by TRPC channels. *Investig. Ophthalmol. Vis. Sci.* **2007**, *48*, 5767–5772. [CrossRef]
152. Bhutto, I.; Lutty, G. Understanding age-related macular degeneration (AMD): Relationships between the photoreceptor/retinal pigment epithelium/Bruch's membrane/choriocapillaris complex. *Mol. Asp. Med.* **2012**, *33*, 295–317. [CrossRef] [PubMed]
153. Coscas, G.; Yamashiro, K.; Coscas, F.; De Benedetto, U.; Tsujikawa, A.; Miyake, M.; Gemmy Cheung, C.M.; Wong, T.Y.; Yoshimura, N. Comparison of exudative age-related macular degeneration subtypes in Japanese and French patients: Multicenter diagnosis with multimodal imaging. *Am. J. Ophthalmol.* **2014**, *158*, 309–318.e2. [CrossRef] [PubMed]
154. Sohrab, M.; Wu, K.; Fawzi, A.A. A pilot study of morphometric analysis of choroidal vasculature *in vivo*, using en face optical coherence tomography. *PLoS ONE* **2012**, *7*, e48631. [CrossRef] [PubMed]
155. McLeod, D.S.; Grebe, R.; Bhutto, I.; Merges, C.; Baba, T.; Lutty, G.A. Relationship between RPE and choriocapillaris in age-related macular degeneration. *Investig. Ophthalmol. Vis. Sci.* **2009**, *50*, 4982–4991. [CrossRef] [PubMed]
156. Metelo, A.M.; Noonan, H.; Iliopoulos, O. HIF2α inhibitors for the treatment of VHL disease. *Oncotarget* **2015**, *6*, 23036–23037. [CrossRef] [PubMed]
157. Metelo, A.M.; Noonan, H.R.; Li, X.; Jin, Y.; Baker, R.; Kamentsky, L.; Zhang, Y.; van Rooijen, E.; Shin, J.; Carpenter, A.E.; et al. Pharmacological HIF2α inhibition improves VHL disease-associated phenotypes in zebrafish model. *J. Clin. Investig.* **2015**, *125*, 1987–1997. [CrossRef] [PubMed]
158. Datta, S.; Cano, M.; Ebrahimi, K.; Wang, L.; Handa, J.T. The impact of oxidative stress and inflammation on RPE degeneration in non-neovascular AMD. *Prog. Retin. Eye Res.* **2017**, *60*, 201–218. [CrossRef]
159. Brantley, M.A.; Osborn, M.P.; Sanders, B.J.; Rezaei, K.A.; Lu, P.; Li, C.; Milne, G.L.; Cai, J.; Sternberg, P.; Sternberg, P.; et al. Plasma biomarkers of oxidative stress and genetic variants in age-related macular degeneration. *Am. J. Ophthalmol.* **2012**, *153*, 460–467.e1. [CrossRef]
160. Morrow, J.D.; Frei, B.; Longmire, A.W.; Gaziano, M.; Lynch, S.M.; Shyr, Y.; Strauss, W.E.; Oates, J.A.; Roberts, L.J., II. Increase in circulating products of lipid peroxidation (F_2-isoprostanes) in smokers. *N. Engl. J. Med.* **1995**, *332*, 1198–1203. [CrossRef]

161. Jones, D.P.; Mody, V.C., Jr.; Carlson, J.L.; Lynn, M.J.; Sternberg, P. Redox analysis of human plasma allows separation of pro-oxidant events of aging from decline in antioxidant defenses. *Free Radic. Biol. Med.* **2002**, *33*, 1290–1300. [CrossRef]
162. Moriarty, S.E.; Shah, J.H.; Lynn, M.; Jiang, S.; Openo, K.; Jones, D.P.; Sternberg, P. Oxidation of glutathione and cysteine in human plasma associated with smoking. *Free Radic. Biol. Med.* **2003**, *35*, 1582–1588. [CrossRef] [PubMed]
163. Merle, B.M.J.; Benlian, P.; Puche, N.; Bassols, A.; Delcourt, C.; Souied, E.H. Circulating omega-3 fatty acids and neovascular age-related macular degeneration. *Investig. Ophthalmol. Vis. Sci.* **2014**, *55*, 2010–2019. [CrossRef]
164. Merle, B.M.J.; Delyfer, M.-N.; Korobelnik, J.-F.; Rougier, M.-B.; Malet, F.; Féart, C.; Le Goff, M.; Peuchant, E.; Letenneur, L.; Dartigues, J.-F.; et al. High concentrations of plasma n3 fatty acids are associated with decreased risk for late age-related macular degeneration. *J. Nutr.* **2013**, *143*, 505–511. [CrossRef] [PubMed]
165. Johnson, S.R.; Lange, B.M. Open-access metabolomics databases for natural product research: Present capabilities and future potential. *Front. Bioeng. Biotechnol.* **2015**, *3*, 22. [CrossRef] [PubMed]
166. Tolleson, W.H. Human melanocyte biology, toxicology, and pathology. *J. Environ. Sci. Health Part C Environ. Carcinog. Ecotoxicol. Rev.* **2005**, *23*, 105–161. [CrossRef]
167. Chen, G.; Walmsley, S.; Cheung, G.C.; Chen, L.; Cheng, C.-Y.; Beuerman, R.W.; Wong, T.Y.; Zhou, L.; Choi, H. Customized consensus spectral library building for untargeted quantitative metabolomics analysis with data independent acquisition mass spectrometry and MetaboDIA workflow. *Anal. Chem.* **2017**, *89*, 4897–4906. [CrossRef]
168. Pushpoth, S.; Fitzpatrick, M.; Young, S.; Yang, Y.; Talks, J.; Wallace, G. Metabolomic analysis in patients with age related macular degeneration. *Investig. Ophthalmol. Vis. Sci.* **2013**, *54*, 3662.
169. Pushpoth, S.; Fitzpatrick, M.; Talks, J.S.; Young, S.; Yang, Y.C.; Wallace, G.R. Metabolomic analysis of urine in patients with age related macular degeneration. *Investig. Ophthalmol. Vis. Sci.* **2015**, *56*, 368.
170. Wong, C.W.; Yanagi, Y.; Lee, W.-K.; Ogura, Y.; Yeo, I.; Yin Wong, T.; Ming Gemmy Cheung, C. Age-related macular degeneration and polypoidal choroidal vasculopathy in Asians. *Prog. Retin. Eye Res.* **2016**, *53*, 107–139. [CrossRef]
171. Li, M.; Zhang, X.; Liao, N.; Ye, B.; Peng, Y.; Ji, Y.; Wen, F. Analysis of the serum lipid profile in polypoidal choroidal vasculopathy. *Sci. Rep.* **2016**, *6*, 38342. [CrossRef]
172. Schoumacher, M.; Lambert, V.; Leenders, J.; Roblain, Q.; Govaerts, B.; Rakic, J.-M.; Noël, A.; De Tullio, P. NMR-based metabolomics for new target discovery and personalized medicine: Application to age-related macular degeneration (AMD). In *EUROMAR (European Magnetic Resonance Meeting)*; Université de Nantes: Nantes, France, 2018.
173. Schoumacher, M.; De Tullio, P.; Lambert, V.; Hansen, S.; Leenders, J.; Govaerts, B.; Pirotte, B.; Rakic, J.-M.; Noël, A. From metabolomics to identification of a new therapeutic approach for age-related macular degeneration (AMD). In *30ièmes Journées Franco-Belges de Pharmacochimie*; Université d'Orléans: Amboise, France, 2016.
174. Kowalczuk, L.; Matet, A.; Dor, M.; Bararpour, N.; Daruich, A.; Dirani, A.; Behar-Cohen, F.; Thomas, A.; Turck, N. Proteome and metabolome of subretinal fluid in central serous chorioretinopathy and rhegmatogenous retinal detachment: A pilot case study. *Transl. Vis. Sci. Technol.* **2018**, *7*, 3. [CrossRef] [PubMed]
175. Wang, Y.; Wang, M.; Zhang, X.; Zhang, Q.; Nie, J.; Zhang, M.; Liu, X.; Ma, L. The association between the lipids levels in blood and risk of age-related macular degeneration. *Nutrients* **2016**, *8*, 663. [CrossRef] [PubMed]
176. Mishima, S.; Gasset, A.; Klyce, S.D.; Baum, J.L. Determination of tear volume and tear flow. *Investig. Ophthalmol. Vis. Sci.* **1966**, *5*, 264–276.
177. Zhou, L.; Beuerman, R.W.; Foo, Y.; Liu, S.; Ang, L.P. Characterisation of human tear proteins using high-resolution mass spectrometry. *Ann. Acad. Med. Singap.* **2006**, *35*, 400–407. [PubMed]
178. Zhou, L.; Zhao, S.Z.; Koh, S.K.; Chen, L.; Vaz, C.; Tanavde, V.; Li, X.R.; Beuerman, R.W. In-depth analysis of the human tear proteome. *J. Proteom.* **2012**, *75*, 3877–3885. [CrossRef] [PubMed]
179. Lam, S.M.; Tong, L.; Duan, X.; Petznick, A.; Wenk, M.R.; Shui, G. Extensive characterization of human tear fluid collected using different techniques unravels the presence of novel lipid amphiphiles. *J. Lipid Res.* **2014**, *55*, 289–298. [CrossRef] [PubMed]

180. Chen, L.; Zhou, L.; Chan, E.C.Y.; Neo, J.; Beuerman, R.W. Characterization of the human tear metabolome by LC-MS/MS. *J. Proteome Res.* **2011**, *10*, 4876–4882. [CrossRef]
181. Hagan, S.; Martin, E.; Enríquez-de-Salamanca, A. Tear fluid biomarkers in ocular and systemic disease: Potential use for predictive, preventive and personalised medicine. *EPMA J.* **2016**, *7*, 15. [CrossRef]
182. Pieragostino, D.; D'Alessandro, M.; di Ioia, M.; Di Ilio, C.; Sacchetta, P.; Del Boccio, P. Unraveling the molecular repertoire of tears as a source of biomarkers: Beyond ocular diseases. *Proteom. Clin. Appl.* **2015**, *9*, 169–186. [CrossRef]
183. Borchman, D.; Foulks, G.N.; Yappert, M.C.; Tang, D.; Ho, D. V Spectroscopic evaluation of human tear lipids. *Chem. Phys. Lipids* **2007**, *147*, 87–102. [CrossRef]
184. Saville, J.T.; Zhao, Z.; Willcox, M.D.P.; Blanksby, S.J.; Mitchell, T.W. Detection and quantification of tear phospholipids and cholesterol in contact lens deposits: The effect of contact lens material and lens care solution. *Investig. Ophthalmol. Vis. Sci.* **2010**, *51*, 2843–2851. [CrossRef] [PubMed]
185. Lam, S.M.; Tong, L.; Reux, B.; Lear, M.J.; Wenk, M.R.; Shui, G. Rapid and sensitive profiling of tear wax ester species using high performance liquid chromatography coupled with tandem mass spectrometry. *J. Chromatogr. A* **2013**, *1308*, 166–171. [CrossRef] [PubMed]
186. Rantamäki, A.H.; Seppänen-Laakso, T.; Oresic, M.; Jauhiainen, M.; Holopainen, J.M. Human tear fluid lipidome: From composition to function. *PLoS ONE* **2011**, *6*, e19553. [CrossRef] [PubMed]
187. Dean, A.W.; Glasgow, B.J. Mass spectrometric identification of phospholipids in human tears and tear lipocalin. *Investig. Ophthalmol. Vis. Sci.* **2012**, *53*, 1773–1782. [CrossRef] [PubMed]
188. Butovich, I.A. On the lipid composition of human meibum and tears: Comparative analysis of nonpolar lipids. *Investig. Ophthalmol. Vis. Sci.* **2008**, *49*, 3779–3789. [CrossRef] [PubMed]
189. Butovich, I.A. Tear film lipids. *Exp. Eye Res.* **2013**, *117*, 4–27. [CrossRef] [PubMed]
190. Bleau, G.; Bodley, F.H.; Longpré, J.; Chapdelaine, A.; Roberts, K.D. Cholesterol sulfate. I. Occurrence and possible biological functions as an amphipathic lipid in the membrane of the human erythrocyte. *Biochim. Biophys. Acta* **1974**, *352*, 1–9. [CrossRef]
191. Romano, A.; Rolant, F. A non-invasive method of blood glucose evaluation by tear glucose measurement, for the detection and control of diabetic states. *Metab. Pediatr. Syst. Ophthalmol.* **1988**, *11*, 78–80.
192. Baca, J.T.; Finegold, D.N.; Asher, S.A. Tear glucose analysis for the noninvasive detection and monitoring of diabetes mellitus. *Ocul. Surf.* **2007**, *5*, 280–293. [CrossRef]
193. Daum, K.M.; Hill, R.M. Human tear glucose. *Investig. Ophthalmol. Vis. Sci.* **1982**, *22*, 509–514.
194. Yan, Q.; Peng, B.; Su, G.; Cohan, B.E.; Major, T.C.; Meyerhoff, M.E. Measurement of tear glucose levels with amperometric glucose biosensor/capillary tube configuration. *Anal. Chem.* **2011**, *83*, 8341–8346. [CrossRef]
195. Peng, B.; Lu, J.; Balijepalli, A.S.; Major, T.C.; Cohan, B.E.; Meyerhoff, M.E. Evaluation of enzyme-based tear glucose electrochemical sensors over a wide range of blood glucose concentrations. *Biosens. Bioelectron.* **2013**, *49*, 204–209. [CrossRef] [PubMed]
196. Baca, J.T.; Taormina, C.R.; Feingold, E.; Finegold, D.N.; Grabowski, J.J.; Asher, S.A. Mass spectral determination of fasting tear glucose concentrations in nondiabetic volunteers. *Clin. Chem.* **2007**, *53*, 1370–1372. [CrossRef] [PubMed]
197. Gonzalez, A.N. Measuring Glucose Levels in Tears as an Alternative to Blood Glucose Levels, Bard College, 2016, Vol. Paper 262. Available online: https://digitalcommons.bard.edu/senproj_s2016/262 (accessed on 21 December 2018).
198. Park, M.; Jung, H.; Jeong, Y.; Jeong, K.-H. Plasmonic Schirmer strip for human tear-based gouty arthritis diagnosis using surface-enhanced raman scattering. *ACS Nano* **2017**, *11*, 438–443. [CrossRef] [PubMed]
199. Dame, Z.T.; Aziat, F.; Mandal, R.; Krishnamurthy, R.; Bouatra, S.; Borzouie, S.; Guo, A.C.; Sajed, T.; Deng, L.; Lin, H.; et al. The human saliva metabolome. *Metabolomics* **2015**, *11*, 1864–1883. [CrossRef]
200. Tsuruoka, M.; Hara, J.; Hirayama, A.; Sugimoto, M.; Soga, T.; Shankle, W.R.; Tomita, M. Capillary electrophoresis-mass spectrometry-based metabolome analysis of serum and saliva from neurodegenerative dementia patients. *Electrophoresis* **2013**, *34*, 2865–2872. [CrossRef] [PubMed]
201. Figueira, J.; Jonsson, P.; Nordin Adolfsson, A.; Adolfsson, R.; Nyberg, L. NMR analysis of the human saliva metabolome distinguishes dementia patients from matched controls. *Mol. Biosyst.* **2016**, *12*, 2562–2571. [CrossRef] [PubMed]

202. Haines, N.R.; Manoharan, N.; Olson, J.L.; D'alessandro, A.; Reisz, J.A. Metabolomics analysis of human vitreous in diabetic retinopathy and rhegmatogenous retinal detachment. *J. Proteome Res.* **2018**, *17*, 2421–2427. [CrossRef]
203. Paris, L.P.; Johnson, C.H.; Aguilar, E.; Usui, Y.; Cho, K.; Hoang, L.T.; Feitelberg, D.; Benton, H.P.; Westenskow, P.D.; Kurihara, T.; et al. Global metabolomics reveals metabolic dysregulation in ischemic retinopathy. *Metabolomics* **2016**, *12*, 15. [CrossRef]
204. Schwartzman, M.L.; Iserovich, P.; Gotlinger, K.; Bellner, L.; Dunn, M.W.; Sartore, M.; Pertile, M.G.; Leonardi, A.; Sathe, S.; Beaton, A.; et al. Profile of lipid and protein autacoids in diabetic vitreous correlates with the progression of diabetic retinopathy. *Diabetes* **2010**, *59*, 1780–1788. [CrossRef]
205. Al-Shabrawey, M.; Mussell, R.; Kahook, K.; Tawfik, A.; Eladl, M.; Sarthy, V.; Nussbaum, J.; El-Marakby, A.; Park, S.Y.; Gurel, Z.; et al. Increased expression and activity of 12-lipoxygenase in oxygen-induced ischemic retinopathy and proliferative diabetic retinopathy: Implications in retinal neovascularization. *Diabetes* **2011**, *60*, 614–624. [CrossRef] [PubMed]
206. Barba, I.; Garcia-Ramírez, M.; Hernández, C.; Alonso, M.A.; Masmiquel, L.; García-Dorado, D.; Simó, R. Metabolic fingerprints of proliferative diabetic retinopathy: An ^{1}H-NMR–based metabonomic approach using vitreous humor. *Investig. Ophthalmol. Vis. Sci.* **2010**, *51*, 4416–4421. [CrossRef] [PubMed]
207. Young, S.P.; Nessim, M.; Falciani, F.; Trevino, V.; Banerjee, S.P.; Scott, R.A.H.; Murray, P.I.; Wallace, G.R. Metabolomic analysis of human vitreous humor differentiates ocular inflammatory disease. *Mol. Vis.* **2009**, *15*, 1210–1217. [PubMed]
208. Li, M.; Li, H.; Jiang, P.; Liu, X.; Xu, D.; Wang, F. Investigating the pathological processes of rhegmatogenous retinal detachment and proliferative vitreoretinopathy with metabolomics analysis. *Mol. Biosyst.* **2014**, *10*, 1055–1062. [CrossRef] [PubMed]
209. Yu, M.; Wu, Z.; Zhang, Z.; Huang, X.; Zhang, Q. Metabolomic analysis of human vitreous in rhegmatogenous retinal detachment associated with choroidal detachment. *Investig. Ophthalmol. Vis. Sci.* **2015**, *56*, 5706–5713. [CrossRef]
210. Mains, J.; Tan, L.E.; Zhang, T.; Young, L.; Shi, R.; Wilson, C. Species variation in small molecule components of animal vitreous. *Investig. Ophthalmol. Vis. Sci.* **2012**, *53*, 4778–4786. [CrossRef] [PubMed]
211. Agudo-Barriuso, M.; Lahoz, A.; Nadal-Nicolás, F.M.; Sobrado-Calvo, P.; Piquer-Gil, M.; Díaz-Llopis, M.; Vidal-Sanz, M.; Mullor, J.L. Metabolomic changes in the rat retina after optic nerve crush. *Investig. Ophthalmol. Vis. Sci.* **2013**, *54*, 4249–4259. [CrossRef]
212. Edwards, G.; Aribindi, K.; Guerra, Y.; Bhattacharya, S.K. Sphingolipids and ceramides of mouse aqueous humor: Comparative profiles from normotensive and hypertensive DBA/2J mice. *Biochimie* **2014**, *105*, 99–109. [CrossRef]
213. Mayordomo-Febrer, A.; Lopez-Murcia, M.; Morales-Tatay, J.; Monleon-Salvado, D.; Pinazo-Durán, M.D. Metabolomics of the aqueous humor in the rat glaucoma model induced by a series of intracamerular sodium hyaluronate injection. *Exp. Eye Res.* **2015**, *131*, 84–92. [CrossRef]
214. Edwards, G.; Aribindi, K.; Guerra, Y.; Lee, R.K.; Bhattacharya, S.K. Phospholipid profiles of control and glaucomatous human aqueous humor. *Biochimie* **2014**, *101*, 232–247. [CrossRef]
215. Aribindi, K.; Guerra, Y.; Piqueras, M.D.C.; Banta, J.T.; Lee, R.K.; Bhattacharya, S.K. Cholesterol and glycosphingolipids of human trabecular meshwork and aqueous humor: Comparative profiles from control and glaucomatous donors. *Curr. Eye Res.* **2013**, *38*, 1017–1026. [CrossRef] [PubMed]
216. Aljohani, A.J.; Edwards, G.; Guerra, Y.; Dubovy, S.; Miller, D.; Lee, R.K.; Bhattacharya, S.K. Human trabecular meshwork sphingolipid and ceramide profiles and potential latent fungal commensalism. *Investig. Ophthalmol. Vis. Sci.* **2014**, *55*, 3413–3422. [CrossRef] [PubMed]
217. Barbas-Bernardos, C.; Armitage, E.G.; García, A.; Mérida, S.; Navea, A.; Bosch-Morell, F.; Barbas, C. Looking into aqueous humor through metabolomics spectacles—Exploring its metabolic characteristics in relation to myopia. *J. Pharm. Biomed. Anal.* **2016**, *127*, 18–25. [CrossRef] [PubMed]
218. Ji, Y.; Rao, J.; Rong, X.; Lou, S.; Zheng, Z.; Lu, Y. Metabolic characterization of human aqueous humor in relation to high myopia. *Exp. Eye Res.* **2017**, *159*, 147–155. [CrossRef] [PubMed]
219. Snytnikova, O.A.; Khlichkina, A.A.; Yanshole, L.V.; Yanshole, V.V.; Iskakov, I.A.; Egorova, E.V.; Stepakov, D.A.; Novoselov, V.P.; Tsentalovich, Y.P. Metabolomics of the human aqueous humor. *Metabolomics* **2017**, *13*. [CrossRef]

220. Lakowicz, J.R. Fluorophores. In *Principles of Fluorescence Spectroscopy*; Springer: Boston, MA, USA, 2006; pp. 63–95.
221. Chance, B. Pyridine nucleotide as an indicator of the oxygen requirements for energy-linked functions of mitochondria. *Circ. Res.* **1976**, *38*, I31–I38.
222. Lakowicz, J.R.; Szmacinski, H.; Nowaczyk, K.; Johnson, M.L. Fluorescence lifetime imaging of free and protein-bound NADH. *Proc. Natl. Acad. Sci. USA* **1992**, *89*, 1271–1275. [CrossRef]
223. Sharick, J.T.; Favreau, P.F.; Gillette, A.A.; Sdao, S.M.; Merrins, M.J.; Skala, M.C. Protein-bound NAD(P)H lifetime is sensitive to multiple fates of glucose carbon. *Sci. Rep.* **2018**, *8*. [CrossRef]
224. Lakowicz, J.R. Fluorescence-Lifetime Imaging Microscopy. In *Principles of Fluorescence Spectroscopy*; Springer: Boston, MA, USA, 2006; pp. 741–755. ISBN 9780387463124.
225. Becker, W. Fluorescence lifetime imaging—Techniques and applications. *J. Microsc.* **2012**, *247*, 119–136. [CrossRef]
226. Schweitzer, D.; Klemm, M.; Quick, S.; Deutsch, L.; Jentsch, S.; Hammer, M.; Dawczynski, J.; Kloos, C.H.; Mueller, U.A. Detection of early metabolic alterations in the ocular fundus of diabetic patients by time-resolved autofluorescence of endogenous fluorophores. In *Clinical and Biomedical Spectroscopy and Imaging II*; Ramanujam, N., Popp, J., Eds.; International Society for Optics and Photonics: Bellingham WA, USA, 2011; Volume SPIE 8087, p. 80871G.
227. Blacker, T.S.; Mann, Z.F.; Gale, J.E.; Ziegler, M.; Bain, A.J.; Szabadkai, G.; Duchen, M.R. Separating NADH and NADPH fluorescence in live cells and tissues using FLIM. *Nat. Commun.* **2014**, *5*, 3936. [CrossRef]
228. Patterson, G.H.; Piston, D.W.; Barisas, B.G. Förster distances between green fluorescent protein pairs. *Anal. Biochem.* **2000**, *284*, 438–440. [CrossRef] [PubMed]
229. Richards-Kortum, R.; Drezek, R.; Sokolov, K.; Pavlova, I.; Pollen, M. Survey of Endogenous Biological Fluorophores. In *Handbook of Biomedical Fluorescence*; Mycek, M.A., Pogue, B.W., Eds.; Marcel Dekker, Inc.: New York, NY, USA; Basel, Switzerland, 2003; pp. 237–264.
230. Urayama, P.; Mycek, M.-A. Fluorescence Lifetime Imaging Microscopy of Endogenous Biological Fluorescence. In *Handbook of Biomedical Fluorescence*; Mycek, M.-A., Pogue, B.W., Eds.; Macel Dekker, Inc.: New York, NY, USA; Basel, Switzerland, 2003; pp. 211–236. ISBN 0-8247-0955-1.
231. Schweitzer, D.; Schenke, S.; Hammer, M.; Schweitzer, F.; Jentsch, S.; Birckner, E.; Becker, W.; Bergmann, A.A. Towards metabolic mapping of the human retina. *Microsc. Res. Tech.* **2007**, *70*, 410–419. [CrossRef] [PubMed]
232. Marcu, L. Fluorescence lifetime techniques in medical applications. *Ann. Biomed. Eng.* **2012**, *40*, 304–331. [CrossRef] [PubMed]
233. Miura, Y.; Huettmann, G.; Orzekowsky-Schroeder, R.; Steven, P.; Szaszák, M.; Koop, N.; Brinkmann, R. Two-photon microscopy and fluorescence lifetime imaging of retinal pigment epithelial cells under oxidative stress. *Investig. Ophthalmol. Vis. Sci.* **2013**, *54*, 3366–3377. [CrossRef] [PubMed]
234. Schweitzer, D.; Gaillard, E.R.; Dillon, J.; Mullins, R.F.; Russell, S.; Hoffmann, B.; Peters, S.; Hammer, M.; Biskup, C. Time-resolved autofluorescence imaging of human donor retina tissue from donors with significant extramacular drusen. *Investig. Ophthalmol. Vis. Sci.* **2012**, *53*, 3376–3386. [CrossRef]
235. Schweitzer, D.; Hammer, M.; Schweitzer, F.; Anders, R.; Doebbecke, T.; Schenke, S.; Gaillard, E.R.; Gaillard, E.R. *In vivo* measurement of time-resolved autofluorescence at the human fundus. *J. Biomed. Opt.* **2004**, *9*, 1214–1222. [CrossRef] [PubMed]
236. Delori, F.C.; Dorey, C.K.; Staurenghi, G.; Arend, O.; Goger, D.G.; Weiter, J.J. In vivo fluorescence of the ocular fundus exhibits retinal pigment epithelium lipofuscin characteristics. *Investig. Ophthalmol. Vis. Sci.* **1995**, *36*, 718–729.
237. Pau, H.; Degen, J.; Schmidtke, H.-H. Different regional changes of fluorescence spectra of clear human lenses and nuclear cataracts. *Graefe's Arch. Clin. Exp. Ophthalmol.* **1993**, *231*, 656–661. [CrossRef]
238. Klemm, M.; Schweitzer, D.; Peters, S.; Sauer, L.; Hammer, M.; Haueisen, J. FLIMX: A software package to determine and analyze the fluorescence lifetime in time-resolved fluorescence data from the human eye. *PLoS ONE* **2015**, *10*, e0131640. [CrossRef]
239. Delori, F.C.; Webb, R.H.; Sliney, D.H. Maximum permissible exposures for ocular safety (ANSI 2000), with emphasis on ophthalmic devices. *J. Opt. Soc. Am. A* **2007**, *24*, 1250–1265. [CrossRef]
240. Dysli, C.; Quellec, G.; Abegg, M.; Menke, M.N.; Wolf-Schnurrbusch, U.; Kowal, J.; Blatz, J.; La Schiazza, O.; Leichtle, A.B.; Wolf, S.; et al. Quantitative analysis of fluorescence lifetime measurements of the macula using the fluorescence lifetime imaging ophthalmoscope in healthy subjects. *Investig. Ophthalmol. Vis. Sci.* **2014**, *55*, 2106–2113. [CrossRef] [PubMed]

241. Dysli, C.; Dysli, M.; Zinkernagel, M.S.; Enzmann, V. Effect of pharmacologically induced retinal degeneration on retinal autofluorescence lifetimes in mice. *Exp. Eye Res.* **2016**, *153*, 178–185. [CrossRef] [PubMed]
242. Dysli, C.; Fink, R.; Wolf, S.; Zinkernagel, M.S. Fluorescence lifetimes of drusen in age-related macular degeneration. *Investig. Ophthalmol. Vis. Sci.* **2017**, *58*, 4856–4862. [CrossRef]
243. Sauer, L.; Gensure, R.H.; Andersen, K.M.; Kreilkamp, L.; Hageman, G.S.; Hammer, M.; Bernstein, P.S. Patterns of fundus autofluorescence lifetimes in eyes of individuals with nonexudative age-related macular degeneration. *Investig. Ophthalmol. Vis. Sci.* **2018**, *59*, AMD65–AMD77. [CrossRef] [PubMed]
244. Spaide, R.F.; Curcio, C.A. Drusen characterization with multimodal imaging. *Retina* **2010**, *30*, 1441–1454. [CrossRef] [PubMed]
245. Dysli, C.; Wolf, S.; Berezin, M.Y.; Sauer, L.; Hammer, M.; Zinkernagel, M.S. Fluorescence lifetime imaging ophthalmoscopy. *Prog. Retin. Eye Res.* **2017**, *60*, 120–143. [CrossRef] [PubMed]
246. Sauer, L.; Andersen, K.M.; Dysli, C.; Zinkernagel, M.S.; Bernstein, P.S.; Hammer, M. Review of clinical approaches in fluorescence lifetime imaging ophthalmoscopy. *J. Biomed. Opt.* **2018**, *23*, 1–20. [CrossRef]
247. Gray, N.; Lewis, M.R.; Plumb, R.S.; Wilson, I.D.; Nicholson, J.K. High-throughput microbore UPLC−MS metabolic phenotyping of urine for large-Scale epidemiology studies. *J. Proteome Res.* **2015**, *14*, 2714–2721. [CrossRef]
248. Marko, C.K.; Laíns, I.; Husain, D.; Miller, J.W. AMD biomarkers identified by metabolomics. *Retin. Phys.* **2018**, *15*, 22–24.
249. Rieckmann, J.C.; Geiger, R.; Hornburg, D.; Wolf, T.; Kveler, K.; Jarrossay, D.; Sallusto, F.; Shen-Orr, S.S.; Lanzavecchia, A.; Mann, M.; et al. Social network architecture of human immune cells unveiled by quantitative proteomics. *Nat. Immunol.* **2017**, *18*, 583–593. [CrossRef]
250. Szmacinski, H.; Hegde, K.; Zeng, H.-H.; Eslami, K.; Puche, A.; Lakowicz, J.R.; Lengyel, I.; Thompson, R.B. Towards early detection of age-related macular degeneration with tetracyclines and FLIM. In *Proceedings of SPIE*; Vo-Dinh, T., Mahadevan-Jansen, A., Grundfest, W.S., Eds.; SPIE: Bellingham, WA, USA, 2018; Volume 10484.
251. Merle, B.M.J.; Colijn, J.M.; Cougnard-Grégoire, A.; de Koning-Backus, A.P.M.; Delyfer, M.-N.; Kiefte-de Jong, J.C.; Meester-Smoor, M.; Féart, C.; Verzijden, T.; Samieri, C.; et al. Mediterranean diet and incidence of advanced age-related macular degeneration: The EYE-RISK Consortium. *Ophthalmology* **2018**. [CrossRef] [PubMed]
252. Lien, E.L.; Hammond, B.R. Nutritional influences on visual development and function. *Prog. Retin. Eye Res.* **2011**, *30*, 188–203. [CrossRef] [PubMed]
253. Hollyfield, J.G.; Bonilha, V.L.; Rayborn, M.E.; Yang, X.; Shadrach, K.G.; Lu, L.; Ufret, R.L.; Salomon, R.G.; Perez, V.L. Oxidative damage-induced inflammation initiates age-related macular degeneration. *Nat. Med.* **2008**, *14*, 194–198. [CrossRef] [PubMed]
254. Gu, X.; Meer, S.G.; Miyagi, M.; Rayborn, M.E.; Hollyfield, J.G.; Crabb, J.W.; Salomon, R.G. Carboxyethylpyrrole protein adducts and autoantibodies, biomarkers for age-related macular degeneration. *J. Biol. Chem.* **2003**, *278*, 42027–42035. [CrossRef]
255. Rinninella, E.; Mele, M.; Merendino, N.; Cintoni, M.; Anselmi, G.; Caporossi, A.; Gasbarrini, A.; Minnella, A.; Rinninella, E.; Mele, M.C.; et al. The role of diet, micronutrients and the gut microbiota in age-related macular degeneration: New perspectives from the gut–retina axis. *Nutrients* **2018**, *10*, 1677. [CrossRef] [PubMed]
256. Biarnés, M.; Vassiliev, V.; Nogoceke, E.; Emri, E.; Rodríguez-Bocanegra, E.; Ferraro, L.; Garcia, M.; Fauser, S.; Monés, J.; Lengyel, I.; et al. Precision medicine for age-related macular degeneration: Current developments and prospects. *Expert Rev. Precis. Med. Drug Dev.* **2018**, *3*, 249–263. [CrossRef]
257. Lorés-Motta, L.; de Jong, E.K.; den Hollander, A.I. Exploring the use of molecular biomarkers for precision medicine in age-related macular degeneration. *Mol. Diagn. Ther.* **2018**, *22*, 315–343. [CrossRef]
258. Park, K.S.; Xu, C.L.; Cui, X.; Tsang, S.H. Reprogramming the metabolome rescues retinal degeneration. *Cell. Mol. Life Sci.* **2018**, *75*, 1559–1566. [CrossRef]

metabolites

MDPI

Article

Metabolomic Profiling of Cerebral Palsy Brain Tissue Reveals Novel Central Biomarkers and Biochemical Pathways Associated with the Disease: A Pilot Study

Zeynep Alpay Savasan [1,2,*], Ali Yilmaz [3], Zafer Ugur [3], Buket Aydas [4], Ray O. Bahado-Singh [1,2] and Stewart F. Graham [2,3]

1 Department of Obstetrics and Gynecology, Maternal Fetal Medicine Division, Beaumont Health System, 3811 W. 13 Mile Road, Royal Oak, MI 48073, USA; Ray.Bahado-Singh@beaumont.org

2 Oakland University-William Beaumont School of Medicine, Beaumont Health, 3811 W. 13 Mile Road, Royal Oak, MI 48073, USA; Stewart.Graham@beaumont.org

3 Beaumont Research Institute, Beaumont Health, 3811 W. 13 Mile Road, Royal Oak, MI 48073, USA; Ali.Yilmaz@beaumont.org (A.Y.); Zafer.Ugur@beaumont.org (Z.U.)

4 Departments of Mathematics and Computer Sciences, Albion College, 611 E. Porter St., Albion, MI 49224, USA; baydas@albion.edu

* Correspondence: Zeynep.AlpaySavasan@beaumont.org; Tel.: +1-248-712-4595

Received: 9 January 2019; Accepted: 31 January 2019; Published: 2 February 2019

Abstract: Cerebral palsy (CP) is one of the most common causes of motor disability in childhood, with complex and heterogeneous etiopathophysiology and clinical presentation. Understanding the metabolic processes associated with the disease may aid in the discovery of preventive measures and therapy. Tissue samples (caudate nucleus) were obtained from post-mortem CP cases (n = 9) and age- and gender-matched control subjects (n = 11). We employed a targeted metabolomics approach using both ^{1}H NMR and direct injection liquid chromatography-tandem mass spectrometry (DI/LC-MS/MS). We accurately identified and quantified 55 metabolites using ^{1}H NMR and 186 using DI/LC-MS/MS. Among the 222 detected metabolites, 27 showed significant concentration changes between CP cases and controls. Glycerophospholipids and urea were the most commonly selected metabolites used to develop predictive models capable of discriminating between CP and controls. Metabolomics enrichment analysis identified folate, propanoate, and androgen/estrogen metabolism as the top three significantly perturbed pathways. We report for the first time the metabolomic profiling of post-mortem brain tissue from patients who died from cerebral palsy. These findings could help to further investigate the complex etiopathophysiology of CP while identifying predictive, central biomarkers of CP.

Keywords: cerebral palsy; metabolomics; ^{1}H NMR; targeted mass spectrometry; metabolic pathways

PACS: J0101

1. Introduction

Cerebral palsy (CP) is the most common cause of severe neurodisability in children [1]. Although the main underlying causal factor is considered to be birth asphyxia, the pathophysiology of the disease is still not well understood. There are other causal factors that occur later in life that are hypothesized to be involved in the development of CP [2–4]. Congenital malformations are rarely identified [5]. Genetic predispositions with exposure to environmental factors can lead to CP. Common cerebral lesions seen in CP include destructive injuries, predominantly in the white matter in preterm infants and in the gray matter and the brainstem nuclei in full-term newborns [4]. The effect of these lesions, especially on the immature brain, could alter the series of developmental events [6]. Alteration in cell morphology or function and cell death observed in hypoxic ischemia or in inflammatory conditions leading to excessive production of proinflammatory cytokines [7,8], oxidative stress [9], maternal growth factor deprivation [10], extracellular matrix modifications [10], and excessive release of glutamate [11] have been shown to trigger the excitotoxic cascade and predispose the development of CP [12–15].

Cerebral palsy is a heterogeneous condition with multiple causes; clinical types and patterns of neuropathology on brain imaging; multiple associated developmental pathologies, such as intellectual disability, autism, epilepsy, and visual impairment; and, more recently, multiple rare pathogenic genetic mutations [2,16–18]. This is a clinical spectrum with many causal pathways and many types and degrees of disability [12]. These various pathways and etiologies have each resulted in a non-specific non-progressive disorder of posture and movement control. Thus, CP should be considered as a descriptive term for affected individuals, with each case requiring a detailed consideration of the underlying etiology. The described feature of this condition is one of the challenges for researchers due to the possibility of various underlying etiologies and confounders [12]. To our knowledge, there is currently no method available for predicting those at greatest risk of developing the disease. Moreover, only two strategies have succeeded in decreasing CP in 2-year-old children, which include the use of hypothermia in full-term newborns with moderate neonatal encephalopathy [19,20] and the administration of magnesium sulfate to mothers in preterm labor [21,22].

Among the new omics, metabolomics has the huge potential to advance our understanding of many complex diseases by uniquely detecting rapid biochemical pathway alterations and uncovering multiple biomarker panels, especially in various neurological disorders [23–31]. Over the past decade, the search for useful biomarkers to accurately predict brain pathology has become a growing area of interest. Biomarkers such as neuroimaging markers showing corticospinal tract integrity, metabolite ratios in brain regions, and brain volumes [32,33], multiorgan injury markers [34], and inflammatory markers [35,36] were studied as prediction models, however the description of metabolomic alterations or the identification of clinically approved biomarkers in CP has not been reported. There is accumulating evidence that metabolomic profiling of post-mortem brain tissue helps in understanding the pathophysiology of neurologic, neurodegenerative, and psychiatric disorders [23,24,28,37–43]. Thus, this study aims to biochemically profile post-mortem brain tissue from patients who died from CP and compare those with age-, and gender-matched controls. We believe that this approach will help us to identify central biomarkers of the disease while uncovering previously unreported biochemical pathways associated with the disease.

2. Results

Using ^{1}H NMR and direct injection liquid chromatography-tandem mass spectrometry (DI/LC-MS/MS), we biochemically profiled post-mortem human brain tissue from people who died from CP and compared them with age- and gender-matched controls. We accurately identified and quantified 55 metabolites using ^{1}H NMR and 186 using DI/LC-MS/MS. Figure 1 represents a labelled 1D ^{1}H NMR spectrum acquired from an extract of caudate nucleus harvested from a person who died from CP.

Due to the complementarity between the two techniques, there was a certain degree of observed overlap in terms of the metabolites measured (19 metabolites). To account for this, we took the average value for the individual metabolites and used this concentration value in our analyses, leaving us with 222 metabolites. Principal component analysis (PCA) was performed on the data to check for any intrinsic variation and subsequently remove any potential outliers ($p < 0.05$) based on the Hotelling's T^2 plot. No outliers were detected. Univariate analysis of the data revealed that of the 222 metabolites, 27 of them were at statistically, significantly different concentrations between CP and control tissue (Table 1; $p < 0.05$; $q < 0.3$). A full list of the 222 measured metabolites is available in Table S1 in the Supplementary Materials.

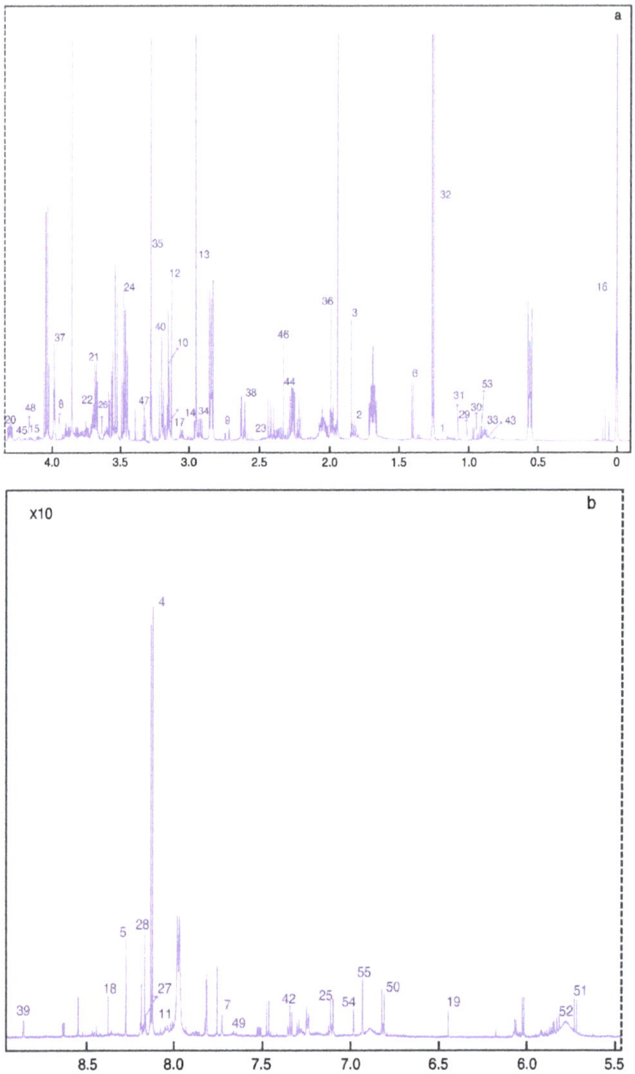

Figure 1. Typical (**a**) aliphatic and (**b**) aromatic region of 600 MHz ^{1}H-NMR spectra of brain tissue extract, the metabolites are listed as follows. 1: 3-Hydroxybutyrate; 2: 4-Aminobutyrate; 3: Acetate; 4: Adenine; 5: Adenosine; 6: Alanine; 7: Anserine; 8: Ascorbate; 9: Aspartate; 10: Carnitine; 11: Carnosine; 12: Choline; 13: Creatine; 14: Creatine phosphate; 15: Creatinine; 16: DSS; 17: Ethanolamine; 18: Formate; 19: Fumarate; 20: Glucose; 21: Glutamate; 22: Glutamine; 23: Glutathione; 24: Glycine; 25: Histamine; 26: Homocitrulline; 27: Hypoxanthine; 28: Inosine; 29: Isobutyrate; 30: Isoleucine; 31: Isopropanol; 32: Lactate; 33: Leucine; 34: Lysine; 35: Methanol; 36: Methionine; 37: Myo-inositol; 38: N-Acetylaspartate; 39: Niacinamide; 40: O-Acetylcholine; 41: O-Phosphocholine; 42: Phenylalanine; 43: Propylene glycol; 44: Pyruvate; 45: sn-Glycero-3-phosphocholine; 46: Succinate; 47: Taurine; 48: Threonine; 49: Tryptophan; 50: Tyrosine; 51: Uracil; 52: Urea; 53: Valine; 54: π-Methylhistidine; 55: τ-Methylhistidine.

Table 1. Statistically significant metabolite concentrations (µM; $p < 0.05$; $q < 0.05$) for CP vs control PM brain extracts. *t*-test values were calculated as a default and values with (W) were calculated using the Wilcoxon–Mann–Whitney test.

HMDB	Compound ID	Mean (SD) of Control (µM)	Mean (SD) of CP (µM)	*p*-Value	*q*-Value (FDR)	Fold Change
HMDB00294	Urea	59.236 (37.499)	184.144 (14.774)	0.0074 (W)	0.299	−3.11
HMDB00148	L-Glutamic acid	499.627 (15.680)	6.767 (13.764)	0.0106 (W)	0.299	73.83
HMDB13456	PC(o-22:2(13Z,16Z)/22:3(10Z,13Z,16Z))	1.187 (0.902)	0.335 (0.379)	0.0125 (W)	0.299	3.54
HMDB08276	PC(20:0/20:2(11Z,14Z))	0.265 (0.190)	0.051 (0.110)	0.0166 (W)	0.299	5.16
HMDB13450	PC(o-22:0/22:6(4Z,7Z,10Z,13Z,16Z,19Z))	0.847 (0.710)	0.231 (0.404)	0.0166 (W)	0.299	3.66
HMDB00195	Inosine	8.082 (4.627)	14.333 (6.338)	0.0201	0.299	−1.77
HMDB13333	3-Hydroxy-9-hexadecenoylcarnitine	0.061 (0.062)	0.129 (0.076)	0.0204 (W)	0.299	-2.13
HMDB10379	LysoPC(14:0)	5.237 (1.153)	4.151 (0.665)	0.0224	0.299	1.26
HMDB13433	PC(o-18:1(9Z)/22:0)	1.334 (0.714)	0.638 (0.487)	0.023	0.299	2.09
HMDB13453	PC(o-22:1(13Z)/22:3(10Z,13Z,16Z))	0.281 (0.180)	0.133 (0.069)	0.0248	0.299	2.12
HMDB07991	PC(16:0/22:6(4Z,7Z,10Z,13Z,16Z,19Z))	55.251 (4.352)	19.532 (5.971)	0.0249	0.299	2.83
HMDB08055	PC(18:0/22:5(4Z,7Z,10Z,13Z,16Z))	9.151 (6.281)	3.871 (2.773)	0.0249	0.299	2.36
HMDB06083	Troxerutin	188.555 (18.953)	432.889 (25.759)	0.0250 (W)	0.299	−2.3
HMDB08048	PC(18:0/20:4(5Z,8Z,11Z,14Z))	114.082 (59.935)	56.311 (43.130)	0.0264	0.299	2.03
HMDB00142	Formic acid	4.718 (2.078)	7.489 (3.055)	0.0269	0.299	−1.59
HMDB08057	PC(18:0/22:6(4Z,7Z,10Z,13Z,16Z,19Z))	23.314 (15.829)	11.438 (6.380)	0.0275 (W)	0.299	2.04
HMDB07892	PC(14:0/22:6(4Z,7Z,10Z,13Z,16Z,19Z))	0.405 (0.338)	0.139 (0.090)	0.028	0.299	2.91
HMDB0029205	LysoPC(26:0)	0.227 (0.197)	0.456 (0.235)	0.0293	0.299	−2.01
HMDB07874	PC(14:0/18:2(9Z,12Z))	3.462 (3.478)	0.558 (0.715)	0.0297 (W)	0.299	6.21
HMDB03334	Symmetric dimethylarginine	0.638 (0.399)	1.405 (0.802)	0.0310 (W)	0.299	−2.2
HMDB10394	LysoPC(20:3(8Z,11Z,14Z))	1.213 (0.902)	0.492 (0.500)	0.0310 (W)	0.299	2.46
HMDB08288	PC(20:0/22:6(4Z,7Z,10Z,13Z,16Z,19Z))	0.367 (0.230)	0.186 (0.100)	0.0332	0.299	1.98
HMDB11151	PC(O-16:0/18:2(9Z,12Z))	10.915 (6.853)	5.759 (2.592)	0.0381	0.299	1.9
HMDB13469	SM(d18:0/24:1(15Z)(OH))	1.353 (0.764)	2.168 (1.131)	0.0402 (W)	0.299	−1.6
HMDB13458	PC(o-24:0/18:3(6Z,9Z,12Z))	0.909 (0.441)	0.536 (0.290)	0.0428	0.299	1.7
HMDB08138	PC(18:2(9Z,12Z)/18:2(9Z,12Z))	189.522 (12.500)	60.640 (6.755)	0.0465 (W)	0.299	3.13
HMDB13411	PC(o-16:1(9Z)/16:1(9Z))	0.720 (0.496)	0.362 (0.212)	0.048	0.299	1.99

Those compounds highlighted in bold are considered statistically, significantly different. (W)-data were non-normally distributed and the *p*-value was calculated by the Wilcoxon–Mann–Whitney test.

Having confirmed that there were significant differences between CP and control brains, we wanted to investigate a number of machine learning techniques to identify which method worked best for accurately discriminating between CP cases and controls. We used the variable importance functions *varimp* in h2o and *varImp* in caret R packages to rank the models' features in each of the predictive algorithms. Feature predictors were estimated using a model-based approach. In other words, a feature was considered important if it contributed to the model performance. We extracted 20 important predictors from each of the models used for predicting CP. From these 20 features, the top metabolites were chosen and used to generate the specific predictive model. These were also compared across the different machine learning approaches (Table 2).

Table 2. List of panels of metabolites used in different artificial intelligence methods. LR: logistic regression; SVM: support vector machine; PLS-DA: partial least square-discriminant analysis, RF: random forest; PAM: prediction analysis for microarrays; DL: deep learning.

Models	Selected Features
LR	PC ae C44:5, Urea
SVM	PC ae C44:5, Urea, C9
PLS-DA	PC ae C44:5, Urea, C9, PC aa C40:6, PC ae C40:1, PC ae C44:6
RF	PC ae C44:5, Urea, C9, PC aa C40:6, PC ae C40:1
PAM	Urea, PC ae C44:5, PC ae C44:6, C9, PC aa C40:6, PC ae C40:1
DL	C9, PC ae C40:1, Urea, PC ae C44:6, PC ae C44:5

Table 3 lists the average AUCs, sensitivity values, and specificity values calculated on the holdout test sets. Of all the methods employed, prediction analysis for microarrays (PAM) performed the best in terms of AUC, sensitivity and specificity combined.

Table 3. Results for the various predictive modeling techniques employed.

	LR	SVM	PLS-DA	RF	PAM	DL
AUC (95% CI)	0.861 (0.688–1)	0.925 (0.73–1)	0.929 (0.8–1)	0.899 (0.6–1)	0.93 (0.8–1)	0.937 (0.8–1)
Sensitivity	0.842	0.778	0.870	0.889	0.899	0.833
Specificity	0.909	0.625	0.725	0.850	0.855	0.667

Metabolomics enrichment analysis highlighted six metabolic pathways as significantly disturbed in the CP brain as compared with controls. These include folate metabolism, propanoate metabolism, androgen and estrogen metabolism, androstenedione metabolism, pterine metabolism, and steroid metabolism (Figure 2).

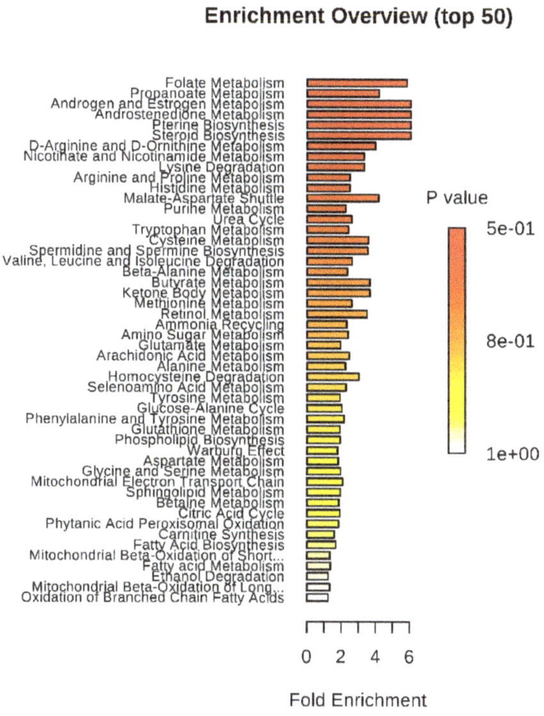

Figure 2. Results of the metabolite pathway enrichment analysis.

3. Discussion

To our knowledge, this is the first study to use targeted and quantitative metabolomics to biochemically profile post-mortem brain tissue from people who died from CP and compared them with age- and gender-matched controls. Our univariate analysis of the concentration data highlighted 27 metabolites to be significantly different concentrations between CP and control brains (Table 1).

We achieved consistently good diagnostic performance (AUC > 0.80) using six different Machine Learning approaches. PAM analysis, following cross validation, yielded an AUC (95% CI) = 0.930 (0.8–1) with a sensitivity and specificity of 0.899 and 0.855, respectively. LR had the smallest AUC among all the algorithms used. This was probably due to its sensitivity and not being the most ideal method for nonlinear analysis. When we looked at all the variables used as predictors in all of the models, we identified glycerophospholipids (PC ae C44:5, PC ae C44:5, PC ae C44:6, 40:1, 40:6) and urea to be the common denominators.

In our univariate analysis, we found that glutamate was included in the top significantly different metabolites in CP brains. Glutamic acid, known as a key molecule in cellular metabolism, is the most abundant fast excitatory neurotransmitter in the nervous system [44]. Glutamic acid is believed to be involved in cognitive functions such as learning and memory in the brain due to its function in synapsis [44]. In brain injury or disease, excess glutamate can accumulate outside the cells. This process causes calcium ions to enter cells, leading to neuronal damage and eventual cell death, known as excitotoxicity [45]. Excitotoxicity due to glutamate occurs as part of the ischemic cascade and is associated with stroke and diseases like amyotrophic lateral sclerosis, lathyrism, and Alzheimer's disease [46–48]. A fundamental process that leads to perinatal brain damage with hypoxic-ischemic injury is believed to be the damage to neurons with excitotoxicity [49].

The potential sources of cellular glutamate available for release during ischemia include astrocytes, oligodendrocytes, axons, and cells from neighboring structures such as the choroid plexus. Of these sources, ischemic glutamate release from astrocytes has been well characterized in gray matter [50] as well as periventricular white matter which is a lesion associated with chronic neurologic morbidity, especially CP seen in premature neonates [51]. Moreover, in animal models, prenatal magnesium sulfate use had prevented local glutamate level elevation and neurologic impairment after an excitotoxic brain lesion [52]. This effect was more significant in males compared to females [52]. It is not surprising that our results supported the previous reports on the importance of glutamate metabolism in lesions associated with CP.

Glycerophospholipids or phosphoglycerides are the most significant metabolites identified in CP in our machine learning techniques. Glycerophospholipids function in signal induction and transport. They provide the precursors for prostanglandins and leukotrienes [53] for biological responses [54]. They are also involved in apoptosis, modulation of the activities of transporters, and membrane-bound enzymes [55–57]. Marked alterations in neural membrane glycerophospholipid composition have been reported to occur in neurological disorders such as Alzheimer's disease, depression, and anxiety [54,58,59]. These alterations result in changes in membrane fluidity and permeability. These processes along with the accumulation of lipid peroxides and compromised energy metabolism may be responsible for the neurodegeneration observed in CP [60,61]. Umbilical cord metabolomic profiles in neonates with perinatal asphyxia who have substantial risk to develop CP showed significant alterations in amino acids, acylcarnitines, and glycerophospholipids [62] similar to our findings in the brain tissue of patients with CP.

Machine learning techniques also identified urea as a good predictive variable across all of our models. In neurodegenerative disorders such as Huntington's disease, changes in urea levels were identified in post-mortem brain tissues [63]. Widespread elevation of urea has also been reported in brain tissues with Alzheimer's disease [64], suggesting that urea cycle disruption could also be a unifying pathogenic feature of neurodegenerative diseases. Excessive levels of urea and its nitrogenous precursor ammonia are neurotoxic, as evidenced by uremic encephalopathy and the urea cycle disorders. Urea cycle disorders are genetic disorders caused by a mutation that results in a deficiency of enzymes in the urea cycle [65]. These enzymes are responsible for removing ammonia from the blood stream. In urea cycle disorders, nitrogen accumulates, resulting in hyperammonemia that can cause irreversible brain damage, with manifestations ranging from lethargy and abnormal behavior such as disordered sleep and neurological posturing through to acute psychosis, seizure, coma, and death [66]. Similarly, uremic encephalopathy typically occurs in patients with renal failure, which can lead to symptoms ranging from mild fatigue and generalized weakness to seizure and coma [67]. There have been no previous reports showing an association between urea cycle abnormalities and CP. Argininemia, which is a rare urea cycle defect disorder, has been reported in a small case series of young children leading to progressive spastic tetraplegia, poor physical growth, and mental retardation with seizures mimicking CP [68]. Our study is the first showing altered urea concentration in the post-mortem CP brain tissue supporting previous studies about other neurological disorders.

The results of the pathway enrichment analysis highlighted folic acid metabolism as the most perturbed biochemical pathway. Methylation cycle and folate metabolism are important in neurotransmitter regulation, nerve myelination, and DNA synthesis. Thus, folate metabolites play a critical role in cognitive function and neuromuscular stability. A previous study showed a possible protective effect of prenatal folic acid supplementation on CP development [69]. There is evidence that children with CP show dysregulation of methylation capacity and folate metabolism despite adequate levels of folate and vitamin B12 [70]. Maintenance of methylation activity is crucial for RNA and DNA synthesis and subsequent growth and development as well as maintaining neurodevelopment. Interestingly, there is a cerebral folate deficiency syndrome described in children with developmental delay and deceleration of head growth, psychomotor retardation, and hypotonia.

One-third of these children develop ataxia, spasticity, dyskinesia, speech difficulties, and seizures similar to children with CP [71]. In mouse models, folate deficiency has been demonstrated to decrease neurotransmitter acethylcholine activity, which in turn significantly decreases cognitive performance [72]. Furthermore, low serum folate concentrations were also found in patients with Alzheimer's disease and dementia [73]. There is also evidence of the beneficiary effect of folate therapy on both EEG patterns and neuropsychological performance in patients with neuropathy and cerebral atrophy [74].

Additionally, our pathway enrichment analysis identified propionate metabolism as being significantly perturbed in CP brains. Propionate is the most common short-chain fatty acid produced by the human gut microbiota in response to indigestible carbohydrates such as fiber in the diet. Propionate and other short-chain fatty acids are produced in the body during normal cellular metabolism following enteric bacterial fermentation of dietary carbohydrates and proteins [75]. Propionate-producing enteric bacteria, including unique *Clostridial*, *Desulfovibrio*, and *Bacteriodetes* species, have been isolated from patients with regressive autism spectrum disorders [76,77]. Propionate is also present naturally in a variety of foods and is a common food preservative in refined wheat and dairy products. Under normal circumstances, these short-chain fatty acids are primarily metabolized in the liver. However, if there are genetic and/or acquired aberrations in metabolism [78], higher than normal levels of short-chain fatty acids can be present in the circulating blood, and can cross the gut–blood and blood–brain barriers. They can concentrate intracellularly, particularly in acidotic conditions, where they may have deleterious effects on brain development and function [79]. This could be important in the context of neurological disorders, since propionate is known to affect cell signaling, neurotransmitter synthesis and release, mitochondrial function/CoA sequestration, lipid metabolism, immune function, gap junction modulation, and gene expression [79–84], all of which have been implicated in a variety of neurological disorders including autism spectrum diseases [79,85]. Intracerebroventricular infusions with propionate produced short bouts of behavioral and electrophysiological effects, coupled with biochemical and neuropathological alterations in adult rats, consistent with those seen in autism disorder [86–89]. A recent study showed infusions with propionate or butyrate altered the brain acylcarnitine and phospholipid profiles [90], which are known to affect membrane fluidity, peroxisomal function, gap junction coupling capacity, signaling, and neuroinflammation [79], supporting our findings as earlier defined in CP brain tissues.

Finally, we found that the sex steroid metabolism pathway was significantly altered in the brain tissue of patients with CP. Although there is paucity of data on the effect of sex steroids in the development of CP, estradiol has been shown to have a dose-dependent protection on oxygen-induced apoptotic cell death in oligodendrocytes in animal models [83]. This may suggest a possible role for estrogens in the prevention of neonatal oxygen-induced white matter injury [91]. Although sex steroid levels were low for both genders after birth, our preliminary finding should be investigated more deeply to identify the correlation with the better survival rates of female premature babies compared to males [92]. Estrogen could be effective in modulating glutamate-induced neurotoxicity [85]. However, the mechanism underlying estrogen's neuroprotective effect is not fully clarified [93]. Moreover, as previously mentioned, there may be a gender-specific neuroprotective effect of magnesium sulfate in the premature brain [52]. When plasma levels of androgens were analyzed in male subjects with autism compared to males with mental retardation and control subjects, androgenic hormone levels were not different among the groups, except that the DHEAS levels were higher in mentally retarded patients with CP compared to age-matched mentally retarded patients without CP or controls [94].

In our study, the number of cases and controls was small due to the difficulties in obtaining the post-mortem brain tissues from patients with CP. Clinical information for both cases and controls was limited. The age and gender for cases and controls were matched with the best available samples in the NIH NeuroBioBank. The biopsy specimens were obtained from the same but one anatomical location of each brain to be analyzed.

4. Materials and Methods

4.1. Tissue Samples

Only a limited number of specimens and tissue was available for this pilot study. Tissue samples (caudate nucleus) were obtained from post-mortem CP cases (n = 9) and age- and gender-matched control subjects (n = 11). Tissues were obtained from the Harvard University Tissue Resource Center, the University of Maryland Brain and Tissue Bank, and the University of Miami Miller School of Medicine, which are all Brain and Tissue Repositories of the NIH NeuroBioBank. This study was approved by the Beaumont Health System's Human Investigation Committee (HIC No.: 2018-387). The methods were carried out in accordance with the approved guidelines. Details such as age, gender, race, and post-mortem delay can be found in Table S2 in the Supplementary Materials.

4.2. Sample Preparation

Samples were stored at −80 °C prior to preparation. Subsequently, samples were lyophilized and milled to a fine powder under liquid nitrogen to limit the amount of heat production. For ^{1}H NMR, 50 mg samples were extracted in 50% methanol/water (1 g/mL) in a sterile 2 mL Eppendorf tube. The samples were mixed for 20 min and sonicated for 20 min, and the protein was removed by centrifugation at 13,000× g at 4 °C for 30 min. Supernatants were collected, dried under vacuum using a Savant DNA SpeedVac (Thermo Scientific, Waltham, MA USA), and reconstituted in 285 µL of 50 mM potassium phosphate buffer (pH 7.0), 30 µL of sodium 2,2-dimethyl-2-silapentane-5-sulfonate (DSS), and 35 µL of D_2O [95]. A 200 µL portion of the reconstituted sample was transferred to a 3 mm Bruker NMR tube for analysis. All samples were housed at 4 °C in a thermostatically controlled SampleJet autosampler (Bruker-Biospin, Billerica, MA, USA) and heated to room temperature over 3 minutes prior to analysis by NMR.

For analysis by targeted mass spectrometry, the tissue samples were analyzed using the commercially available AbsoluteIDQ p180 (Biocrates, Innsbruck, Austria) kit. In brief, 10 mg (±3 mg) of milled tissue were extracted in 300 µL of solvent (85% ethanol and 15% phosphate buffered saline solution). The samples were shaken at 700 rpm for 10 min, followed by sonication for 20 min, and centrifuged at 13,000× g for 20 min. The supernatant was collected and 10 µL were used for analysis with the kit. A 10 µL portion of blank, 3 zero samples, 7 calibration standards, and 3 quality control samples were loaded onto the filters in the upper 96-well plate and dried under nitrogen using a positive pressure processor (Waters Technologies Corporation, Milford, MA, USA). Subsequently, 50 µL of phenylisothiocyanate derivatization solution were added to each well and left at room temperature for 20 min. The plate was subsequently dried under nitrogen for 60 min, followed by the addition of 300 µL of methanol containing 5 mM ammonium acetate for the extraction of metabolites. The plate was shaken at 700 rpm for 30 min and the extracts filtered to the lower collection plate using the positive pressure processor. Eluates were diluted with water for the analysis of the metabolites with the workflow using ultra-pressure liquid chromatography mass spectrometry (UPLC-MS) and diluted with running solvent for flow injection analysis (FIA)-MS (for lipids).

4.3. Data Collection and Metabolic Profiling

Using a randomized running order, all 1D ^{1}H NMR data were recorded at 300 (±0.5) K on a Bruker ASCEND HD 600 MHz spectrometer (Bruker-Biospin, Billerica, MA, USA) coupled with a 5 mm TCI cryoprobe. For each sample, 256 transients were collected as 64k data points with a spectral width of 12 kHz (20 ppm), using a pulse sequence called CPP WaterSupp (Bruker pulse program: pusenoesypr1d) developed by Mercier et al. [96] and an inter-pulse delay of 9.65 s. The data collection protocol included a 180-s temperature equilibration period, fast 3D shimming using the z-axis profile of the ^{2}H NMR solvent signal, receiver gain adjustment, and acquisition. The free induction decay signal was zero filled to 128k and exponentially multiplied with a 0.1 Hz line broadening factor. The zero and first order phase constants were manually optimized after Fourier transformation and a polynomial baseline correction of the free induction decay (FID; degree 5) was applied for precise quantitation. All spectra were processed and analyzed using Chenomx NMR Suite (v8.0, Chenomx, Edmonton, AB, Canada).

As previously noted, targeted MS analysis was carried out using AbsoluteIDQ p180 kit (Biocrates Life Sciences AG, Innsbruck, Austria). Data was acquired using a Xevo TQ-S mass spectrometer coupled to an Acquity I Class UPLC system (Waters Technologies Corporation, Milford, MA, USA) as per the manufacturer's instructions. The system allows for the accurate quantification of up to 188 endogenous metabolites including amino acids, acylcarnitines, biogenic amines, glycerophospholipids, sphingolipids, and sugars. Sample registration and the automated calculation of metabolite concentrations and export of data were carried out with Biocrates MetIDQ software. We accurately identified and quantified 59 metabolites using ^{1}H NMR and 173 using DI/LC-MS/MS. Some overlap was observed (22 metabolites) between the two platforms and as such, we reported the average values for each individual metabolite measured using both analytical platforms.

4.4. Statistical Analysis

Using MetaboAnalyst (v4.0) [97], the data were analyzed using a two-tailed Student's *t*-test to determine the statistical significance between the metabolite concentration in CP and corresponding controls ($p < 0.05$, FDR < 0.3).

We selected a representative set of six artificial intelligence algorithms, which have been applied for problems of data classification in the bioinformatics field. These included logistic regression (LR), prediction analysis for microarrays (PAM), partial least square-discriminant analysis (PLS-DA), deep learning (DL), random forest (RF), and support vector machine (SVM).

Using publicly available toolboxes in R, important parameters for each model were optimized so that the best prediction performance could be achieved [98–103]. In order to assess model performance of each approach or algorithm, the data were split into training and testing sets (80% and 20% respectively). In an attempt of avoiding sampling bias, the splitting process was repeated ten times and the AUC values were averaged out. Sensitivity and specificity values were calculated at 95% confidence intervals.

4.5. Metabolite Pathway Enrichment Analysis

Metabolite set enrichment analysis (MSEA) was completed using MetaboAnalyst (v4.0) [97]. Metabolite names were converted to Human Metabolome Database (HMDB) identifiers. The raw data was subjected to sum normalization and autoscaling. The pathway-associated metabolite set was the chosen metabolite library, and all compounds in this library were used. Pathways with a raw p value < 0.01 were considered to be significantly altered upon CP.

5. Conclusions

We report for the first time a targeted, quantitative metabolomic approach for profiling post-mortem human brain tissue from patients with CP. Metabolomic analysis provided new insights into the dysregulated brain metabolism associated with CP. The metabolites and associated biochemical pathways identified herein could potentially facilitate the understanding of the underlying complex pathophysiology associated with CP as well as possible central biomarkers for early detection and prediction of CP. There is a need for future studies to confirm our current preliminary data in more accessible biomatrices.

Supplementary Materials: The following are available online at http://www.mdpi.com/2218-1989/9/2/27/s1, Table S1: Metabolite Concentrations (μM) for CP vs. Control PM Brain Extracts. Those compounds highlighted in bold are considered statistically, significantly different ($p < 0.05$; $q < 0.05$). *t*-test values were calculated as a default and values with (W) were calculated using the Wilcoxon Mann Whitney test. Table S2. A list of the available demographic information. PM-post-mortem.

Author Contributions: Designing research studies (S.F.G., Z.A.S.), conducting experiments (A.Y., Z.U.), analyzing data (A.Y., B.A., S.F.G., Z.U.), drafting the manuscript (A.Y., S.F.G., Z.A.S., Z.U.), and writing the manuscript. All authors contributed to the editing of the manuscript (A.Y., S.F.G., Z.A.S., Z.U., R.O.B.-S., B.A.).

Funding: This research received no external funding.

Acknowledgments: The authors would like to acknowledge the NIH NeuroBioBank for graciously providing the specimens and enabling them to conduct their experiments. In addition, this work was partly funded by the generous contribution made by the Fred A. and Barbara M. Erb Foundation.

Conflicts of Interest: The authors declare no conflict of interest.

References

1. Blair, E.; Watson, L. Epidemiology of cerebral palsy. *Semin. Fetal Neonatal Med.* **2006**, *11*, 117–125. [CrossRef] [PubMed]
2. Wimalasundera, N.; Stevenson, V.L. Cerebral palsy. *Pract. Neurol.* **2016**, *16*, 184–194. [CrossRef] [PubMed]
3. Johnston, M.V.; Hoon, A.H., Jr. Cerebral palsy. *Neuromol. Med.* **2006**, *8*, 435–450. [CrossRef]
4. Drougia, A.; Giapros, V.; Krallis, N.; Theocharis, P.; Nikaki, A.; Tzoufi, M.; Andronikou, S. Incidence and risk factors for cerebral palsy in infants with perinatal problems: A 15-year review. *Early Hum. Dev.* **2007**, *83*, 541–547. [CrossRef] [PubMed]
5. Nelson, K.B.; Blair, E. Prenatal factors in singletons with cerebral palsy born at or near term. *N. Engl. J. Med.* **2015**, *373*, 946–953. [CrossRef]
6. Boyle, A.K.; Rinaldi, S.F.; Norman, J.E.; Stock, S.J. Preterm birth: Inflammation, fetal injury and treatment strategies. *J. Reprod. Immunol.* **2017**, *119*, 62–66. [CrossRef] [PubMed]
7. MacLennan, A.H.; Thompson, S.C.; Gecz, J. Cerebral palsy: Causes, pathways, and the role of genetic variants. *Am. J. Obstet. Gynecol.* **2015**, *213*, 779–788. [CrossRef]
8. Keogh, J.M.; Badawi, N. The origins of cerebral palsy. *Curr. Opin. Neurol.* **2006**, *19*, 129–134. [CrossRef]
9. Tonni, G.; Leoncini, S.; Signorini, C.; Ciccoli, L.; De Felice, C. Pathology of perinatal brain damage: Background and oxidative stress markers. *Arch. Gynecol. Obstet.* **2014**, *290*, 13–20. [CrossRef]
10. Korzeniewski, S.J.; Slaughter, J.; Lenski, M.; Haak, P.; Paneth, N. The complex aetiology of cerebral palsy. *Nat. Rev. Neurol.* **2018**, *14*, 528–543. [CrossRef]
11. Spedding, M.; Gressens, P. Neurotrophins and cytokines in neuronal plasticity. *Novartis Found. Symp.* **2008**, *289*, 222–233; discussion 233–240. [PubMed]
12. Marret, S.; Vanhulle, C.; Laquerriere, A. Pathophysiology of cerebral palsy. *Handb. Clin. Neurol.* **2013**, *111*, 169–176. [PubMed]
13. Denihan, N.M.; Boylan, G.B.; Murray, D.M. Metabolomic profiling in perinatal asphyxia: A promising new field. *Biomed. Res. Int.* **2015**, *2015*, 254076. [CrossRef] [PubMed]
14. Yli, B.M.; Kjellmer, I. Pathophysiology of foetal oxygenation and cell damage during labour. *Best Pract. Res. Clin. Obstet. Gynaecol.* **2016**, *30*, 9–21. [CrossRef] [PubMed]
15. Hoon, A.H., Jr.; Vasconcellos Faria, A. Pathogenesis, neuroimaging and management in children with cerebral palsy born preterm. *Dev. Disabil. Res. Rev.* **2010**, *16*, 302–312. [CrossRef] [PubMed]
16. Novak, C.M.; Ozen, M.; Burd, I. Perinatal brain injury: Mechanisms, prevention, and outcomes. *Clin. Perinatol.* **2018**, *45*, 357–375. [CrossRef]
17. Fahey, M.C.; Maclennan, A.H.; Kretzschmar, D.; Gecz, J.; Kruer, M.C. The genetic basis of cerebral palsy. *Dev. Med. Child Neurol.* **2017**, *59*, 462–469. [CrossRef]
18. Pascal, A.; Govaert, P.; Oostra, A.; Naulaers, G.; Ortibus, E.; Van den Broeck, C. Neurodevelopmental outcome in very preterm and very-low-birthweight infants born over the past decade: A meta-analytic review. *Dev. Med. Child Neurol.* **2018**, *60*, 342–355. [CrossRef]
19. Shankaran, S. Prevention, diagnosis, and treatment of cerebral palsy in near-term and term infants. *Clin. Obstet. Gynecol.* **2008**, *51*, 829–839. [CrossRef]
20. Rizzotti, A.; Bas, J.; Cuestas, E. Efficacy and securyty of therapeutic hypothermia for hypoxic ischemic encephalopathy: A meta-analysis. *Rev. Fac. Cienc. Med.* **2010**, *67*, 15–23.
21. Shepherd, E.; Salam, R.A.; Middleton, P.; Han, S.; Makrides, M.; McIntyre, S.; Badawi, N.; Crowther, C.A. Neonatal interventions for preventing cerebral palsy: An overview of cochrane systematic reviews. *Cochrane Database Syst. Rev.* **2018**, *6*, Cd012409. [CrossRef] [PubMed]

22. Chollat, C.; Sentilhes, L.; Marret, S. Protection of brain development by antenatal magnesium sulphate for infants born preterm. *Dev. Med. Child Neurol.* **2019**, *61*, 25–30. [CrossRef] [PubMed]
23. Graham, S.F.; Rey, N.L.; Yilmaz, A.; Kumar, P.; Madaj, Z.; Maddens, M.; Bahado-Singh, R.O.; Becker, K.; Schulz, E.; Meyerdirk, L.K.; et al. Biochemical profiling of the brain and blood metabolome in a mouse model of prodromal parkinson's disease reveal distinct metabolic profiles. *J. Proteome Res.* **2018**, *17*, 2460–2469. [CrossRef] [PubMed]
24. Graham, S.F.; Chevallier, O.P.; Kumar, P.; Turkoglu, O.; Bahado-Singh, R.O. Metabolomic profiling of brain from infants who died from sudden infant death syndrome reveals novel predictive biomarkers. *J. Perinatol.* **2017**, *37*, 91–97. [CrossRef] [PubMed]
25. Yilmaz, A.; Geddes, T.; Han, B.; Bahado-Singh, R.O.; Wilson, G.D.; Imam, K.; Maddens, M.; Graham, S.F. Diagnostic biomarkers of alzheimer's disease as identified in saliva using 1h nmr-based metabolomics. *J. Alzheimers Dis.* **2017**, *58*, 355–359. [CrossRef] [PubMed]
26. Pan, X.; Elliott, C.T.; McGuinness, B.; Passmore, P.; Kehoe, P.G.; Holscher, C.; McClean, P.L.; Graham, S.F.; Green, B.D. Metabolomic profiling of bile acids in clinical and experimental samples of alzheimer's disease. *Metabolites* **2017**, *7*, 28. [CrossRef] [PubMed]
27. Bahado-Singh, R.O.; Graham, S.F.; Turkoglu, O.; Beauchamp, K.; Bjorndahl, T.C.; Han, B.; Mandal, R.; Pantane, J.; Kowalenko, T.; Wishart, D.S.; et al. Identification of candidate biomarkers of brain damage in a mouse model of closed head injury: A metabolomic pilot study. *Metabolomics* **2016**, *12*, 42. [CrossRef]
28. Graham, S.F.; Kumar, P.; Bahado-Singh, R.O.; Robinson, A.; Mann, D.; Green, B.D. Novel metabolite biomarkers of huntington's disease (hd) as detected by high resolution mass spectrometry. *J. Proteome Res.* **2016**, *15*, 1592–1601. [CrossRef]
29. Fiandaca, M.S.; Zhong, X.; Cheema, A.K.; Orquiza, M.H.; Chidambaram, S.; Tan, M.T.; Gresenz, C.R.; FitzGerald, K.T.; Nalls, M.A.; Singleton, A.B.; et al. Plasma 24-metabolite panel predicts preclinical transition to clinical stages of alzheimer's disease. *Front. Neurol.* **2015**, *6*, 237. [CrossRef]
30. Mapstone, M.; Cheema, A.K.; Fiandaca, M.S.; Zhong, X.; Mhyre, T.R.; MacArthur, L.H.; Hall, W.J.; Fisher, S.G.; Peterson, D.R.; Haley, J.M.; et al. Plasma phospholipids identify antecedent memory impairment in older adults. *Nat. Med.* **2014**, *20*, 415–418. [CrossRef]
31. Varma, V.R.; Oommen, A.M.; Varma, S.; Casanova, R.; An, Y.; Andrews, R.M.; O'Brien, R.; Pletnikova, O.; Troncoso, J.C.; Toledo, J.; et al. Brain and blood metabolite signatures of pathology and progression in alzheimer disease: A targeted metabolomics study. *PLoS Med.* **2018**, *15*, e1002482. [CrossRef] [PubMed]
32. Parikh, N.A. Advanced neuroimaging and its role in predicting neurodevelopmental outcomes in very preterm infants. *Semin. Perinatol.* **2016**, *40*, 530–541. [CrossRef]
33. Jaspers, E.; Byblow, W.D.; Feys, H.; Wenderoth, N. The corticospinal tract: A biomarker to categorize upper limb functional potential in unilateral cerebral palsy. *Front. Pediatr.* **2015**, *3*, 112. [CrossRef] [PubMed]
34. Aslam, S.; Molloy, E.J. Biomarkers of multiorgan injury in neonatal encephalopathy. *Biomark. Med.* **2015**, *9*, 267–275. [CrossRef] [PubMed]
35. Shalak, L.F.; Perlman, J.M. Infection markers and early signs of neonatal encephalopathy in the term infant. *Ment. Retard. Dev. Disabil. Res. Rev.* **2002**, *8*, 14–19. [CrossRef] [PubMed]
36. Jin, C.; Londono, I.; Mallard, C.; Lodygensky, G.A. New means to assess neonatal inflammatory brain injury. *J. Neuroinflamm.* **2015**, *12*, 180. [CrossRef] [PubMed]
37. Graham, S.F.; Chevallier, O.P.; Kumar, P.; Türkoğlu, O.; Bahado-Singh, R.O. High resolution metabolomic analysis of asd human brain uncovers novel biomarkers of disease. *Metabolomics* **2016**, *12*, 62. [CrossRef]
38. Graham, S.F.; Chevallier, O.P.; Roberts, D.; Hölscher, C.; Elliott, C.T.; Green, B.D. Investigation of the human brain metabolome to identify potential markers for early diagnosis and therapeutic targets of alzheimer's disease. *Anal. Chem.* **2013**, *85*, 1803–1811. [CrossRef] [PubMed]
39. Graham, S.F.; Holscher, C.; Green, B.D. Metabolic signatures of human alzheimer's disease (ad): 1h nmr analysis of the polar metabolome of post-mortem brain tissue. *Metabolomics* **2014**, *10*, 744–753. [CrossRef]
40. Graham, S.F.; Holscher, C.; McClean, P.; Elliott, C.T.; Green, B.D. 1 h nmr metabolomics investigation of an alzheimer's disease (ad) mouse model pinpoints important biochemical disturbances in brain and plasma. *Metabolomics* **2013**, *9*, 974–983. [CrossRef]

41. Graham, S.F.; Kumar, P.K.; Bjorndahl, T.; Han, B.; Yilmaz, A.; Sherman, E.; Bahado-Singh, R.O.; Wishart, D.; Mann, D.; Green, B.D. Metabolic signatures of huntington's disease (hd): (1)h nmr analysis of the polar metabolome in post-mortem human brain. *Biochim. Biophys. Acta* **2016**, *1862*, 1675–1684. [CrossRef] [PubMed]
42. Graham, S.F.; Pan, X.; Yilmaz, A.; Macias, S.; Robinson, A.; Mann, D.; Green, B.D. Targeted biochemical profiling of brain from huntington's disease patients reveals novel metabolic pathways of interest. *Biochim. Biophys. Acta* **2018**, *1864*, 2430–2437. [CrossRef] [PubMed]
43. Graham, S.F.; Turkoglu, O.; Kumar, P.; Yilmaz, A.; Bjorndahl, T.C.; Han, B.; Mandal, R.; Wishart, D.S.; Bahado-Singh, R.O. Targeted metabolic profiling of post-mortem brain from infants who died from sudden infant death syndrome. *J. Proteome Res.* **2017**, *16*, 2587–2596. [CrossRef] [PubMed]
44. Cotman, C.W.; Foster, A.; Lanthorn, T. An overview of glutamate as a neurotransmitter. *Adv. Biochem. Psychopharmacol.* **1981**, *27*, 1–27. [PubMed]
45. Beal, M.F. Role of excitotoxicity in human neurological disease. *Curr. Opin. Neurobiol.* **1992**, *2*, 657–662. [CrossRef]
46. Battaglia, G.; Bruno, V. Metabotropic glutamate receptor involvement in the pathophysiology of amyotrophic lateral sclerosis: New potential drug targets for therapeutic applications. *Curr. Opin. Pharmacol.* **2018**, *38*, 65–71. [CrossRef] [PubMed]
47. Wang, R.; Reddy, P.H. Role of glutamate and nmda receptors in alzheimer's disease. *J. Alzheimers Dis.* **2017**, *57*, 1041–1048. [CrossRef]
48. Walker, J.E. Glutamate, gaba, and cns disease: A review. *Neurochem. Res.* **1983**, *8*, 521–550. [CrossRef]
49. Choi, D.W.; Rothman, S.M. The role of glutamate neurotoxicity in hypoxic-ischemic neuronal death. *Annu. Rev. Neurosci.* **1990**, *13*, 171–182. [CrossRef]
50. Anderson, C.M.; Swanson, R.A. Astrocyte glutamate transport: Review of properties, regulation, and physiological functions. *Glia* **2000**, *32*, 1–14. [CrossRef]
51. Back, S.A.; Craig, A.; Kayton, R.J.; Luo, N.L.; Meshul, C.K.; Allcock, N.; Fern, R. Hypoxia-ischemia preferentially triggers glutamate depletion from oligodendroglia and axons in perinatal cerebral white matter. *J. Cereb. Blood Flow Metab.* **2007**, *27*, 334–347. [CrossRef] [PubMed]
52. Daher, I.; Le Dieu-Lugon, B.; Dourmap, N.; Lecuyer, M.; Ramet, L.; Gomila, C.; Ausseil, J.; Marret, S.; Leroux, P.; Roy, V.; et al. Magnesium sulfate prevents neurochemical and long-term behavioral consequences of neonatal excitotoxic lesions: Comparison between male and female mice. *J. Neuropathol. Exp. Neurol.* **2017**, *76*, 883–897. [CrossRef] [PubMed]
53. Hermansson, M.; Hokynar, K.; Somerharju, P. Mechanisms of glycerophospholipid homeostasis in mammalian cells. *Prog. Lipid Res.* **2011**, *50*, 240–257. [CrossRef] [PubMed]
54. Farooqui, A.A.; Horrocks, L.A.; Farooqui, T. Glycerophospholipids in brain: Their metabolism, incorporation into membranes, functions, and involvement in neurological disorders. *Chem. Phys. Lipids* **2000**, *106*, 1–29. [CrossRef]
55. Fuchs, B.; Schiller, J.; Cross, M.A. Apoptosis-associated changes in the glycerophospholipid composition of hematopoietic progenitor cells monitored by 31p nmr spectroscopy and maldi-tof mass spectrometry. *Chem. Phys. Lipids* **2007**, *150*, 229–238. [CrossRef]
56. Farooqui, A.A.; Horrocks, L.A.; Farooqui, T. Interactions between neural membrane glycerophospholipid and sphingolipid mediators: A recipe for neural cell survival or suicide. *J. Neurosci. Res.* **2007**, *85*, 1834–1850. [CrossRef]
57. Yang, Y.; Lee, M.; Fairn, G.D. Phospholipid subcellular localization and dynamics. *J. Biol. Chem.* **2018**, *293*, 6230–6240. [CrossRef]
58. Frisardi, V.; Panza, F.; Seripa, D.; Farooqui, T.; Farooqui, A.A. Glycerophospholipids and glycerophospholipid-derived lipid mediators: A complex meshwork in alzheimer's disease pathology. *Prog. Lipid Res.* **2011**, *50*, 313–330. [CrossRef]
59. Muller, C.P.; Reichel, M.; Muhle, C.; Rhein, C.; Gulbins, E.; Kornhuber, J. Brain membrane lipids in major depression and anxiety disorders. *Biochim. Biophys. Acta* **2015**, *1851*, 1052–1065. [CrossRef]
60. Villamil-Ortiz, J.G.; Barrera-Ocampo, A.; Arias-Londono, J.D.; Villegas, A.; Lopera, F.; Cardona-Gomez, G.P. Differential pattern of phospholipid profile in the temporal cortex from e280a-familiar and sporadic alzheimer's disease brains. *J. Alzheimers Dis.* **2018**, *61*, 209–219. [CrossRef]

61. Zhang, J.; Zhang, X.; Wang, L.; Yang, C. High performance liquid chromatography-mass spectrometry (lc-ms) based quantitative lipidomics study of ganglioside-nana-3 plasma to establish its association with parkinson's disease patients. *Med Sci. Monit. Int. Med J. Exp. Clin. Res.* **2017**, *23*, 5345–5353. [CrossRef]
62. Walsh, B.H.; Broadhurst, D.I.; Mandal, R.; Wishart, D.S.; Boylan, G.B.; Kenny, L.C.; Murray, D.M. The metabolomic profile of umbilical cord blood in neonatal hypoxic ischaemic encephalopathy. *PLoS ONE* **2012**, *7*, e50520. [CrossRef] [PubMed]
63. Handley, R.R.; Reid, S.J.; Brauning, R.; Maclean, P.; Mears, E.R.; Fourie, I.; Patassini, S.; Cooper, G.J.S.; Rudiger, S.R.; McLaughlan, C.J.; et al. Brain urea increase is an early huntington's disease pathogenic event observed in a prodromal transgenic sheep model and hd cases. *Proc. Natl. Acad. Sci. USA* **2017**, *114*, E11293–E11302. [CrossRef] [PubMed]
64. Xu, J.; Begley, P.; Church, S.J.; Patassini, S.; Hollywood, K.A.; Jullig, M.; Curtis, M.A.; Waldvogel, H.J.; Faull, R.L.; Unwin, R.D.; et al. Graded perturbations of metabolism in multiple regions of human brain in alzheimer's disease: Snapshot of a pervasive metabolic disorder. *Biochim. Biophys. Acta* **2016**, *1862*, 1084–1092. [CrossRef] [PubMed]
65. Summar, M.L.; Mew, N.A. Inborn errors of metabolism with hyperammonemia: Urea cycle defects and related disorders. *Pediatr. Clin. N. Am.* **2018**, *65*, 231–246. [CrossRef] [PubMed]
66. Stone, W.L.; Jaishankar, G.B. *Urea Cycle Disorders*; StatPearls Publishing: Treasure Island, FL, USA, 2018.
67. Baluarte, J.H. Neurological complications of renal disease. *Semin. Pediatr. Neurol.* **2017**, *24*, 25–32. [CrossRef] [PubMed]
68. Wu, T.; Li, X.; Ding, Y.; Liu, Y.; Song, J.; Wang, Q.; Li, M.; Qin, Y.; Yang, Y. Seven patients of argininemia with spastic tetraplegia as the first and major symptom and prenatal diagnosis of two fetuses with high risk. *Zhonghua Er Ke Za Zhi = Chin. J. Pediatr.* **2015**, *53*, 425–430.
69. Gao, J.; Zhao, B.; He, L.; Sun, M.; Yu, X.; Wang, L. Risk of cerebral palsy in chinese children: A n:M matched case control study. *J. Paediatr. Child Health* **2017**, *53*, 464–469. [CrossRef]
70. Schoendorfer, N.C.; Obeid, R.; Moxon-Lester, L.; Sharp, N.; Vitetta, L.; Boyd, R.N.; Davies, P.S. Methylation capacity in children with severe cerebral palsy. *Eur. J. Clin. Investig.* **2012**, *42*, 768–776. [CrossRef]
71. Nabiuni, M.; Rasouli, J.; Parivar, K.; Kochesfehani, H.M.; Irian, S.; Miyan, J.A. In vitro effects of fetal rat cerebrospinal fluid on viability and neuronal differentiation of pc12 cells. *Fluids Barriers CNS* **2012**, *9*, 8. [CrossRef]
72. Chan, A.; Tchantchou, F.; Graves, V.; Rozen, R.; Shea, T.B. Dietary and genetic compromise in folate availability reduces acetylcholine, cognitive performance and increases aggression: Critical role of s-adenosyl methionine. *J. Nutr. Health Aging* **2008**, *12*, 252–261. [CrossRef] [PubMed]
73. Lovati, C.; Galimberti, D.; Pomati, S.; Capiluppi, E.; Dolci, A.; Scapellato, L.; Rosa, S.; Mailland, E.; Suardelli, M.; Vanotti, A.; et al. Serum folate concentrations in patients with cortical and subcortical dementias. *Neurosci. Lett.* **2007**, *420*, 213–216. [CrossRef] [PubMed]
74. Botez, M.I.; Peyronnard, J.M.; Berube, L.; Labrecque, R. Relapsing neuropathy, cerebral atrophy and folate deficiency. A close association. *Appl. Neurophysiol.* **1979**, *42*, 171–183. [PubMed]
75. Mortensen, P.B.; Clausen, M.R. Short-chain fatty acids in the human colon: Relation to gastrointestinal health and disease. *Scand. J. Gastroenterol. Suppl.* **1996**, *216*, 132–148. [CrossRef] [PubMed]
76. Finegold, S.M.; Molitoris, D.; Song, Y.; Liu, C.; Vaisanen, M.L.; Bolte, E.; McTeague, M.; Sandler, R.; Wexler, H.; Marlowe, E.M.; et al. Gastrointestinal microflora studies in late-onset autism. *Clin. Infect. Dis.* **2002**, *35*, S6–S16. [CrossRef] [PubMed]
77. Finegold, S.M.; Dowd, S.E.; Gontcharova, V.; Liu, C.; Henley, K.E.; Wolcott, R.D.; Youn, E.; Summanen, P.H.; Granpeesheh, D.; Dixon, D.; et al. Pyrosequencing study of fecal microflora of autistic and control children. *Anaerobe* **2010**, *16*, 444–453. [CrossRef] [PubMed]
78. Conn, A.R.; Fell, D.I.; Steele, R.D. Characterization of alpha-keto acid transport across blood-brain barrier in rats. *Am. J. Physiol.* **1983**, *245*, E253–E260. [CrossRef]
79. Thomas, R.H.; Meeking, M.M.; Mepham, J.R.; Tichenoff, L.; Possmayer, F.; Liu, S.; MacFabe, D.F. The enteric bacterial metabolite propionic acid alters brain and plasma phospholipid molecular species: Further development of a rodent model of autism spectrum disorders. *J. Neuroinflamm.* **2012**, *9*, 153. [CrossRef]
80. Koh, A.; Molinaro, A.; Stahlman, M.; Khan, M.T.; Schmidt, C.; Manneras-Holm, L.; Wu, H.; Carreras, A.; Jeong, H.; Olofsson, L.E.; et al. Microbially produced imidazole propionate impairs insulin signaling through mtorc1. *Cell* **2018**, *175*, 947–961. [CrossRef]

81. Morland, C.; Froland, A.S.; Pettersen, M.N.; Storm-Mathisen, J.; Gundersen, V.; Rise, F.; Hassel, B. Propionate enters gabaergic neurons, inhibits gaba transaminase, causes gaba accumulation and lethargy in a model of propionic acidemia. *Biochem. J.* **2018**, *475*, 749–758. [CrossRef]
82. Hoyles, L.; Snelling, T.; Umlai, U.K.; Nicholson, J.K.; Carding, S.R.; Glen, R.C.; McArthur, S. Microbiome-host systems interactions: Protective effects of propionate upon the blood-brain barrier. *Microbiome* **2018**, *6*, 55. [CrossRef] [PubMed]
83. van den Berge, M.; Jonker, M.R.; Miller-Larsson, A.; Postma, D.S.; Heijink, I.H. Effects of fluticasone propionate and budesonide on the expression of immune defense genes in bronchial epithelial cells. *Pulm. Pharmacol. Ther.* **2018**, *50*, 47–56. [CrossRef] [PubMed]
84. Pluciennik, F.; Verrecchia, F.; Bastide, B.; Herve, J.C.; Joffre, M.; Deleze, J. Reversible interruption of gap junctional communication by testosterone propionate in cultured sertoli cells and cardiac myocytes. *J. Membr. Biol.* **1996**, *149*, 169–177. [CrossRef] [PubMed]
85. Frye, R.E.; Rossignol, D.A. Mitochondrial dysfunction can connect the diverse medical symptoms associated with autism spectrum disorders. *Pediatr. Res.* **2011**, *69*, 41r–47r. [CrossRef] [PubMed]
86. MacFabe, D.F.; Cain, D.P.; Rodriguez-Capote, K.; Franklin, A.E.; Hoffman, J.E.; Boon, F.; Taylor, A.R.; Kavaliers, M.; Ossenkopp, K.P. Neurobiological effects of intraventricular propionic acid in rats: Possible role of short chain fatty acids on the pathogenesis and characteristics of autism spectrum disorders. *Behav. Brain Res.* **2007**, *176*, 149–169. [CrossRef] [PubMed]
87. MacFabe, D.F.; Cain, N.E.; Boon, F.; Ossenkopp, K.P.; Cain, D.P. Effects of the enteric bacterial metabolic product propionic acid on object-directed behavior, social behavior, cognition, and neuroinflammation in adolescent rats: Relevance to autism spectrum disorder. *Behav. Brain Res.* **2011**, *217*, 47–54. [CrossRef] [PubMed]
88. Shultz, S.R.; Macfabe, D.F.; Martin, S.; Jackson, J.; Taylor, R.; Boon, F.; Ossenkopp, K.P.; Cain, D.P. Intracerebroventricular injections of the enteric bacterial metabolic product propionic acid impair cognition and sensorimotor ability in the long-evans rat: Further development of a rodent model of autism. *Behav. Brain Res.* **2009**, *200*, 33–41. [CrossRef]
89. Shultz, S.R.; MacFabe, D.F.; Ossenkopp, K.P.; Scratch, S.; Whelan, J.; Taylor, R.; Cain, D.P. Intracerebroventricular injection of propionic acid, an enteric bacterial metabolic end-product, impairs social behavior in the rat: Implications for an animal model of autism. *Neuropharmacology* **2008**, *54*, 901–911. [CrossRef]
90. Broeder, C.E.; Brenner, M.; Hofman, Z.; Paijmans, I.J.; Thomas, E.L.; Wilmore, J.H. The metabolic consequences of low and moderate intensity exercise with or without feeding in lean and borderline obese males. *Int. J. Obes.* **1991**, *15*, 95–104.
91. Gerstner, B.; Lee, J.; DeSilva, T.M.; Jensen, F.E.; Volpe, J.J.; Rosenberg, P.A. 17beta-estradiol protects against hypoxic/ischemic white matter damage in the neonatal rat brain. *J. Neurosci. Res.* **2009**, *87*, 2078–2086. [CrossRef]
92. Zisk, J.L.; Genen, L.H.; Kirkby, S.; Webb, D.; Greenspan, J.; Dysart, K. Do premature female infants really do better than their male counterparts? *Am. J. Perinatol.* **2011**, *28*, 241–246. [CrossRef] [PubMed]
93. Lan, Y.L.; Zhao, J.; Li, S. Estrogen receptors' neuroprotective effect against glutamate-induced neurotoxicity. *Neurol. Sci.* **2014**, *35*, 1657–1662. [CrossRef] [PubMed]
94. Tordjman, S.; Anderson, G.M.; McBride, P.A.; Hertzig, M.E.; Snow, M.E.; Hall, L.M.; Ferrari, P.; Cohen, D.J. Plasma androgens in autism. *J. Autism Dev. Disord.* **1995**, *25*, 295–304. [CrossRef] [PubMed]
95. Ravanbakhsh, S.; Liu, P.; Bjordahl, T.C.; Mandal, R.; Grant, J.R.; Wilson, M.; Eisner, R.; Sinelnikov, I.; Hu, X.; Luchinat, C.; et al. Accurate, fully-automated nmr spectral profiling for metabolomics. *PLoS ONE* **2015**, *10*, e0124219. [CrossRef] [PubMed]
96. Mercier, P.; Lewis, M.J.; Chang, D.; Baker, D.; Wishart, D.S. Towards automatic metabolomic profiling of high-resolution one-dimensional proton nmr spectra. *J. Biomol. NMR* **2011**, *49*, 307–323. [CrossRef] [PubMed]
97. Chong, J.; Soufan, O.; Li, C.; Caraus, I.; Li, S.; Bourque, G.; Wishart, D.S.; Xia, J. Metaboanalyst 4.0: Towards more transparent and integrative metabolomics analysis. *Nucleic Acids Res.* **2018**, *46*, W486–W494. [CrossRef] [PubMed]
98. Min, S.; Lee, B.; Yoon, S. Deep learning in bioinformatics. *Brief. Bioinform.* **2017**, *18*, 851–869. [CrossRef] [PubMed]

99. Alakwaa, F.M.; Chaudhary, K.; Garmire, L.X. Deep learning accurately predicts estrogen receptor status in breast cancer metabolomics data. *J. Proteome Res.* **2018**, *17*, 337–347. [CrossRef] [PubMed]
100. Kuhn, M. Building predictive models in r using the caret package. *J. Stat. Softw.* **2008**, *28*, 26. [CrossRef]
101. Angermueller, C.; Parnamaa, T.; Parts, L.; Stegle, O. Deep learning for computational biology. *Mol. Syst. Biol.* **2016**, *12*, 878. [CrossRef] [PubMed]
102. Zou, H.; Hastie, T. Regularization and variable selection via the elastic net. *J. R. Stat. Soc. Ser. B Stat. Methodol.* **2005**, *67*, 301–320. [CrossRef]
103. Faul, F.; Erdfelder, E.; Buchner, A.; Lang, A.G. Statistical power analyses using g*power 3.1: Tests for correlation and regression analyses. *Behav. Res. Methods* **2009**, *41*, 1149–1160. [CrossRef] [PubMed]

metabolites

MDPI

Article

Palmitate and Stearate are Increased in the Plasma in a 6-OHDA Model of Parkinson's Disease

Anuri Shah [1,2,†], Pei Han [1,3,†], Mung-Yee Wong [2], Raymond Chuen-Chung Chang [2,4,*] and Cristina Legido-Quigley [1,5,*]

1 Institute of Pharmaceutical Science, Faculty of Life Sciences and Medicine, King's College London, London SE19NH, UK; anuri.shah@hku.hk (A.S.); hanpei0126@hotmail.com (P.H.)
2 Laboratory of Neurodegenerative Diseases, School of Biomedical Sciences, LKS Faculty of Medicine, The University of Hong Kong, Hong Kong, China; mungyee.wong@gmail.com
3 Institute of Materia Medica, Chinese Academy of Medical Sciences & Peking Union Medical College, Beijing 100006, China
4 State Key Laboratory of Brain and Cognitive Sciences, The University of Hong Kong, Hong Kong, China
5 Steno Diabetes Center Copenhagen, DK-2820 Gentofte, Denmark
* Correspondence: rccchang@hku.hk (R.C.-C.C.); cristina.legido.quigley@regionh.dk (C.L.-Q.); Tel.: +852-39179127 (R.C.-C.C.); +45-30913083 (C.L.-Q.)
† These authors contributed equally to this work.

Received: 30 December 2018; Accepted: 9 February 2019; Published: 13 February 2019

Abstract: Introduction: Parkinson's disease (PD) is the second most common neurodegenerative disorder, without any widely available curative therapy. Metabolomics is a powerful tool which can be used to identify unexpected pathway-related disease progression and pathophysiological mechanisms. In this study, metabolomics in brain, plasma and liver was investigated in an experimental PD model, to discover small molecules that are associated with dopaminergic cell loss. Methods: Sprague Dawley (SD) rats were injected unilaterally with 6-hydroxydopamine (6-OHDA) or saline for the vehicle control group into the medial forebrain bundle (MFB) to induce loss of dopaminergic neurons in the substantia nigra pars compacta. Plasma, midbrain and liver samples were collected for metabolic profiling. Multivariate and univariate analyses revealed metabolites that were altered in the PD group. Results: In plasma, palmitic acid ($q = 3.72 \times 10^{-2}$, FC = 1.81) and stearic acid ($q = 3.84 \times 10^{-2}$, FC = 2.15), were found to be increased in the PD group. Palmitic acid ($q = 3.5 \times 10^{-2}$) and stearic acid ($q = 2.7 \times 10^{-2}$) correlated with test scores indicative of motor dysfunction. Monopalmitin ($q = 4.8 \times 10^{-2}$, FC = −11.7), monostearin ($q = 3.72 \times 10^{-2}$, FC = −15.1) and myo-inositol ($q = 3.81 \times 10^{-2}$, FC = −3.32), were reduced in the midbrain. The liver did not have altered levels of these molecules. Conclusion: Our results show that saturated free fatty acids, their monoglycerides and myo-inositol metabolism in the midbrain and enteric circulation are associated with 6-OHDA-induced PD pathology.

Keywords: Parkinson's disease; 6-OHDA; GC-MS; plasma; midbrain; fatty acid metabolism; myo-inositol

1. Introduction

Parkinson's disease (PD) affects approximately 1% of the population above the age of 50 years, worldwide [1]. Costs of treatment per capita in the U.K. alone can be up to £13,804 annually [2]. Aging populations are generally at greatest risk [3]. Up to 18 genetic loci have been demonstrated to contribute to familial cases of the disease [4]. It is estimated that the incidence of PD worldwide will double within the next decade [1]. All these factors combined are increasingly directing research and development not only towards novel therapies, but also technologies for better diagnosis and disease management. Diagnosis of PD is primarily based on clinical symptoms. However, the error rate is high [5]. Of these, the main features used for clinical diagnosis are heavily relied on symptoms such as bradykinesia, rigidity, tremors, postural instability and freezing [6].

Metabolomics is a well-defined approach used for biomarker discovery and investigating disease mechanisms. Burté et al. undertook metabolomic profiling of the serum samples of early stage PD patients and found an increase in metabolites of the fatty acid beta oxidation pathways [7]. In one longitudinal study, it was found that a combination of plasma and CSF xanthine and fatty acid metabolites showed significant changes between baseline and the study end point (two years from baseline) [8]. Markers that were highly correlated with a change in UPDRS scores, indicative of PD progression, included benzoate in the CSF and phenylcarnitine and aspartylphenylalanine in the plasma. Several studies on urine samples of PD patients showed an increase in amino acid metabolism, including phenylalanine [9], histidine, glycine and tryptophan/kynurenine [10]. A study by Ohman et al., using NMR based metabolomics on CSF, demonstrated the role of the amino acid alanine, energy metabolism (creatinine) and glucose metabolism (mannose) in distinguishing PD patients from controls [11]. Metabolic profiling of CSF has also been useful in differentiating newly diagnosed PD patients from controls. Trupp et al. revealed an increase in levels of the amino acids, alanine and methionine, and a reduction of saturated and unsaturated fatty acids in the CSF of newly diagnosed PD patients [12]. A differential role of glutathione metabolism has also shown a link to PD. While one study found increased glutathione in the plasma [5], another study revealed lower oxidized glutathione levels in the CSF [13], further suggesting the involvement of free radicals in PD.

A wide range of in vivo transgenic- and toxin-based models of PD are available. Studies on the mesencephalon of MPTP-induced PD in mice have revealed the role of altered energy [14], ceramide and sphingolipid metabolism [15]. A range of lipid species, including lysophosphatidylcholines and phosphatidylcholines are shown to be involved in 6-OHDA toxicity in rat midbrains [16]. An MPTP-treated goldfish model also indicated the role of phosphocholine metabolism, along with amino acids such as leucine, valine and glutamine [17]. Additionally, cardiolipins and mitochondria-associated phospholipids were detected in mesencephalon and plasma samples of rotenone-treated rats [18].

The aim of this study is to discover metabolic pathways in different tissues in a unilateral PD model. Plasma and liver profiling are often used as an indicator of metabolic changes, whereas in this case the midbrain is the site of primary pathological changes in the brain.

2. Results

2.1. Validation of the 6-OHDA Model

The apomorphine-induced rotation test was used to monitor the intensity of the ipsilateral lesion, while the cylinder test was used to assess motor dysfunction presented by asymmetry in the forepaw use of the rats. It was observed that the number of contralateral rotations in 30 min was significantly higher in rats that received 6-OHDA, with an average of 220 rotations (Figure 1A). Rats that received 6-OHDA injection showed an inability to use the contralateral limb while rearing, compared to their sham counterpart. This was evidenced by a 75% dependence on the ipsilateral limb while rearing (Figure 1B).

After behavioural testing, immunohistochemistry was done to confirm loss of dopaminergic neurons in the SNpc by counting TH immuno-positive neurons. It was observed that rats in the sham group had a similar TH count on both sides (Figure 1C), while the 6-OHDA group showed only 28% TH density on the ipsilateral side compared to the contralateral side (Figure 1C).

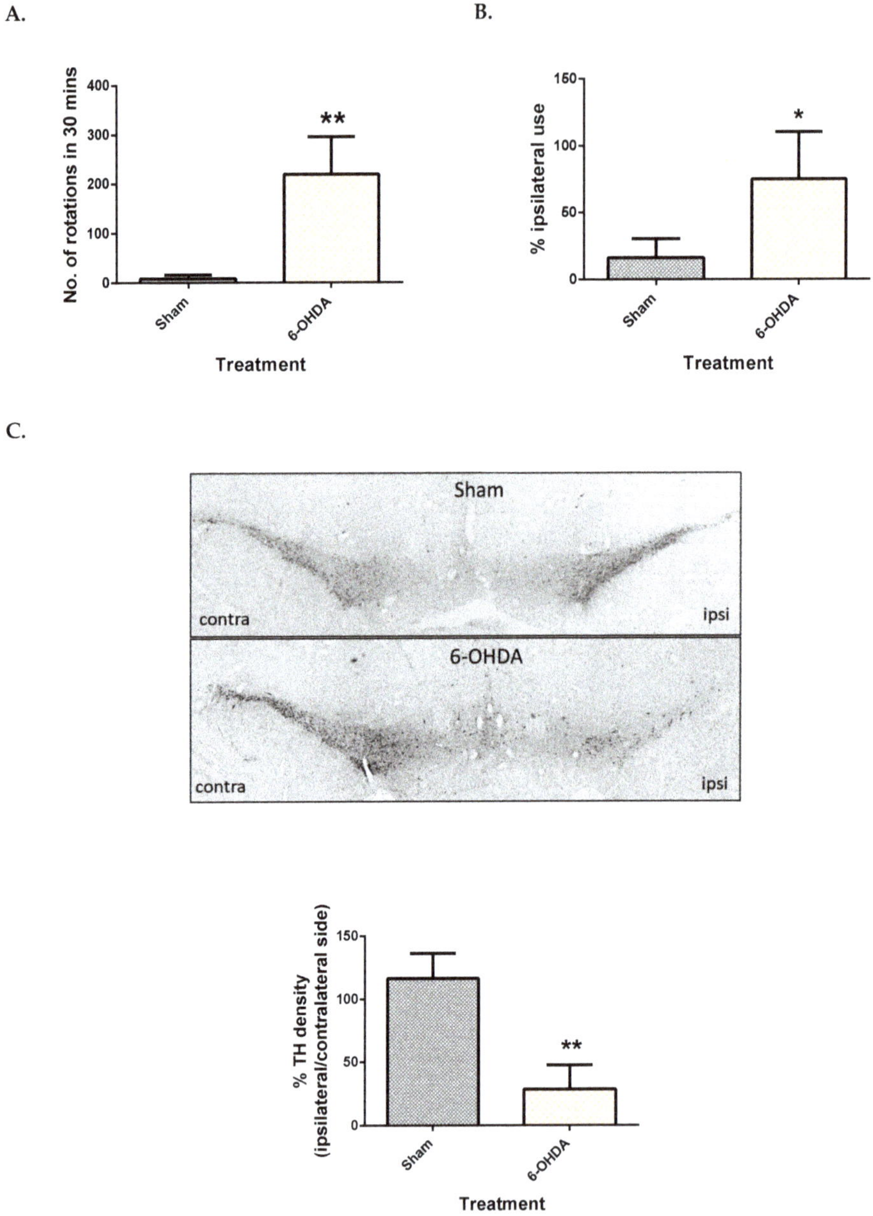

Figure 1. Comparison of motor function and dopaminergic cell loss between control and 6-OHDA groups. Apomorphine induced rotation test (**A**), Cylinder test (**B**) and TH density (**C**) between the sham and 6-OHDA group. Contra = contralateral and ipsi = ipsilateral. Data represent mean ± S.D of at least 5 rats in each group. (* indicates p value < 0.05 and ** $p < 0.01$, using Mann–Whitney test.).

2.2. *Metabolomic Method Validation and Feature Selection*

The metabolomics workflow is described in Supplementary Figure S2. To assess the reproducibility of our analysis, PCA plots were used. QC samples in both the plasma and the midbrain plots clustered together showing good repeatability (supplementary Figure S3). Up to 1500 metabolic features in the plasma and 2500 metabolic features in the mesencephalon regions were obtained.

OPLS-DA multivariate analysis showed a significant separation between the sham and 6-OHDA groups in the plasma and midbrain (supplementary Figure S4). Corresponding S-plots then revealed 16 metabolic features (4 from plasma tissues, and 12 from midbrain) altered between the 6-OHDA and sham groups. After comparing to NIST library, 13 were identified with similarity index >85%, two were identified as sugars and one remained unknown.

2.3. *Metabolite Levels in the Plasma, Brain and Liver*

Five features (two from the plasma, and three from the brain) showed significant difference between the groups after Benjamini–Hochberg correction. The two plasma metabolite features, which were identified as palmitic acid and stearic acid (similarity index >90%), were significantly upregulated in 6-OHDA group (Figure 2), compared to the sham ($q = 3.72 \times 10^{-2}$ for palmitate and $q = 3.84 \times 10^{-2}$ for stearate). Post-hoc power analysis yielded a statistical power of 93.2% for palmitic acid and 86.5% for stearic acid.

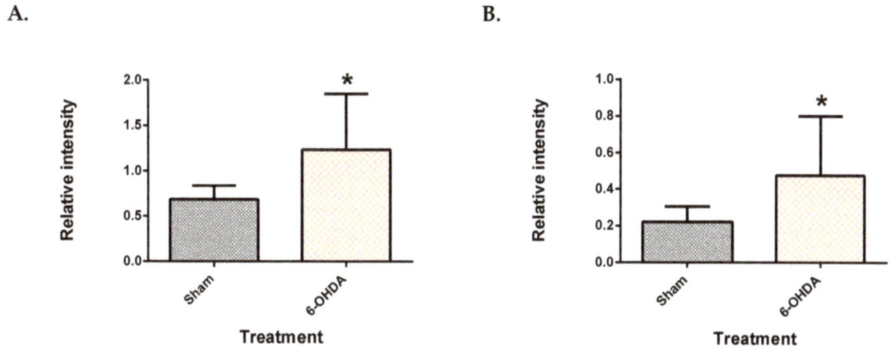

Figure 2. Saturated free fatty acids in the plasma. Palmitic acid (**A**) and stearic acid (**B**) were upregulated in the plasma of 6-OHDA-lesioned rats. Data represent mean ± S.D of at least 5 animals in each group. (* indicates q value < 0.05 using Mann–Whitney test, followed by Benjamini–Hochberg correction.).

From the mesencephalon, all three metabolite features presented lower levels in the 6-OHDA group compared to the sham (Figure 3). These were identified as monopalmitin ($q = 4.8 \times 10^{-2}$), monostearin ($q = 3.72 \times 10^{-2}$) and myo-inositol ($q = 3.81 \times 10^{-2}$). Monopalmitin and monostearin had a similarity index of more than 90%, while myo-inositol was 88%. The myo-inositol pure standard confirmed the identity of myo-inositol (supplementary Figure S5). Post-hoc power analysis revealed that the two monoglycerides showed a statistical power below 80% while myo-inositol had a statistical power of 97.4%.

The same univariate approach was applied to liver palmitic acid, stearic acid, monopalmitin and monostearin levels. The levels of these four metabolites remained unchanged in the liver, between the 6-OHDA and the sham groups (Supplementary Figure S6). A summary of all metabolite changes has been illustrated in Figures 4 and 5.

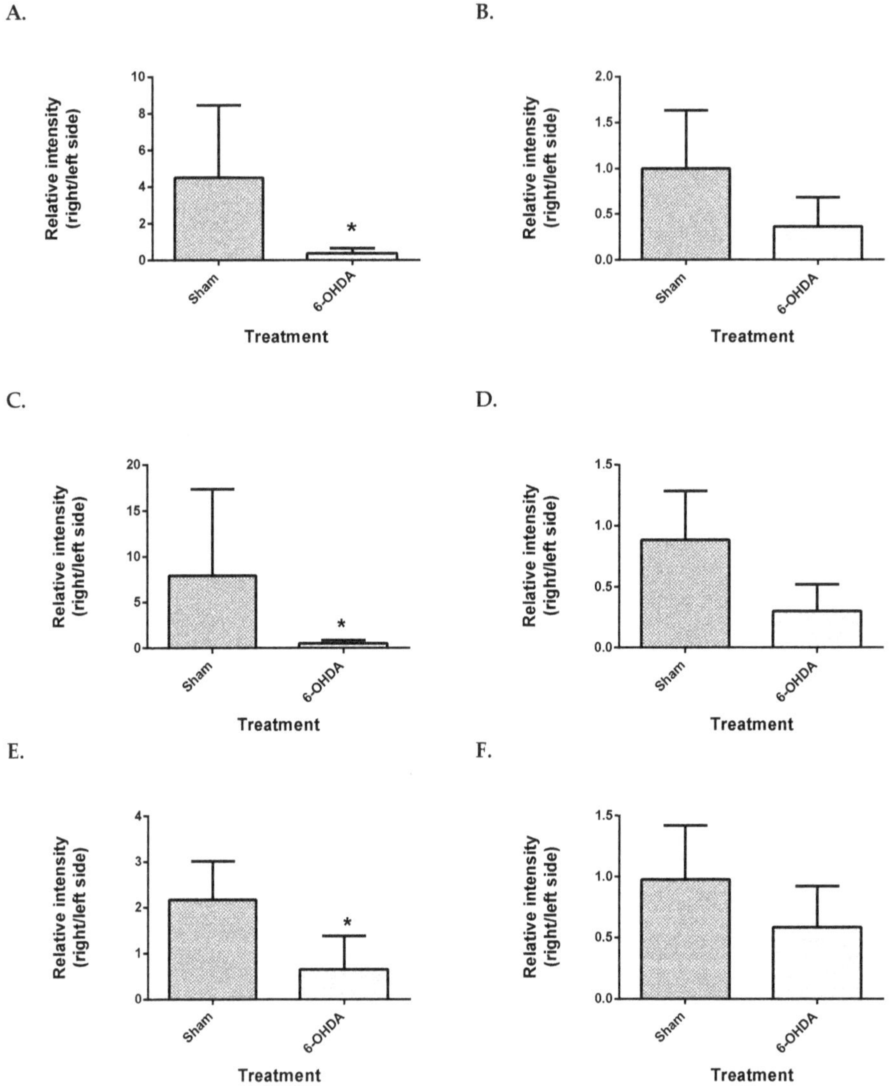

Figure 3. Comparison of brain metabolite changes between the mesencephalon and cerebellum. Midbrain monopalmitin (**A**), monostearin (**C**) myo-inositol (**E**) were significantly altered while cerebellar monopalmitin (**B**), monostearin (**D**), myo-inositol (**F**) and) were unchanged. Data represent mean ± S.D of at least 5 animals in each group. (* indicates *q* value < 0.05 using Mann–Whitney test, followed by Benjamini–Hochberg correction).

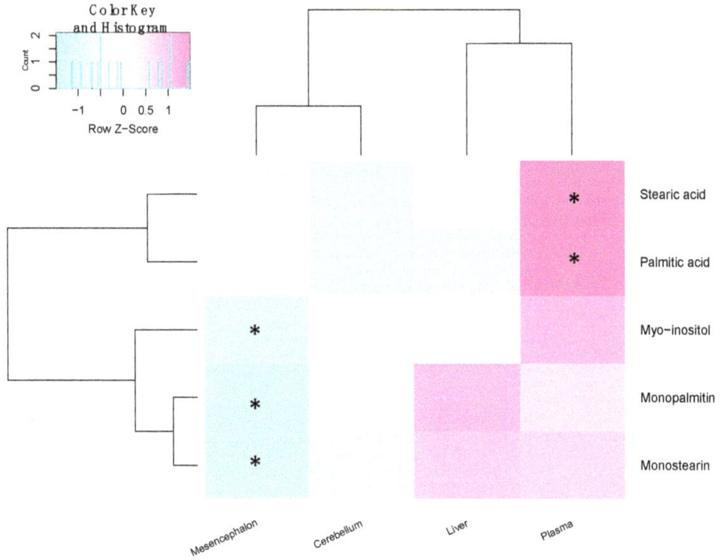

Figure 4. Heat map showing the fold change of metabolites between the sham and 6-OHDA groups within the different tissues. Similarly changed metabolites are clustered together, while tissues with similar changes in metabolites are near to each other. Significantly changed metabolites between the sham and 6-OHDA treated rats are marked for each tissue. (* indicates $p < 0.05$ after Benjamini–Hochberg correction, in that tissue.

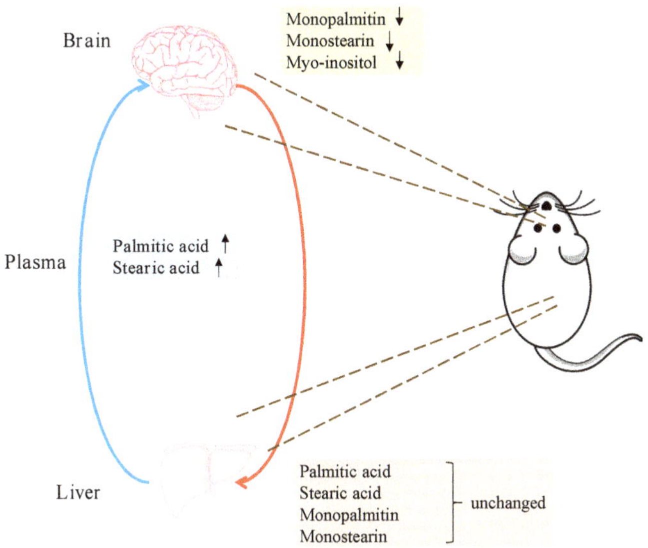

Figure 5. Summary of region-specific metabolite changes. Arrows indicate metabolites significantly increased or decreased in the midbrain and plasma of the 6-OHDA-treated animals. Levels of these metabolites remained unchanged in the liver.

2.4. *Correlation of Plasma and Midbrain Features with Motor Dysfunction*

Spearman's correlation was done to gauge the relationship between the levels of all five features and the motor impairment examined by the behaviour tests Table 1. The plasma metabolites were found to be highly correlated with motor dysfunction. Palmitic acid showed a strong positive correlation ($r = 0.674$, $q = 0.035$) with rigidity in the contralateral forelimb movement, as examined by the cylinder test. Stearic acid also had a high positive correlation with the cylinder test ($r = 0.649$, $p = 0.027$).

Table 1. Summary of correlation results of plasma and midbrain metabolites with the cylinder test.

Metabolite	Site	Spearman's Correlation Coefficient	*q*-Value
Palmitic acid	Plasma	0.674	**0.035**
Stearic acid	Plasma	0.649	**0.027**
Monopalmitin	Midbrain	−0.578	0.07
Monostearin	Midbrain	−0.439	0.205
Myo-inositol	Midbrain	−0.205	0.438

3. Discussion

The aim of this study was to elucidate metabolite changes in an in vivo model of PD. To confirm the effectiveness of our model, we performed behaviour tests and immunohistochemistry to examine the loss of dopaminergic neurons in the SNpc. Unilateral lesions of 6-OHDA successfully resulted in the manifestation of motor symptoms, as observed by the cylinder test, and the apomorphine-induced rotation test indicating the intensity of the lesions. Additionally, a significant loss of dopaminergic neurons was observed on the ipsilateral side, as measured by counting TH immuno-reactive positive cells. All these results proved the validity of a "hemi-parkinsonian" model.

Palmitic acid and stearic acid were significantly increased in the plasma, while an imbalance of their monoglyceride forms in the midbrain was also observed. Fatty acids can be transported through the blood–brain barrier via two major routes, either passive diffusion [19], or facilitated by transporters [20]. The transport of these two upregulated fatty acids in the context of PD needs to be further studied. One study has shown a decrease of palmitic and linoleic acid in the human plasma [12]. Moreover, the significant correlation between plasma palmitate and stearate and the cylinder test is of particular interest, suggesting an association with symptoms. This association of the fatty acids with motor symptoms is indicative of the severity of the 6-OHDA lesion and thus damage induced by it. Owing to a lack of biomarkers for PD, diagnosis currently relies heavily on symptoms. It is worthwhile to investigate the clinical potential of these fatty acids in PD, further. Moreover, longitudinal studies will aid in assessing whether palmitate and stearate levels change as neuronal damage progresses in this model. Finding a biomarker that correlates with worsening of symptoms is ideal for tracking disease progression.

Saturated free fatty acids are released into the blood by two major pathways. Lipolysis [21] is the breakdown of fats in adipose tissue to release triglycerides and free fatty acids into the blood, whereas de novo lipogenesis (DNL) occurs when saturated fatty acids are synthesized from glucose and its metabolites in the liver are subsequently released into the plasma to target tissues in need. There is also evidence to show that palmitic acid may be the major product of DNL [22]. However, our results show that there was no increase of palmitic or stearic acid in the liver tissue. This observation can be attributed to a swift clearance of the liver fatty acids by the plasma. In addition, oxidation of fatty acids, which takes place in the mitochondria, is an important pathway providing energy [23]. Mitochondrial dysfunction has also been implicated in PD [24]. The increased levels of stearic acid and palmitic acid in plasma could, therefore, be a consequence of impaired mitochondria in hepatic or extrahepatic tissues.

Saturated free fatty acids have known effects in the context of neuronal conditions. In one study, a diet rich in palmitic acid (30% palmitic acid of total fat) fed to mice resulted in reduced hippocampal neurogenesis [25]. Additionally, a diet supplemented with palmitate (2.2% *w*/*w*) induced endoplasmic reticulum (ER) stress in murine hippocampi and cortices. This study also assessed effects of increasing concentrations of palmitate, of up to 500 μM, on human neuroblastoma SH-SY5Y cells, which are commonly used for PD studies. It was found that 100 μM and higher concentrations of palmitic acid led to an upregulation of the ER stress-associated pro-apoptotic signalling machinery CHOP [26]. Additionally, 0.2 mM palmitic and stearic acid induced hyperphosphorylation of tau in rat primary cortical neurons, which was facilitated by astrocyte-induced oxidative stress [27]. Hyperphosphorylation of tau is a hallmark of Alzheimer's disease. An increased uptake of labelled palmitate into the brain was also observed in patients with metabolic syndrome compared to that of control subjects. Elevated levels of free fatty acids in the plasma have also been linked to metabolic syndrome [28–32]. This is particularly important given that PD affects an aging population, many of which suffer from metabolic syndrome as well.

On the other hand, monoglycerides of palmitic acid and stearic acid had a significant difference between the lesioned and non-lesioned sides of the mesencephalon, with no corresponding changes in the cerebellum. The cerebellum was used as a control region because this region remains unaffected by 6-OHDA lesions into the MFB. Monoglycerides are an intermediate product formed during the breakdown of triglycerides by lipolysis [33]. Whether the imbalance of these metabolic features can be stem from the imbalance of dopaminergic neurons must be further assessed. Additionally, the corresponding levels of these metabolites in the striatal terminals will give a more holistic idea about their association with dopaminergic loss; 1-monopalmitin and 1-monostearin have been shown to be altered in the CSF of patients with inflammatory demyelinating disease such as multiple sclerosis, a disorder affecting nerve fibres [34].

In this study, myo-inositol also showed an imbalance in the mesencephalon. Studies on the basal ganglia of patients with a *PINK1* mutation have reported an increase in myo-inositol by using MR spectroscopy [35]. Myo-inositol is purported to be a marker of glial cell death and neuroinflammation [36]. It is noteworthy that monoglycerides and myo-inositol are also downstream products of the IP_3-DAG signalling pathway [37]. This pathway plays a role in facilitating release of Ca^{2+}, which is important in cellular growth and synaptic plasticity [37].

Our results are in line with some of the findings from other studies. Lu et al. studied the changes in metabolites in the brains of goldfish treated with MPTP. ^{1}H NMR-based metabolomics revealed an increase of myo-inositol and linoleic acid in the PD brain [17]. In another independent study, a'paraquat-treated *Drosophila* model was used to elucidate changes of metabolites. It was shown that myo-inositol, 1-monopalmitin, 1-monostearin and the fatty acids palmitate and oleate were increased in the heads of the paraquat-treated flies [38]. In these studies, however, specific changes in the midbrain only were not determined. Furthermore, it must be noted that the findings from this study must be further validated for specificity to dopaminergic loss. This can be confirmed using a lesion that spares dopaminergic neurons but selectively damages surrounding neurons, such as an excitotoxic lesion of the striatum.

4. Materials and Methods

Twenty-six, four to six weeks old male Sprague-Dawley rats were purchased from the Laboratory Animal Unit at The University of Hong Kong. All experimental procedures were in accordance with the Committee on the Use of Live Animals in Teaching and Research of The University of Hong Kong (3491-14). The animals weighed 200 g at the beginning of treatment and were housed in pairs, in a temperature-controlled room with a 12-h dark/light cycle and free access to food and water.

All chemicals and reagents were obtained from Sigma Aldrich (United Kingdom), unless stated otherwise.

4.1. Stereotactic Injection of 6-OHDA

The rats were randomly divided into sham (n = 13) and 6-OHDA (n = 13) groups. Fresh stock solution (3 μg/μL) of 6-hydroxydopamine hydrobromide was prepared in saline (0.9% *w*/*v* NaCl) containing 0.2 mg/mL ascorbic acid. Rats were anaesthetized with 60 mg/kg pentobarbital (Alfansan International, Netherlands). 12 μg (in 4 μL) of 6-OHDA or vehicle was introduced into the right medial forebrain bundle (MFB) of the rat, using a Hamilton syringe connected to a 33G needle, at the rate of 1 μL/min. The coordinates of the injection site were: ML = −1.2, AP = −4 and DV = +7.5 (below dura), with the nose bar position at 4.5, based on the atlas by Paxinos and Watson. These coordinates were slightly modified from the study by Torres et al. [39]. Sham rats were injected with the same volume (4 μL) of vehicle. The needle was left in place for five minutes before retracting, and the incision was sutured. Body temperature and heart rate of the animals was measured constantly throughout the procedure.

4.2. Behavioural Assessment

At two weeks post-surgery, behaviour assessment for motor function was carried out in an isolated room, in the following order:

Cylinder test: The protocol used by Schallert et al. [40] was modified slightly. Rats were placed in a transparent acrylic cylinder for a total of three minutes and recorded. During every rear, the use of ipsilateral, contralateral or both forelimbs was counted, for a minimum of three and a maximum of ten rears or three minutes, whichever was first. The cylinder was cleaned with 70% ethanol between each use. Results were expressed as % trials with ipsilateral use only.

Apomorphine-induced rotation test: Rats were injected subcutaneously with 0.3 mg/kg of apomorphine hydrochloride dissolved in saline. Five minutes after injection, each rat was placed in a cylinder and recorded for 40 min. The number of contralateral rotations in 30 min was measured. Minimum four rotations per minute was considered as acceptable criteria for a successful model. Only the rats that were successfully lesioned based on this criterion were used for further studies.

4.3. Immunohistochemistry

After behavioural assessment, the mesencephalon of five rats from each group were harvested for immunohistochemical staining of tyrosine hydroxylase as follows:

Tissue processing and frozen-sectioning: Rats were overdosed with 150–200 mg/kg pentobarbital, followed by intra-cardiac transfusion of ice-cold saline and subsequently, freshly prepared ice-cold 4% paraformaldehyde (PFA) in 0.1 M phosphate-buffer. The substantia nigra *pars compacta* (SNpc) was then dissected out, post-fixed and soaked in increasing concentrations of sucrose. The tissue was then snap-frozen and stored in −80 °C until use. Thin slices of 15 μM were the sectioned using on a cryostat (Leica, Germany) and mounted. Every 6th section of the mesencephalon was collected, dried and stored at 4 °C until use.

DAB staining and imaging: Sectioned tissues were washed with 0.1 M PBS thrice, followed by blocking of endogenous peroxidase activity with 30% hydrogen peroxide in methanol for 30 min. Tissues were then incubated with the anti-tyrosine hydroxylase biotin-conjugated antibody (1:400, Cell Signaling Technologies, Danvers, MA, USA), in a humid slide chamber at 4 °C overnight. Following anti-biotin secondary antibody (1:400, Dako, USA) incubation, slides were stained using 3,3′-diaminobenzidine (DAB) solution from the ABC staining kit (Invitrogen, USA) according to the manufacturer's protocol. Brain sections were then counter stained with hematoxylin, dehydrated with ethanol and toluene and mounted. Slides were observed at a magnification of 5× using Brightfield microscopy (Zeiss Axioplasm, Germany), and stitched using the Image Composite Editor software (Microsoft, Albuquerque, NM, USA). Cell counting was then done using ImageJ (National Institute of Health, Bethesda, MD, USA). Cell counts were expressed as a % of the right side to the left side.

4.4. Tissue Harvest

After behavioural assessment, rats were asphyxiated using CO_2 and tissues were harvested for metabolomics analysis as follows:

Plasma extraction: the plasma was extracted form a total of 10 rats for each group. Briefly, one millilitre of blood was drawn by intra-cardiac transfusion, using an EDTA-buffer coated syringe. The needle was taken off and blood transferred to Eppendorf tubes and shaken. Samples were kept on ice, and then centrifuged at 4500 rpm in a 4 °C Eppendorf centrifuge for 15 min. The supernatants were collected, and samples were stored at −80 °C until use.

Microdissection of brain: The brain of eight rats from each group were harvested and briefly rinsed in ice cold 0.1 M phosphate-buffered saline (PBS) to remove any excess blood. It was then slit down the middle to divide the right and left sides and the cerebellum and entire mesencephalon tissues were separated according to our previous protocol [41], snap frozen in liquid N_2 and stored in −80 °C until use.

Liver: A part of the liver from eight rats in each group was cut and briefly rinsed in ice cold 0.1 M PBS to remove any excess blood. It was then snap frozen in liquid N_2 and stored in −80 °C until use.

4.5. Sample Extraction for Metabolomics

Plasma sample extraction: In vial dual extraction (IVDE) was slightly modified based on our previous protocol [42]. Briefly, 20 μL of LC-MS grade water was added to 40 μL plasma, followed by 80 μL of LC-MS grade methanol containing 10 μg/mL of succinic- d4- acid as internal standard (IS). After vortex, 400 μL of LC-MS grade methyl tertiary butyl ether (MTBE) with 10 μg/mL of tripentadecanoin as internal standard was added and then the samples were mixed thoroughly. Following a final addition of 100 μL LC-MS grade water, samples were centrifuged at 3000× *g* for 10 min at 4 °C to give a clear separation of MTBE (upper) and aqueous (lower) phases with protein aggregated at the bottom. The aqueous and MTBE layers were collected and stored until analysis at −20 °C and −80 °C, respectively.

Brain and liver sample extraction: IVDE was slightly modified based on our previous protocol [43]. Prior to homogenisation, 5 μL of methanol and 5 μL of IS (50 μg/mL succinic-d4 acid in 80% methanol) was added per milligram of tissue. The tissue was then homogenised using a Tissuelyzer (Qiagen, Germany) for ten cycles of 30 s at 25 Hz. Subsequently, 80 μL of homogenate was diluted with 120 μL of methanol. The subsequent extraction procedure was similar to plasma extraction, with addition of 40 μL of water, 1000 μL of MTBE containing tripentadecanoin (10 μg/mL) and thorough vortex. After addition of 160 μL of water, samples were then centrifuged at 3000× *g* for 10 min at 4 °C. The aqueous and MTBE layers were then separated and stored.

4.6. Derivatization of Tissues for GC-MS Analysis

Roughly 20 μL of plasma/brain/liver sample was dried under a stream of N_2. For plasma, 50 μL of O-methoxyamine-HCL (MOX) in pyridine (20 mg/mL) was added to the residue and maintained at 70 °C for 30 min. Samples were then dried again and reconstituted in a 1:1 (*v*/*v*) solution of acetonitrile and the derivatizing agent BSTFA (1% TMCS). The derivatization process was operated at 70 °C for an hour. For brain and liver tissues, the residues were reconstituted in a 1:1 (*v*/*v*) solution of acetonitrile and the derivatizing agent BSTFA (1% TMCS) directly and incubated at 37 °C for an hour. All the resulting derivatized samples were transferred to amber HPLC vials with inserts for GC-MS analysis.

4.7. GC-MS Analysis

GC-MS analysis was carried out on a Shimadzu GC-2010 Plus gas chromatograph equipped with a GCMS-QP2010 SE single quadruple mass spectrometer (Shimadzu, Japan). Sample (0.5 μL) was injected on a BP5MS (5% phenyl polysilphenylene-siloxane) capillary column (length 30 m, thickness 0.25 mm, diameter 0.25 mm) in the split mode with a split ratio of 1:60. The gradient temperature started from 60 °C and was held for 1 min, followed by a linear increase of 10 °C/min to 320 °C. Then, it was kept at 320 °C for 4 min. The carrier gas (helium) flow rate was set at 40 cm/s. Mass spectra analysis was performed using electron impact ionisation of 70 eV with an ion-source temperature of 200 °C, an interface temperature of 320 °C and an injection temperature of 280 °C. Data were collected between m/z 50–600 Da in a SCAN mode. Quality control (QC) samples made from pooled corresponding samples were injected periodically.

4.8. Data Processing and Metabolite Identification

The raw data generated was converted to mzXML using the GC-MS Postrun Analysis software (Shimadzu, Japan). The mzXML files were further processed for peak picking (signal to noise threshold = 5) and retention time correction using the XCMS package in R. The pre-processed metabolomics data were normalized separately in R which resulted in the best clustering of QC samples in principal component analysis (PCA) score plot. All semi-quantification was done using raw peak areas of selected features normalized to that of internal standard. Metabolite identification was done by comparing the GC-MS fragmentation mass spectra to those found in the National Institute of Standards and Technology (NIST) database. Identification of metabolites was confirmed by comparing the retention time and mass spectrum to pure standards if the similarity index was less than 90%.

4.9. Statistical Analysis

SIMCA (Umetrics, Sweden) was used for multivariate statistical analysis. Principle component analysis (PCA) was performed on the plasma and midbrain samples combined to assess reproducibility of the data. Orthogonal partial least squares-discriminant analysis (OPLS-DA) plots were then built for both the plasma and mesencephalon tissues after Pareto scaling and excluding features with VIP values <1. Corresponding S-plots were used for feature selection. Since the lesion for mesencephalon tissues was unilateral, the metabolite levels on the lesioned side were normalized to the intact side and these ratios were subsequently used for OPLS-DA plots.

Following feature selection, univariate analysis was performed on semi-quantified data using GraphPad Prism (GraphPad, USA). Mann-Whitney test followed by Benjamini and Hochberg correction (q value < 0.05) were used to find statistically significant features. Significant midbrain features were also measured in the cerebellum. Spearman's correlation test, followed by Benjamini and Hochberg correction was performed to assess correlation with the behaviour assessment. Post-hoc power analysis was performed in R with package "pwr". Heat map cluster analysis was performed in R with package "ggplot". All data is expressed at mean ± S.D.

5. Conclusions

In summary, this study used metabolomics to elucidate changes in 6-OHDA-induced parkinsonism. Two saturated free fatty acids, palmitic and stearic acid, were increased in the plasma of rats that underwent 6-OHDA injection. Monopalmitin, monostearin and myo-inositol showed an asymmetric distribution between the ipsilateral and contralateral mesencephalon. Changes of the midbrain metabolites may be associated with neuronal loss elicited by 6-OHDA while palmitic acid and stearic acid showed a high correlation with behaviour tests, indicating a possible association with disease severity.

Supplementary Materials: The following are available online at http://www.mdpi.com/2218-1989/9/2/31/s1.

Author Contributions: Conceptualization, A.S., R.C.-C.C. and C.L.-Q.; Data curation, A.S., P.H. and M.-Y.W.; Funding acquisition, R.C.-C.C.; Investigation, A.S. and P.H.; Methodology, A.S., P.H., R.C.-C.C. and C.L.-Q.; Project administration, R.C.-C.C. and C.L.-Q.; Resources, R.C.-C.C. and C.L.-Q.; Supervision, R.C.-C.C. and C.L.-Q.; Writing—original draft, A.S.; Writing—review & editing, A.S., P.H., R.C.-C.C. and C.L.-Q.

Funding: This study was partly funded by the Health Medical Research Fund (02131496) from the Food and Health Bureau of Hong Kong S.A.R Government to RCCC.

Acknowledgments: We thank Sarah Salvage and Atsuko Hikima for generously permitting use of their Zeiss microscope. We also thank Mathew Arno from the Genomics Centre at KCL for permitting use of the Tissuelyser. AS and MYW are supported by the Postgraduate Scholarship from The University of Hong Kong. PH is sponsored by the China Scholarship Council.

Conflicts of Interest: The authors declare that they have no conflict of interest.

References

1. Dorsey, E.R.; Constantinescu, R.; Thompson, J.P.; Biglan, K.M.; Holloway, R.G.; Kieburtz, K.; Marshall, F.J.; Ravina, B.M.; Schifitto, G.; Siderowf, A.; et al. Projected Number of People with Parkinson Disease in the Most Populous Nations, 2005 through 2030. *Neurology* **2007**, *68*, 384–386. [CrossRef] [PubMed]
2. McCrone, P.; Allcock, L.M.; Burn, D.J. Predicting the Cost of Parkinson's Disease. *Mov. Disord.* **2007**, *22*, 804–812. [CrossRef] [PubMed]
3. Kowal, S.L.; Dall, T.M.; Chakrabarti, R.; Storm, M.V.; Jain, A. The Current and Projected Economic Burden of Parkinson's Disease in the United States. *Mov. Disord.* **2013**, *28*, 311–318. [CrossRef] [PubMed]
4. Lill, C.M. Genetics of Parkinson's Disease. *Mol. Cell. Probes* **2016**, *30*, 386–396. [CrossRef] [PubMed]
5. Bogdanov, M.; Matson, W.R.; Wang, L.; Matson, T.; Saunders-Pullman, R.; Bressman, S.S.; Beal, M.F. Metabolomic Profiling to Develop Blood Biomarkers for Parkinson's Disease. *Brain* **2008**, *131*, 389–396. [CrossRef] [PubMed]
6. Jankovic, J. Parkinson's Disease: Clinical Features and Diagnosis. *J. Neurol. Neurosurg. Psychiatry* **2008**, *79*, 368–376. [CrossRef] [PubMed]
7. Burté, F.; Houghton, D.; Lowes, H.; Pyle, A.; Nesbitt, S.; Yarnall, A.; Yu-Wai-Man, P.; Burn, D.J.; Santibanez-Koref, M.; Hudson, G. Metabolic Profiling of Parkinson's Disease and Mild Cognitive Impairment. *Mov. Disord.* **2017**, *32*, 927–932. [CrossRef] [PubMed]
8. Lewitt, P.A.; Lu, M.; Auinger, P. Metabolomic biomarkers as strong correlates of Parkinson disease progression. *Neurology* **2017**, *88*, 862–869. [CrossRef] [PubMed]
9. Hatano, T.; Saiki, S.; Okuzumi, A.; Mohney, R.P.; Hattori, N. Identification of Novel Biomarkers for Parkinson's Disease by Metabolomic Technologies. *J. Neurol. Neurosurg. Psychiatry* **2016**, *87*, 295–301. [CrossRef] [PubMed]
10. Luan, H.; Liu, L.F.; Tang, Z.; Zhang, M.; Chua, K.K.; Song, J.X.; Mok, V.C.T.; Li, M.; Cai, Z. Comprehensive Urinary Metabolomic Profiling and Identification of Potential Noninvasive Marker for Idiopathic Parkinson s Disease. *Sci. Rep.* **2015**, 1–11.
11. Öhman, A.; Forsgren, L. NMR Metabonomics of Cerebrospinal Fluid Distinguishes between Parkinson's Disease and Controls. *Neurosci. Lett.* **2015**, *594*, 36–39. [CrossRef] [PubMed]
12. Trupp, M.; Jonsson, P.; Ohrfelt, A.; Zetterberg, H.; Obudulu, O.; Malm, L.; Wuolikainen, A.; Linder, J.; Moritz, T.; Blennow, K.; et al. Metabolite and Peptide Levels in Plasma and CSF Differentiating Healthy Controls from Patients with Newly Diagnosed Parkinson's Disease. *J. Parkinsons. Dis.* **2014**, *4*, 549–560. [PubMed]
13. Lewitt, P.A.; Li, J.; Lu, M.; Beach, T.G.; Adler, C.H.; Guo, L. 3-Hydroxykynurenine and Other Parkinson's Disease Biomarkers Discovered by Metabolomic Analysis. *Mov. Disord.* **2013**, *28*, 1653–1660. [CrossRef] [PubMed]
14. Poliquin, P.O.; Chen, J.; Cloutier, M.; Trudeau, L.É.; Jolicoeur, M. Metabolomics and In-Silico Analysis Reveal Critical Energy Deregulations in Animal Models of Parkinson's Disease. *PLoS ONE* **2013**, *8*, e69146. [CrossRef] [PubMed]
15. Li, X.Z.; Zhang, S.N.; Lu, F.; Liu, C.F.; Wang, Y.; Bai, Y.; Wang, N.; Liu, S.M. Cerebral Metabonomics Study on Parkinson's Disease Mice Treated with Extract of Acanthopanax Senticosus Harms. *Phytomedicine* **2013**, *20*, 1219–1229. [CrossRef] [PubMed]
16. Farmer, K.; Smith, C.A.; Hayley, S.; Smith, J. Major Alterations of Phosphatidylcholine and Lysophosphotidylcholine Lipids in the Substantia Nigra Using an Early Stage Model of Parkinson's Disease. *Int. J. Mol. Sci.* **2015**, *16*, 18865–18877. [CrossRef]
17. Lu, Z.; Wang, J.; Li, M.; Liu, Q.; Wei, D.; Yang, M.; Kong, L. 1H NMR-Based Metabolomics Study on a Goldfish Model of Parkinson's Disease Induced by 1-Methyl-4-Phenyl-1,2,3,6-Tetrahydropyridine (MPTP). *Chem. Biol. Interact.* **2014**, *223*, 18–26. [CrossRef]

18. Tyurina, Y.Y.; Polimova, A.M.; Maciel, E.; Tyurin, V.A.; Kapralova, V.I.; Winnica, D.E.; Vikulina, A.S.; Rosario, M.; Mccoy, J.; Sanders, L.H.; et al. LC/MS analysis of cardiolipins in substantia nigra and plasma of rotenone-treated rats: implication for mitochondrial dysfunction in Parkinson's disease. *Free Radic. Res.* **2015**, *49*, 681–691. [CrossRef]
19. Hamilton, J.A.; Kamp, F. How Are Free Fatty Acids Transported in Membranes? *Diabetes* **1999**, *48*, 2255–2269. [CrossRef]
20. Spector, R. Fatty Acid Transport through the Blood–Brain Barrier. *J. Neurochem.* **1988**, *50*, 639–643. [CrossRef]
21. Conner, W.E.; Lin, D.S.; Colvis, C. Differential Mobilization of Fatty Acids from Adipose Tissue. *J. Lipid Res.* **1996**, *37*, 290–298. [PubMed]
22. Hellerstein, M.K.; Christiansen, M.; Kaempfer, S.; Kletke, C.; Wu, K.; Reid, J.S.; Mulligan, K.; Hellerstein, N.S.; Shackleton, C.H.L. Measurement of De Novo Hepatic Lipogenesis in Humans Using Stable Isotopes. *J. Clin. Investig.* **1991**, *87*, 1841–1852. [CrossRef] [PubMed]
23. Houten, S.M.; Wanders, R.J.A. A General Introduction to the Biochemistry of Mitochondrial Fatty Acid β-Oxidation. *J. Inherit. Metab. Dis.* **2010**, *33*, 469–477. [CrossRef]
24. Gu, M.; Owen, A.D.; Toffa, S.E.K.; Cooper, J.M.; Dexter, D.T.; Jenner, P.; Marsden, C.D.; Schapira, A.H.V. Mitochondrial Function, GSH and Iron in Neurodegeneration and Lewy Body Diseases. *J. Neurol. Sci.* **1998**, *158*, 24–29. [CrossRef]
25. Park, H.-R.; Kim, J.-Y.; Park, K.-Y.; Lee, J.-W. Lipotoxicity of Palmitic Acid on Neural Progenitor Cells and Hippocampal Neurogenesis. *Toxicol. Res.* **2011**, *27*, 103–110. [CrossRef] [PubMed]
26. Marwarha, G.; Claycombe, K.; Schommer, J.; Collins, D.; Ghribi, O. Palmitate-Induced Endoplasmic Reticulum Stress and Subsequent C/EBP α Homologous Protein Activation Attenuates Leptin and Insulin-like Growth Factor 1 Expression in the Brain. *Cell. Signal.* **2016**, *28*, 1789–1805. [CrossRef]
27. Patil, S.; Chan, C. Palmitic and Stearic Fatty Acids Induce Alzheimer-like Hyperphosphorylation of Tau in Primary Rat Cortical Neurons. *Neurosci. Lett.* **2005**, *384*, 288–293. [CrossRef]
28. Wang, Y.; Qian, Y.; Fang, Q.; Zhong, P.; Li, W.; Wang, L.; Fu, W.; Zhang, Y.; Xu, Z.; Li, X.; et al. Saturated Palmitic Acid Induces Myocardial Inflammatory Injuries through Direct Binding to TLR4 Accessory Protein MD2. *Nat. Commun.* **2017**, *8*, 13997. [CrossRef]
29. Grimsgaard, S.; Jacobsen, B.K.; Bjerve, K.S. Plasma saturated and linoleic fatty acids are Independently Associated with Blood Pressure. *Hypertension* **1999**, *34*, 478–483. [CrossRef]
30. Nestel, P.; Clifton, P.; Noakes, M. Effects of Increasing Dietary Palmitoleic Acid Compared with Palmitic and Oleic Acids on Plasma Lipids of Hypercholesterolemic Men. *J. Lipid Res.* **1994**, *35*, 656–662.
31. King, I.B.; Song, X.; Ma, W.; Wu, J.H.Y.; Wang, Q.; Lemaitre, R.N.; Mukamal, K.J.; Djousse, L.; Biggs, M.L.; Delaney, J.A.; et al. Prospective Association of Fatty Acids in the de Novo Lipogenesis Pathway with Risk of Type 2 Diabetes: The Cardiovascular Health Study 1–5. *Am. J. Clin. Nutr.* **2015**, *101*, 153–163.
32. Mu, Y.M.; Yanase, T.; Nishi, Y.; Tanaka, A.; Saito, M.; Jin, C.H.; Mukasa, C.; Okabe, T.; Nomura, M.; Goto, K.; et al. Saturated FFAs, Palmitic Acid and Stearic Acid, Induce Apoptosis in Human Granulosa Cells. *Endocrinology* **2001**, *142*, 3590–3597. [CrossRef] [PubMed]
33. Frühbeck, G.; Méndez-Giménez, L.; Fernández-Formoso, J.A.; Fernández, S.; Rodríguez, A. Regulation of Adipocyte Lipolysis. *Nutr. Res. Rev.* **2014**, *27*, 63–93. [CrossRef] [PubMed]
34. Park, S.J.; Jeong, I.H.; Kong, B.S.; Lee, J.E.; Kim, K.H.; Lee, D.Y.; Kim, H.J. Disease Type- and Status-Specific Alteration of CSF Metabolome Coordinated with Clinical Parameters in Inflammatory Demyelinating Diseases of CNS. *PLoS ONE* **2016**, *11*, e0166277. [CrossRef] [PubMed]
35. Prestel, J.; Gempel, K.; Hauser, T.K.; Schweitzer, K.; Prokisch, H.; Ahting, U.; Freudenstein, D.; Bueltmann, E.; Naegele, T.; Berg, D.; et al. Clinical and Molecular Characterisation of a Parkinson Family with a Novel PINK1 Mutation. *J. Neurol.* **2008**, *255*, 643–648. [CrossRef] [PubMed]
36. Badar-Goffer, R.S.; Ben-Yoseph, O.; Bachelard, H.S.; Morris, P.G. Neuronal-Glial Metabolism under Depolarizing Conditions. A 13C-N.M.R. Study. *Biochem. J.* **1992**, *282 (Pt 1)*, 225–230. [CrossRef]
37. Berridge, M.J.; Taylor, C.W. Inositol Trisphosphate and Calcium Signaling. *Cold Spring Harb. Symp. Quant. Biol.* **1988**, *53*, 927–933. [CrossRef]
38. Shukla, A.K.; Ratnasekhar, C.; Pragya, P.; Chaouhan, H.S.; Patel, D.K.; Chowdhuri, D.K.; Mudiam, M.K.R. Metabolomic Analysis Provides Insights on Paraquat-Induced Parkinson-Like Symptoms in Drosophila Melanogaster. *Mol. Neurobiol.* **2016**, *53*, 254–269. [CrossRef]

39. Torres, E.M.; Lane, E.L.; Heuer, A.; Smith, G.A.; Murphy, E.; Dunnett, S.B. Increased Efficacy of the 6-Hydroxydopamine Lesion of the Median Forebrain Bundle in Small Rats, by Modification of the Stereotaxic Coordinates. *J. Neurosci. Methods* **2011**, *200*, 29–35. [CrossRef]
40. Schallert, T.; Fleming, S.M.; Leasure, J.L.; Tillerson, J.L.; Bland, S.T. CNS Plasticity and Assessment of Forelimb Sensorimotor Outcome in Unilateral Rat Models of Stroke, Cortical Ablation, Parkinsonism and Spinal Cord Injury. *Neuropharmacology* **2000**, *39*, 777–787. [CrossRef]
41. Chiu, K.; Lau, W.M.; Lau, H.T.; So, K.-F.; Chang, R.C.-C. Micro-Dissection of Rat Brain for RNA or Protein Extraction from Specific Brain Region. *J. Vis. Exp.* **2007**, 269. [CrossRef] [PubMed]
42. Whiley, L.; Godzien, J.; Ruperez, F.J.; Legido-Quigley, C.; Barbas, C. In-Vial Dual Extraction for Direct LC-MS Analysis of Plasma for Comprehensive and Highly Reproducible Metabolic Fingerprinting. *Anal. Chem.* **2012**, *84*, 5992–5999. [CrossRef] [PubMed]
43. Ebshiana, A.A.; Snowden, S.G.; Thambisetty, M.; Parsons, R.; Hye, A.; Legido-Quigley, C. Metabolomic Method: UPLC-q-ToF Polar and Non-Polar Metabolites in the Healthy Rat Cerebellum Using an in-Vial Dual Extraction. *PLoS ONE* **2015**, *10*, e0122883. [CrossRef] [PubMed]

metabolites

MDPI

Review

Imaging Mass Spectrometry: A New Tool to Assess Molecular Underpinnings of Neurodegeneration

Kevin Chen [1,2,3], Dodge Baluya [4], Mehmet Tosun [2,3], Feng Li [5] and Mirjana Maletic-Savatic [2,3,6,*]

1 Department of Biosciences, Rice University, Houston, TX 77030, USA
2 Department of Pediatrics, Baylor College of Medicine, Houston, TX 77030, USA
3 Jan and Dan Duncan Neurological Research Institute at Texas Children's Hospital, Houston, TX 77030, USA
4 Chemical Imaging Research Core at MD Anderson Cancer Center, University of Texas, Houston, TX 77030, USA
5 Center for Drug Discovery and Department of Molecular and Cellular Biology, Baylor College of Medicine, Houston, TX 77030, USA
6 Department of Neuroscience and Program in Developmental Biology, Baylor College of Medicine, Houston, TX 77030, USA
* Correspondence: maletics@bcm.edu; Tel.: +1-832-824-8807

Received: 10 May 2019; Accepted: 26 June 2019; Published: 10 July 2019

Abstract: Neurodegenerative diseases are prevalent and devastating. While extensive research has been done over the past decades, we are still far from comprehensively understanding what causes neurodegeneration and how we can prevent it or reverse it. Recently, systems biology approaches have led to a holistic examination of the interactions between genome, metabolome, and the environment, in order to shed new light on neurodegenerative pathogenesis. One of the new technologies that has emerged to facilitate such studies is imaging mass spectrometry (IMS). With its ability to map a wide range of small molecules with high spatial resolution, coupled with the ability to quantify them at once, without the need for a priori labeling, IMS has taken center stage in current research efforts in elucidating the role of the metabolome in driving neurodegeneration. IMS has already proven to be effective in investigating the lipidome and the proteome of various neurodegenerative diseases, such as Alzheimer's, Parkinson's, Huntington's, multiple sclerosis, and amyotrophic lateral sclerosis. Here, we review the IMS platform for capturing biological snapshots of the metabolic state to shed more light on the molecular mechanisms of the diseased brain.

Keywords: neurodegeneration; metabolomics; biomarkers; imaging mass spectrometry

---

1. Introduction

Imaging mass spectrometry (IMS) has emerged as a powerful molecular imaging technology, enabling us to map, with high molecular specificity and sensitivity, the spatial distribution of small molecules in tissues. As such, has allowed us to accelerate scientific discoveries on the role small molecules and metabolites play in health and disease. These small molecules, such as lipids, sugars, neurotransmitters, amino acids, and xenobiotics, are not readily detected by traditional methods of molecular imaging, such as microscopy or in situ hybridization. IMS allows us to visualize, map, and analyze hundreds of these molecules at the same time, in a single, label-free sample [1–5]. Thus, it differs from conventional mass spectrometry, which has revolutionized the worlds of omics sciences and drug metabolism, disposition, and development, but has been mostly used for studies of analytes that have been extracted from fluids or tissues [6], while lacking the spatial information of the metabolite distribution. IMS also allows correlation between the abundance and localization of specific compounds in tissue samples with the histological images obtained from the same or adjacent tissue sections [7]. Such knowledge, in turn, is important in elucidating the function of small molecules within complex biochemical pathways and their roles in health and disease [8].

Importantly, in contrast to the simplification of complex biochemical pathways in reductionist viewpoints, IMS allows a systems approach that can be very valuable to integrating the spatial component of small molecules into our understanding of their role in cellular and intercellular connections [6,9]. Furthermore, hundreds of metabolites can be analyzed and mapped at one time, without the need for labels, staining, and radioactive trackers, to distinguish the different metabolites of interest [1]. Thus, IMS brings new dimensions to molecular imaging and places it at the forefront of many applications in metabolism and small molecule/drug discovery research [7,10–13]. This review focuses on the existing IMS technologies and presents some of their potential applications in neurodegenerative diseases.

2. Imaging Mass Spectrometry: Advantages and Disadvantages

For a longwhile, the brain has been considered a homogeneous structure in terms of metabolite distribution. After all, all cells depend on energy metabolism to survive. Without accounting for the regional specificity of metabolites [14], their identification and quantification in a given sample has failed to detect spatially distinct metabolic alterations that may have been important for understanding the disease pathology. The recognition of these shortcomings has increasingly led to harnessing the imaging and profiling capabilities of IMS in the field of neuroscience, in order to better understand and profile the metabolic changes in neurodegenerative diseases, as well as to find potential biomarkers for their diagnosis and monitoring [15].

Interestingly, over the past few years, the lipidome has gained much traction in research on neurodegenerative disorders, due to the importance of lipid signaling in various cellular pathways [16]. Irregularities of the lipidome, such as erroneous lipid metabolism and signaling, have not only been tied to diseases that involve large-scale metabolic dysfunction (e.g., diabetes, hypertension, atherosclerosis, diabetes), but also neurodegenerative diseases such as Alzheimer's, Parkinson's, and Huntington's diseases, amyotrophic lateral sclerosis, and multiple sclerosis [17]. The nervous system is home to the most heterogeneous and distinct lipid classes in the entire body [18]. While lipidomics of the brain tissue has contributed much to our knowledge of the lipid content of the brain, IMS has brought forward the possible relevance of certain classes of lipids with certain brain functions. For example, the selective localization by IMS of C20 gangliosides in the molecular layer of the dentate gyrus, a region of the brain central to learning and memory [19], has pointed to a possible role of these lipids as key constituents of neurons, contributing to learning and memory. The relevance of this discovery is still to be elucidated; regardless, new investigations of specific roles of different classes of lipids are now possible with IMS technology.

2.1. *Advantages*

The IMS has gained its prominence because of its ability to detect, relatively quantify, and map small molecules (<2 kDa) and the metabolites within a sample in situ, maintaining high spatial resolution and molecular specificity without the need for chemical labels, staining procedures, and molecular probes. IMS is one of the technologies producing big data and serves as an addition to the histologist's toolbox: a complement to it, not a replacement for it. By integrating microscopy with IMS, the applications become almost limitless, and could be used to answer a variety of biologically and medically relevant questions. The spatial resolution of IMS-generated images approaches the cellular level, which is advantageous for studies of tissues composed of a variety of cell types, or of genetic chimeras, as the metabolic signals of neighboring cells can be different. Thus, a more comprehensive and holistic understanding of the genotypic, phenotypic, and metabolic responses of the tissue to disease pathology or to changes in the environment can be obtained with IMS in efforts to study complex disease biology.

In addition, the distribution of xenobiotics, including drug species, can also be studied with IMS. The effects of exogenous species on metabolism and endogenous metabolites can accelerate finding the biomarkers and molecular links of a given disease [20]. Pharmaceutical research has typically used liquid chromatography in tandem with mass spectrometry (LC-MS) to conduct pharmacokinetic studies in preclinical trials on animal models, but the rise of IMS platforms promises to shorten the preclinical research flow. Traditional LC-MS fails to provide any information on spatial localization of a given drug, due to the excision and homogenization of tissue samples. In contrast, IMS is the most thorough and unbiased way to map the penetration of compounds of interest into tissues and their distribution throughout the body. This is most valuable for establishing pharmacokinetic properties, including accumulation in non-target tissues and excretion routes. Furthermore, IMS can also detect the bio-transformed metabolites of a drug for the unbiased determination of true, biologically active drug compounds and their toxic effects. Overall, IMS is a powerful, yet cost-effective technology that will enable distribution studies to be performed earlier in the drug discovery process, without any requirement for radiolabeled standards [20,21].

2.2. Disadvantages

Although a very valuable tool for metabolite detection, IMS has some caveats, many of which stem from the process of sample preparation [20]. While simultaneous detection of carbohydrates, proteins, lipids, nucleotides, matrix ions, and salt is one of the crowning achievements of IMS, this process also leaves this platform open to ion suppression, which occurs when the chemically-distinct natures of each metabolite impair the overall detection [22]. Namely, each metabolite impairs the desorption and ionization efficiency of every other molecule; thus, the aggregate effect results in more abundant metabolites being selectively ionized over less abundant ones, leading to the depletion of signals from low-abundance metabolites that may be of interest. One possible approach to overcome this pitfall is to remove a particular metabolite class; however, this can affect sample integrity, causing molecular diffusion and adduct formation, which ultimately leads to increased complexity of data [20,22]. Another disadvantage of IMS is sample degradation, which can occur as the tissue is thawed during sample preparation. Certain analytes are not stable at room temperature and may completely degrade, leading to misinterpretation of the data [22]. For matrix-assisted laser desorption ionization (MALDI) IMS, matrix application must be uniform across the entire sample to minimize image artifacts, and this may be difficult to achieve [20,22]. Furthermore, until sample preparation is standardized, with standard procedures for each class of metabolites, IMS data may not be fullyreproducible, due to batch effects. Alternative ionization methods have already proven effective in circumventing some of these problems. For example, secondary ion MS minimizes problems with matrix inconsistencies and analyte diffusion by not requiring matrix application at all [20,22]. Finally, with respect to data processing, IMS requires substantial computing capabilities, as the acquired data can reach upwards of several gigabytes, and corresponding processing times of several hours [20]. In sum, as sample preparation becomes increasingly standardized and ionization and desorption techniques continue to improve, IMS will continue to see a growth in popularity.

3. Ionization Methods

Many steps need to be conducted correctly to ensure optimal IMS results: sample preparation, sample desorption and ionization, mass spectrum analysis, and image production [20]. Depending on the biological class of the compound of interest, the ionization technique can be modified to best suit data acquisition [20,22]. The three ionization techniques most central to the existing IMS platform are MALDI, desorption electrospray ionization (DESI), and secondary ion mass spectrometry (SIMS) [22] (Table 1).

Either positive or negative ionization modes can be employed to trigger and detect particular ion formations. In positive ion mode, the molecules gain protons to become cations, whereas in negative ion mode, the molecules are deprotonated to form anions [20]. The three ionization techniques influence the lateral resolution in IMS, defined as the minimum pixel size needed to produce a detectable signal. For MALDI, lateral resolution is affected by the laser spot size, while for DESI, by the spray area. Lateral resolution is limited by sensitivity of the IMS, and needs to be taken into account when assessing the data. In general, it can be enhanced by oversampling, but, more preferably, it should be measured against a reference material [23,24].

Table 1. Three main ionization techniques for mass spectrometry: matrix-assisted laser desorption/ionization (MALDI), desorption electrospray ionization (DESI), and secondary ion mass spectrometry (SIMS).

Platform	Mechanism	Advantages	Disadvantages
Matrix-assisted laser desorption/ionization (MALDI)	UVlaser used for the desorption and ionization of analytes after application of an UV-absorbing matrix into a gaseous state	• High sensitivity • Tolerates sample contaminants and impurities • Detects a large range of *m*/*z* values • Simple post-ionization analysis due to generating only singly-protonated or singly deprotonated ions (depending on whether positive or negative mode is used) • Low fragmentation rate allows analysis of metabolites of large relative molecular weight	• Matrix application must be uniform to minimize artifacts during post-ionization analysis and imaging • Matrix application needs to be tailored for each tissue type to ensure optimal coverage and thickness • Overcoating of matrix reduces detected signal. Internal standard is needed for calibration purposes
Desorption electrospray ionization (DESI)	Electrically-charged solvent drops sprayed onto sample surface to eject analyte molecules into a gaseous state	• Extremely high spatial resolution (resolving power capable of reaching the micron scale) • Focusing capabilities of primary ion beams are superior to those of lasers	• Lower spatial resolution compared to other methods of ionization
Secondary ion mass spectrometry (SIMS)	High-energy primary ion beam (i.e., gallium and indium ions) facilitates the desorption and ionization of analytes in the form of secondary ions in the gaseous state	• Extremely high spatial resolution (resolving power capable of reaching the micron scale) • Focusing capabilities of primary ion beams are superior to those of lasers	• Significant damage done to sample upon primary ion beam impact • Ineffective at detecting certain types of metabolites • High rates of molecular fragmentation complicates post-ionization imaging and analysis • Inability to analyze metabolites of higher relative molecular weight (e.g., >1000 *m*/*z*)

3.1. MALDI

MALDI is the most prevalent ionization technique, due to its versatility in detecting metabolites ranging from the realm of hundreds of Da to 100 kDa and above [25]. In MALDI, the tissue samples are completely coated with a laser-absorbing matrix, typically an organic compound of low molecular weight, which allows the extraction of the metabolites from the sample, also known as the analytes, upon matrix crystallization after solvent evaporation (Table 2). The sample is then hit with a sufficiently energized laser to desorb and ionize the aggregate of sample and matrix molecules [25]. In the crystallized form, analyte degradation is reduced and the analyte molecules of larger molecular weight are better able to resist fragmentation, due to the energy-absorbing properties of the matrix molecules [22]. After the metabolites are desorbed and ionized into a gaseous form, the analyte ions are guided by ion lenses to be detected by a mass analyzer, generating a spectrum of mass to charge (*m*/*z*) ratios for all detected analytes [25] (Figure 1). MALDI is one of the most popular ionization techniques for IMS, because of the combination of its high sensitivity, capability of detecting a large range of analytes, tolerance for sample contaminants, and simplification of analysis due to the production of only singly-protonated or singly-deprotonated ions in positive and negative mode, respectively, and protection from fragmentation [21,22,25]. In addition to a traditional vacuum MALDI, atmospheric pressure (AP) MALDI can be used to analyze volatile compounds, as samples are not exposed to a high vacuum environment prior to analysis [26]. AP-MALDI has also been used more frequently over the past few years, particularly for imaging lipids [27], sugars, and peptides [28]. As it has a more focused beam, it produces images with higher resolution compared to vacuum MALDI, which is very important for single cell IMS [29].

Table 2. The different types of matrices used in MALDI and their application.

Matrix	Application	References
2-amino-5-nitropyridine (ANP)	Oligonucleotides < 20 Bases, MALDI (−)	[30,31]
80% anthranilic acid + 20% nicotinic acid (80/20 AA/NA)	Oligonucleotides < 20 Bases, MALDI (+ & −)	[32,33]
6-aza-2-thiothymine (6-ATT)	Oligonucleotides and carbohydrates MALDI (−)	[34,35]
3-hydroxypicolinic acid (3-HPA)	Oligonucleotides 1 kDa–30 kDa	[36,37]
α cyano 4 hydroxycinnamic acid (CHCA)	small molecules, peptides/proteins < 6 kDa	[38–40]
2,5 dihydroxybenzoic acid (2,5 DHB)	small molecules, peptides/proteins < 6 kDa, polymers, carbohydrates	[39,41–43]
9-aminoacridine (9-AA)	small molecules, lipids, MALDI (−)	[44]

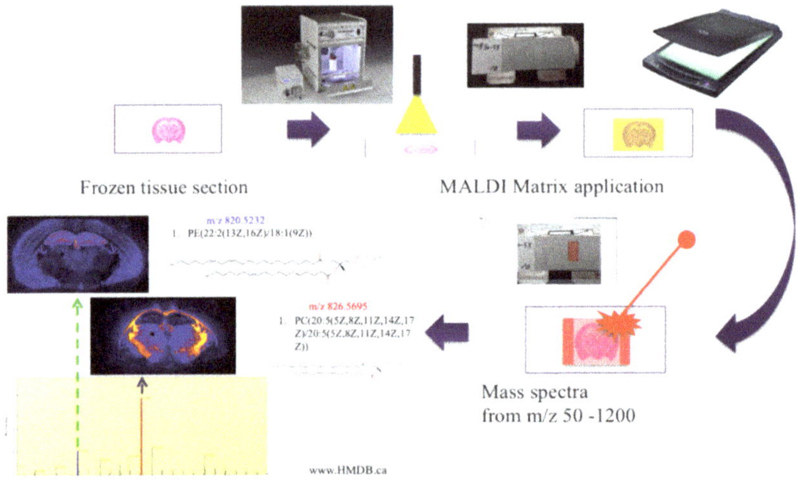

Figure 1. Ion mass spectrometry (IMS) experimental workflow with MALDI. The frozen tissue section must be coated with a matrix and ionized by the laser before metabolite localization within the section can be mapped.

3.2. DESI

With desorption electrospray ionization (DESI), ionization of the sample is possible under atmospheric pressure, unlike the vacuum-induced high pressure environments in which MALDI and SIMS must be carried out [22]. DESI swaps out the laser- and energy-absorbing matrix used in MALDI for electrically charged solvent spray and solvent ion droplets, which are then directed onto the sample surface [45]. The impact of these projectile-like droplet particles provides the energy to eject the analyte molecules into a gaseous state via electrostatic and gaseous forces [46]. After desorption, the singly- or multiply-charged analyte ions are collected through the atmospheric inlet line of a mass spectrometer for subsequent detection [22,46]. By avoiding the rough, high-pressure environment prevalent in most ionization techniques, DESI offers the benefit of conducting sample ionization in a relatively gentle environment [47]. This leads to a higher chance of detecting intact molecular ions rather than fragment ions, resulting in a less complex mass spectrum. Secondly, by not requiring sample preparation procedures such as matrix application, analyses involving DESI ionization avoid problems associatedwith non-homogenous matrix coating or the selection of the optimal matrix and solvent to analyte compatibility [48]. Ionization via DESI can also be preferred for certain classes of molecules; for example, lipids and proteins are generally ionized more effectively by DESI and MALDI, respectively [45,49,50]. The major drawback of DESI, however, is low spatial resolution [50], which limits its widespread use.

3.3. SIMS

Secondary ion mass spectrometry (SIMS) employs the use of a high energy primary ion beam to cause the desorption and ionization of analytes from the sample tissue in the form of secondary ions [22]. One of the most common primary ion sources utilizes high-energy gallium and indium ions [51]. The surplus of energy from these primary ion beams causes high rates of fragmentation to the analytes, producing secondary ions that are accelerated through a mass analyzer until they hit the detector [22]. The main advantage of SIMS is the extremely high spatial resolution, as it can reach a resolving power on the scale of the micron [50]. However, damage to the sample upon impact of the primary ion beam is unavoidable, and, after a certain amount of ionization damage, there is no longer any detectable signal from the emission of the secondary ion [51]. Furthermore, due to its dependence on secondary ion emission, SIMS is ineffective at detecting certain metabolites, specifically hydrophilic metabolites that are present only in low concentrations [52]. Extensive molecular fragmentation often complicates post-ionization analysis, and analytes of high molecular weight are eradicated [22]. However, the development of new primary ion beams with relatively lower energies has been successful in preserving the intactness of higher molecular weight analytes, thereby improving the efficacy of SIMS for a broader range of metabolites of interest [53].

4. IMS Analysis

After the ionization and subsequent detection of the various analytes, an *m*/*z* spectrum corresponding to the entire range of detected metabolites is generated. Software such as *HDImaging* (Waters Corp., Milford, MA, USA) can be used to view and sort these metabolites by selecting *m*/*z* values to visualize the spatial distribution of that particular metabolite, overlaid on the anatomical image of the sample slice. Normalization by total ion current (TIC) is generally done to account for the possible variance in the sample. If searching for a particular metabolite, online databases (such as http://www.hmdb.ca/) can be used to get accurate *m*/*z* value for the metabolite of interest, and then to search for that value in the IMS spectrum, to obtain its spatial distribution. Note that the metabolite of interest is not always present, but this does not necessarily mean that it is not present in the sample. Namely, the concentration might be too low, or a non-ideal sample preparation resulted in destruction of the metabolite. In addition to visualization, quantification of metabolite levels can be performed by selecting the region of interest and extracting the raw data, which can then be further processed with available software, such as *Progenesis QI* (Waters Corp, Milford, MA, USA). In this way, metabolite distributions can be compared between regions of interest and non-interest to illuminate the contribution of given metabolites to the physiological differences of given regions.

5. Neurodegenerative Disorders

Over the past few decades, neurodegenerative diseases have taken the front stage as one of the largest public health concerns. No quantitative biomarkers exist to enable their early diagnosis and commencement of therapy. Furthermore, therapy is non-specific and neurodegeneration progresses, eventually leading to death. The 4.7 million Americans estimated to have Alzheimer's disease (AD) in 2010 has been projected to increase to 13.8 million by 2050 [54] (Figure 2). Worldwide, about 35.6 million people are estimated to be living with dementia, anticipated to increase to much higher rate as low income countries become more developed [55]. In parallel, Parkinson's disease (PD), the second most common neurodegenerative disease, has shown similar increases in prevalence. The 680,000 Americans aged 45 and older who suffered from PD in 2010 is expected to increase to more than 1 million Americans by 2030 [56]. Thus, it is imperative to continue research toward preventing the rising tide of neurodegenerative diseases. Metabolic dysfunction has been reported in most of these diseases as we describe in more detail below, focusing on recent data from IMS studies.

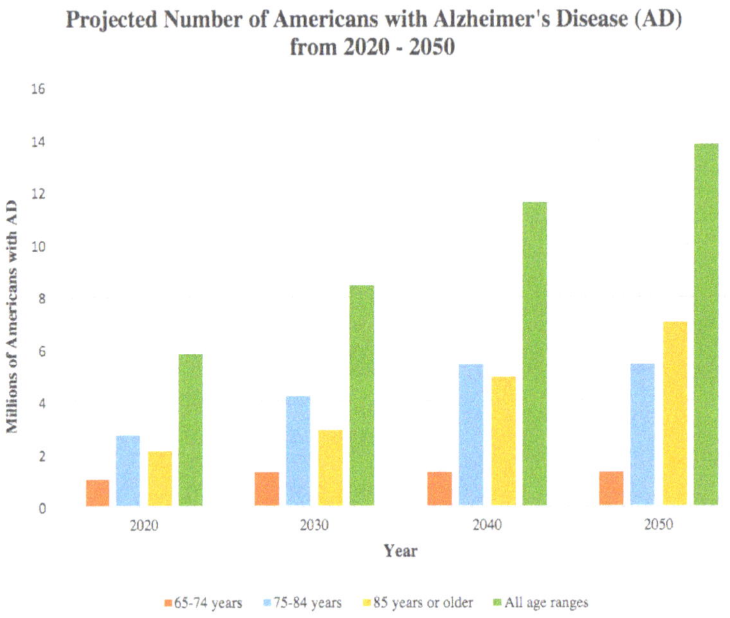

Figure 2. Projected number of Americans with Alzheimer's disease (AD) from 2020–2050. Source: Created from data in Herbert LE, Weuve J, Scherr PA, Evans DA. Alzheimer disease in the United States (2010–2050) estimated using the 2010 Census, *Neurology* 2013; 80(19):1778–1783 [54].

5.1. Alzheimer's Disease (AD)

AD is a progressive neurodegenerative disease that causes gradual loss of memory and other cognitive abilities, as well as emotional and behavioral deficits [17,57]. In addition to the progressive nature of AD, advancing at different rate for each affected individual, its multitude of symptoms can change over time as different brain regions undergo pathology [58]. Common symptoms of AD include loss of explicit memory abilities, difficulties with problem-solving and planning, complications with temporal processing, struggling with writing, alterations in mood and personality, and social withdrawal as difficulties arise with maintaining speech flow during conversations. In the most severe stages of AD, affected individuals become non-ambulatory, and it is their bed-ridden nature that makes them especially susceptible to blood-clots and infectious disease. Eventually, those with AD will succumb to organ failure or aspiration pneumonia, due to chronic infection or accidental ingestion of food into the lungs, respectively.

The complex pathogenesis of AD involves aberrant processing of amyloid-β proteins and the hyperphosphorylation of tau proteins, which aggregate to form amyloid-β plaques, amyloid angiopathy, and tau protein neurofibrillary tangles [57,59]. These aggregations lead to neural degeneration in the hippocampus and entorhinal cortex, the centers of learning and memory in the brain [59,60]. Additionally, the loss of neurons and synaptic activity, oxidative stress, and changes in the activity of reactive glial cells contribute to the gradual decline of cognitive functioning [17,57,59]. Besides protein aggregation, irregular lipid metabolism in the brain has also been implicated in the pathogenesis of AD [61,62]. Lipids are involved in cell signaling pathways and function as the building blocks of cellular membranes; thus, investigations of differential lipid activity in the AD has become a topic of great interest. Furthermore, there are connections to be made between faulty protein and lipid metabolism, as amyloid-β proteins prompt neuronal damage by moderating phospholipase activity [62]. Further, abnormal lipid accumulation may also play a critical role in neurodegeneration [16]. Recent evidence has also implicated impaired adult neurogenesis in the development of the disease, as mouse models of AD established decreased rates of adult neurogenesis [60,63–65], recently confirmed in humans as well [66]. Consequently, it is hypothesized that AD pathogenesis may be related to the effects of aggregates on both mature neuronal death and neural progenitor cell (NPC) activity within the neurogenic niche of the dentate gyrus [67]. Indeed, several computational models have been used to predict the effects of apoptosis on neurogenic potential in both the young and aged brain [65,66,68]. With the recent seminal study on postmortem brains from AD patients [66], it is clear that more research should be done to examine the role of neurogenesis as a possible therapeutic target for AD.

The first studies exploring metabolic alterations in AD used conventional mass spectrometry (MS)-based metabolomics [17]. For example, gas chromatography–mass spectrometry (GC-MS) and ultra-high performance liquid chromatography–mass spectrometry (HLPC-MS) were used to examine transgenic AD mouse models. These studies found that the AD brain exhibited significant metabolomic differences compared to the wild-type mouse control brain [69]. Although the differences existed primarily in the hippocampus and cortex, other regions not traditionally associated with the disease, such as the striatum, cerebellum, and olfactory bulb, were affected as well. These results pointed to AD pathology stemming from myriad dysregulations across several different metabolic pathways [69]. The contributing metabolites to the abnormal mice neurochemical profile were identified as phospholipids, fatty acids, purine and pyrimidine metabolites, sterols, and others [70]. Similar results were obtained when conducting MS analysis of the human AD brain [71]. Through the investigation of seven neural regions, partitioned into the categories of most damaged, moderately damaged, and lightly damaged by AD pathology, the levels of 55 total metabolites were confirmed to be altered in at least one of those regions [71]. The wide range of regions in which these metabolic abnormalities were seen supports the theory of whole-brain degeneration in AD [71], which could also be contributed to by the regional differences in metabolism, as reported in the mouse brain [14].

To correlate the spatial distribution of various biomolecules with the amyloid aggregates within the AD brain, MALDI-IMS found sphingolipid, phospholipid, and lysophospholipid changes associated with individual plaques within tissue samples [72]. These correlative data indicate a possible loop between amyloid aggregates and changes in metabolites that reflect inflammation, oxidative stress, demyelination, and cell death [72] (Table 3). In addition, specific investigations into lipidomic dysfunctions within the human AD brains incorporated the use of MALDI-IMS in both positive and negative ion modes, depending on the relative ease with which lipids were protonated or deprotonated, to identify alterations in different classes of lipids within the hippocampus [73]. In both the control and AD brains, positively and negatively charged ions were readily detected. Although the distribution of the lipid species was consistent across all samples, their relative abundance differed, predominantly in the CA1 region and dentate gyrus of the hippocampus [73].

Besides the neurogenic areas, MALDI-IMS was also conducted on the frontal cortices of postmortem human brain of AD patients, which were further subcategorized into increasing disease severity as AD I–VI, using Braak's histochemical criteria. These studies showed that sulfatide concentration begins to decrease in the early stages of AD in the white and gray matter of the frontal cortex [63]. As these differences were observed only in AD, and not in other neurodegenerative disorders, sulfatide is suspected to play a role in AD pathology, and may be used as a biomarker for AD [61]. Another class of lipids believed to be important for AD pathology are sphingolipids, which include ceramides, sulfatides, and gangliosides [72]. Since their localization is important to validation of this claim, MALDI-IMS was once again utilized to elucidate the sphingolipid spatial profile and its association with the amyloid plaques [72]. IMS revealed that sphingolipids selectively localized to beta-amyloid aggregates within the cortex and hippocampus. Specifically, gangliosides and ceramides were highly localized to beta-amyloid positive plaques, suggesting that differences in sphingolipid concentration may be important to AD pathogenesis [72]. More recently, MALDI-IMS analysis of the lipid signatures in transgenic AD mouse models showed early shifts in lipid homeostasis that commenced early in AD pathology [74]. Namely, the white matter of AD mouse brains contained significantly lower levels of complex gangliosides, such as the GD1 d18:1 species, and higher levels of simple gangliosides in the prodromal phases of AD [74].

Another technique, laser ablation ICP-MS, has also been used to examine the role of iron in AD [75,76]. Unlike MALDI-IMS, ICP-MS vaporizes the sample spot, and the resulting vapors are guided to a sample inlet for ionization by an ICP torch. As such, the ICP-MS breaks all sample molecules into their elemental form. Consequently, this method is excellent for detecting metal ions within samples. The inherent disadvantage of the ICP-MS lies in the analysis of biological components such as lipids, as the technique needs a metal-attached label to its intended target before analysis can be performed. Since iron buildup is implicated in the development of AD, due to iron's capacity to increase oxidative stress, ICP-MS studies were conducted on AD brains, revealing higher levels of iron in the gray matter in the AD frontal cortex [76,77].Recently, higher levels of iron were found within the CA1 region of the hippocampus [75], but this was not accompanied by a corresponding increase in the ferroportin transport protein [75]. Regardless, IMS has provided new data with respect to iron upregulation in AD, further supporting the oxidative stress hypothesis of AD [75,77].

5.2. Parkinson's Disease (PD)

Parkinson's disease (PD) is a progressive neurodegenerative disease characterized by the death of dopaminergic neurons located within the substantia nigra [78]. These neurons form the nigrostriatal dopaminergic pathway; their death results in a dramatic decrease in levels of dopamine in the striatum that results in clinical presentation such as a resting tremor of varying intensity, unstable posture, muscle rigidity, freezing (inability to initiate voluntary movement), and voluntary movement that is characteristically slow and decreased in intensity. PD also affects cognitive abilities and leads to dementia, as neurodegeneration eventually starts to affect the nearby hippocampal and cortical regions [78].

MALDI imaging studies have often been conducted on PD animal models generated through the injection of either 1-methyl-4-phenyl-1,2,3,6-tetra-hydropyridine (MPTP) or 6-hydroxydopamine (6-OHDA) into one hemisphere of the rodent brain, allowing the contralateral side to serve as an internal control [21] (Table 3). MALDI imaging of 6-OHDA mice revealed the unusual presence of collapsin response mediator protein 2 (CRMP-2) in the corpus callosum—this protein is usually found only in the dendrites of hippocampal and cortical CA1 pyramidal cells or Purkinje cerebellar cells [79]. The hyperphosphorylation of certain amino acid residues on CRMP-2 has been associated with AD neural degeneration; this study pointed to the possibility of PD dyshomeostasis involving similar mechanisms, and the potential use of CRMP-2 as biomarkers for PD as well as AD [79–81]. Another MALDI-IMS analysis of MPTP mice localized PEP-19, a calmodulin-binding protein, in the striatum of the control brain, but found a significant reduction in the PD brain [82]. Further, MALDI-IMS was used to examine L-DOPA-induced dyskinesia, a common side effect of PD medications [83–85]. L-DOPA, the precursor of dopamine, is commonly used in the treatment of PD, due to its efficacy in reducing many of the symptoms associated with the neurodegenerative disorder [84]. However, as the medication is taken for years, some patients develop dyskinesia [84]. Interestingly, MALDI-IMS indicated not only a positive correlation between the severity of L-DOPA-induced dyskinesia and the nigral levels of dynorphin B and alpha-neoendorphin, but also that the most significant differences were localized to the lateral substantia nigra [83], suggesting possible mechanisms that might be amenable to targeted treatment.

Table 3. Application of IMS to various neurodegenerative diseases.IMS has been used to study the metabolomics of Alzheimer's disease, Parkinson's disease, Huntington's disease, multiple sclerosis, and amyotrophic lateral sclerosis in both humans and mouse models. Only MALDI-IMS has been used.

Disease	Organism	Findings	Ref.
Alzheimer's Disease (AD)	Humans	Relative abundance of lipid species differed between AD and control brains, predominantly in the CA1 and dentate gyrus regions of the hippocampus.	[73]
-	Humans	Sulfatide concentrations start to decrease during the early stages of AD (determined by Braak's histochemical criteria) in the white and gray matter of the frontal cortex.	[61]
-	Humans	Sphingolipids (e.g., ceramides, sulfatides, and gangliosides) show selective localization to the β-amyloid aggregates within the cortex and hippocampus, with specific localization of gangliosides and ceramides to β-amyloid positive plaques.	[72]
-	Humans	The CA1 region of the hippocampus in the AD brain contains higher levels of iron compared to the control brain, but there is no significant difference in the levels of ferroportin transport protein between AD and control brains.	[75]
-	Mice	Lipid signatures of the AD mouse brain exhibit early shifts in lipid homeostasis: its white matter is composed of higher levels of simple gangliosides and lower levels of complex gangliosides, such as the GD1 d18:1 species.	[74]
Parkinson's Disease (PD)	Mice	Collapsin response mediator protein 2 (CRMP-2) detected in the PD brain though usually only found in hippocampal and cerebellar cells.	[79]
-	Mice	Calmodulin-binding protein (PEP-19), normally localized to the striatum, was significantly downregulated in the stratium of the PD brain.	[82]
-	Mice	Distinct differences in both the levels and localization of various neurotransmitters and amino acid between PD and control brains established.	[85]
-	Mice	Dynorphin B and alpha-neoendrophin nigral levels are positively correlated with the severity of L-DOPA-induced-dyskinesia.	[83]
Huntington's Disease (HD)	Mice	Within the HD myelin layer, sulfatide and triglyceride levels are decreased and sphingomyelin and ceramide-1-phosphate levels are increased in the lamina; within the HD ependymal layer, phosphatidylinositols levels are decreased.	[86]
-	Mice	Efficacy of P42, a 23 amino acid peptide sequence, as a novel therapy for HD was analyzed with IMS to confirm drug delivery, investigate pharmokinetic properties, and observe post-delivery molecular change.	[87]
Multiple Sclerosis (MS)	Humans	Thymosin beta-4 protein localized to active MS lesions that were either chronically demyelinated or only partially remyelinated.	[88]
Amyotrophic Lateral Sclerosis (ALS)	Humans	Truncated ubiquition form (Ubc-174) levels decreased significantly in ALS spinal cords compared to control, which paralleled normal histological distributes of metabolites in the gray matter.	[89]

5.3. *Huntington's Disease (HD)*

Huntington's disease (HD) is a progressive, autosomal-dominant neurodegenerative disease caused by a trinucleotide (CAG) repeat expansion of the gene encoding the huntingtin protein. The resultant mutant protein is believed to be neurotoxic, leading to the destruction of medium spiny neurons within the striatum and resulting in clinical symptoms such as dyskinesia, neuropsychiatric symptoms, and cognitive impairment [90,91]. However, recent evidence also points to a wider extent of neurodegeneration that extends into the cortical areas of the brain, which may explain the diversity of HD clinical presentations [91]. Later stages of HD are often characterized by severe motor impairment and dementia, with judgement, reasoning, and comprehension abilities also experiencing a great loss [91]. As the misfolded proteins accumulate, cellular degradation mechanisms are not able to keep up and eliminate the toxic huntingtin aggregates. However, the exact process by which polyglutamine aggregation causes the selective destruction of neurons remains to be determined [92].

In the past, GC-MS of human HD brains pinpointed urea upregulation as a possible causal factor of HD neurodegeneration. Further, GC-MS studies pointed to a widespread metabolic dysregulation in brain regions that extends outside of canonically damaged regions of the HD brain [93,94]. This was confirmed with direct injection liquid chromatography mass spectrometry (DI/LC-MS/MS) studies on postmortem brains of HD patients, which revealed metabolic differences in both the frontal lobe and striatum [95]. Particularly, the HD brain contained significantly lower endogenous levels of acylcarnitine, the neuroprotective compounds vital for proper energy metabolism, and phospholipids, which are important in both cellular signaling and integrity [95].

Contrary to the findings observed in AD and PD, adult neurogenesis in the HD brain surprisingly exhibits an increase, reported within the subventricular zone, one of two regions where adult neurogenesis occurs [92]. In addition, MALDI-IMS studies of the lipidome indicated significant differences between control and HD brains in the myelin and ependymal layer [86] (Table 3). The HD myelin layer had decreased sulfatides and triglycerides, as well as enrichment of sphingomyelin and ceramide-1-phosphates. The alterations in the HD ependymal layer were mainly attributed to a drastic drop in the concentration of phosphatidylinositols [86].

Finally, in a study evaluating the efficacy of a 23 amino acid peptide sequence known as P42 in exerting a neuroprotective effect on HD mouse models, MALDI-IMS was employed to examine the pharmakokinetics of P42 delivery into the in vivo model, which included the spatial distribution of P42 and its degraded products, the extent to which it was able to reach the target site, and rates of diffusion between various neuronal compartments [87]. With the aid of the imaging technology, the investigators obtained valuable information on the efficacy of drug delivery and targeting of the neurons within the striatum that correlated with the subsequent improvement in performance on behavioral tests and decreases in protein aggregation, leading to an overall improvement of HD symptoms in this model [87].

5.4. Multiple Sclerosis (MS)

Multiple sclerosis (MS) is a complex, chronic, immune-mediated demyelinating disease affecting the central nervous system in which primary inflammation leads to secondary neurodegeneration. It is the most commonly acquired complex degenerative brain disease of young adults, and is among the most frequent causes of disability in early to middle adulthood [96,97]. The disease course varies, and presents with unpredictable symptoms and levels of recovery. Most commonly, patients are diagnosed with Relapsing-Remitting MS (RRMS), which eventually progresses to the Secondary Progressive (SPMS) form and patients rapidly decline. In the pathogenesis of MS, resident microglia and astrocytes, together with infiltrating macrophages and T-and B-lymphocytes, become activated and produce large amounts of inflammatory cytokines, prostaglandins, and other toxins, that lead to demyelination and ultimately to axonal degeneration. The cumulative effects of these mechanisms increase neurodegeneration within the MS brain and eventually lead to the worsening of clinical symptoms, including cognitive deficits, depression, upper motor neuron signs, tremors, fatigue, weakness, and pain [96,97]. The management of MS is plagued by the variability of the clinical course and severity. This poses great difficulty in providing any given individual with an accurate prognosis and customized treatment plan. Unfortunately, there are no biomarkers able to indicate when the next relapse would occur, whether the given therapy would work, and whether the disease process will last several years or several decades. This uncertain aspect of the disease adds to the health care costs and the emotional distress associated with the disorder.

In the past, human MS brain tissue analyzed by electrospray ionization tandem mass spectrometry showed that brains with active MS demonstrated increases in phospholipid levels and decreases in sphingolipid levels in the normal appearing white matter and gray matter. These changes in lipid signatures could result from metabolic dysregulation that causes sphingolipids to be shuttled into synthesizing phospholipids [98]. Additionally, GC-MS showed significant alterations in the abundance of 44 different metabolites in the MS brain, which were traced back to metabolic intermediates integral to biochemical pathways such as bile acid biosynthesis, taurine metabolism, tryptophan and histidine metabolism, linoleic acid, and D-arginine metabolism pathways [99].

To localize metabolic differences reported by GC-MS and other methods, MALDI-IMS was used for the analysis of recurrent inflammatory lesions in post-mortem MS brains to elucidate the spatial distribution of proteins and peptides [88] (Table 3). Specifically, the objective was to characterize the proteins that were only expressed in the intact white and gray matter, as well as the ones preferentially localized to inflammatory lesions. Analysis revealed that thymosin beta-4, a protein involved in cellular migration, proliferation, and differentiation, was localized to active lesions that were chronically demyelinated and lesions that were only partially remyelinated, suggesting a neurorestorative function for the protein to facilitate remyelination through a downregulation of inflammation and upregulation of oligodendrocyte activity in damaged areas. This spatial confirmation of endogenous relevance to the MS disease pathology allowed researchers to conclude that thymosin beta-4 played a neuroprotective and neurorestorative role in the demyelinated CNS [88].

5.5. *Amyotrophic Lateral Sclerosis (ALS)*

Amyotrophic lateral sclerosis (ALS) is a neurodegenerative disease that involves progressive deterioration of motor neurons located in the motor cortex, brainstem, and spinal cord [100]. As both upper and lower motor neurons begin to degenerate, the muscles grow weak from disuse, exhibit fasciculations, and eventually begin to atrophy. Although 10% of ALS diagnosis have been linked to genetic factors, such as a mutation in the copper–zinc superoxide dimutase 1 (SOD1) gene, the disease is generally regarded to be idiopathic, because the overarching biological pathways that lead to ALS are still unknown [101]. Current proposed disease mechanisms for ALS include intricate relationships between cell-damaging gain of function by SOD1, inflammation from microglial activity, intracellular aggregates, defective mitochondrial function, and glutamate-induced excitotoxicity resulting in neurodegeneration and free radical production [102]. However, since no absolute clinical or molecular biomarkers for the disease exist, especially for idiopathic patients that do not carry mutations in SOD1, combined with the similarity of its initial symptoms with other neurological disorders, the accurate diagnosis and prognosis of ALS is difficult [89]. One application of MALDI-IMS towards investigating this neurodegenerative disease focused on post-mortem human spinal cords from both ALS and control patients [89] (Table 3). Control spinal cords had normal metabolite distributions, some of which included histone and thymosin beta-4 proteins. Contrarily, ALS spinal cords contained significantly lower gray matter concentrations of a truncated form of ubiquitin (Ubc-174) in which both C-terminal glycine residues had been removed. The specific localization of the proteins showed that alterations in protease activity validated the hypothesis that proteome dysfunction plays a significant role in ALS pathology [89].

6. Conclusions

With neurodegenerative diseases on the rise all around the world, a lack of understanding of their causative factors continues to contribute to their personal, societal, and financial burden as the aging population grows ever larger. To date, the inability to find biomarkers and cures for the wide spectrum of neurodegenerative disorders can be partially attributed to the lack of analytical technologies and methods that incorporate both the requisite specificity and sensitivity to study the human brain as the primary site of these complex and devastating diseases. Recent advancements in metabolomics have seen the rapid surge of the use of MS to facilitate studies of both the proteome and the lipidome, with the hope that accurate diagnostic and prognostic biomarkers can be identified for early detection and initiation of preventative measures. Furthermore, utilization of IMS has provided spatial localization for those metabolites detected by IMS, solidifying their role in many cases. Thus, IMS, along with other technologies available today, is paving the way for elucidation of metabolic dysregulation and neurodegenerative dyshomeostasis, towardsthe discovery of new targets for precision therapy of these disorders.

Author Contributions: K.C. and M.M.-S. wrote the article. F.L., M.T. and D.B. provided scientific comments. All authors have read and approved the manuscript.

Funding: Publication of this article is funded in part by the NIH grant GM120033-01 (M.M.-S.). The funding bodies did not have any role in the design or conclusions of this study.

Acknowledgments: We thank members of Maletic-Savatic laboratory for helpful comments and suggestions.

Conflicts of Interest: All authors declare that they have no conflicts of interests.

References

1. Watrous, J.D.; Alexandrov, T.; Dorrestein, P.C. The evolving field of imaging mass spectrometry and its impact on future biological research. *J. Mass Spectrom.* **2011**, *46*, 209–222. [CrossRef] [PubMed]
2. Seeley, E.H.; Schwamborn, K.; Caprioli, R.M. Imaging of Intact Tissue Sections: Moving beyond the Microscope. *J. Boil. Chem.* **2011**, *286*, 25459–25466. [CrossRef] [PubMed]
3. Chaurand, P.; Cornett, D.S.; Angel, P.M.; Caprioli, R.M. From whole-body sections down to cellular level, multiscale imaging of phospholipids by MALDI mass spectrometry. *Mol. Cell Proteom.* **2011**, *10*. [CrossRef] [PubMed]
4. Castellino, S.; Groseclose, M.R.; Wagner, D. MALDI imaging mass spectrometry: Bridging biology and chemistry in drug development. *Bioanalysis* **2011**, *3*, 2427–2441. [CrossRef]
5. Caprioli, R.M.; Farmer, T.B.; Gile, J. Molecular Imaging of Biological Samples: Localization of Peptides and Proteins Using MALDI-TOF MS. *Anal. Chem.* **1997**, *69*, 4751–4760. [CrossRef]
6. Arnold, J.M.; Choi, W.T.; Sreekumar, A.; Maletić-Savatić, M. Analytical strategies for studying stem cell metabolism. *Front. Boil.* **2015**, *10*, 141–153. [CrossRef]
7. Miura, D.; Fujimura, Y.; Yamato, M.; Hyodo, F.; Utsumi, H.; Tachibana, H.; Wariishi, H. Ultrahighly Sensitive in Situ Metabolomic Imaging for Visualizing Spatiotemporal Metabolic Behaviors. *Anal. Chem.* **2010**, *82*, 9789–9796. [CrossRef]
8. Kumar, A.; Agarwal, S.; Heyman, J.A.; Matson, S.; Heidtman, M.; Piccirillo, S.; Umansky, L.; Drawid, A.; Jansen, R.; Liu, Y.; et al. Subcellular localization of the yeast proteome. *Genome Res.* **2002**, *16*, 707–719. [CrossRef]
9. Caprioli, R.M. Imaging mass spectrometry: Molecular microscopy for enabling a new age of discovery. *Proteomics* **2014**, *14*, 807–809. [CrossRef]
10. Groseclose, M.R.; Laffan, S.B.; Frazier, K.S.; Hughes-Earle, A.; Castellino, S. Imaging MS in Toxicology: An Investigation of Juvenile Rat Nephrotoxicity Associated with Dabrafenib Administration. *J. Am. Soc. Mass Spectrom.* **2015**, *26*, 887–898. [CrossRef]
11. Groseclose, M.R.; Castellino, S. A Mimetic Tissue Model for the Quantification of Drug Distributions by MALDI Imaging Mass Spectrometry. *Anal. Chem.* **2013**, *85*, 10099–10106. [CrossRef]
12. Nilsson, A.; Peric, A.; Strimfors, M.; Goodwin, R.J.A.; Hayes, M.A.; Andrén, P.E.; Hilgendorf, C. Mass Spectrometry Imaging proves differential absorption profiles of well-characterised permeability markers along the crypt-villus axis. *Sci. Rep.* **2017**, *7*, 6352. [CrossRef] [PubMed]
13. Goodwin, R.; Bunch, J.; McGinnity, D. Mass Spectrometry Imaging in Oncology Drug Discovery. *Adv. Cancer Res.* **2017**, *134*, 133–171.
14. Choi, W.T.; Tosun, M.; Jeong, H.-H.; Karakas, C.; Semerci, F.; Liu, Z.; Maletić-Savatić, M. Metabolomics of mammalian brain reveals regional differences. *BMC Syst. Boil.* **2018**, *12*, 127. [CrossRef] [PubMed]
15. Reyzer, M.L.; Caprioli, R.M. MALDI-MS-based imaging of small molecules and proteins in tissues. *Curr. Opin. Chem. Boil.* **2007**, *11*, 29–35. [CrossRef] [PubMed]
16. Liu, L.; MacKenzie, K.R.; Putluri, N.; Maletić-Savatić, M.; Bellen, H.J. The glia-neuron lactate shuttle and elevated ROS promote lipid synthesis in neurons and lipid droplet accumulation in glia via APOE/D. *Cell Metab.* **2017**, *26*, 719–737. [CrossRef]
17. Botas, A.; Campbell, H.M.; Han, X.; Maletic-Savatic, M. Metabolomics of Neurodegenerative Diseases. *Int. Rev. Neurobiol.* **2015**, *122*, 53–80. [PubMed]
18. Calvano, C.D.; Palmisano, F.; Cataldi, T.R. Understanding neurodegenerative disorders by MS-based lipidomics. *Bioanalysis* **2018**, *10*, 787–790. [CrossRef] [PubMed]
19. Sugiura, Y.; Shimma, S.; Konishi, Y.; Yamada, M.K.; Setou, M. Imaging Mass Spectrometry Technology and Application on Ganglioside Study; Visualization of Age-Dependent Accumulation of C20-Ganglioside Molecular Species in the Mouse Hippocampus. *PLoS ONE* **2008**, *3*, e3232. [CrossRef]

20. Cobice, D.F.; Goodwin, R.J.A.; Andren, P.E.; Nilsson, A.; Mackay, C.L.; Andrew, R. Future technology insight: Mass spectrometry imaging as a tool in drug research and development. *Br. J. Pharmacol.* **2015**, *172*, 3266–3283. [CrossRef]
21. Michno, W.; Wehrli, P.M.; Blennow, K.; Zetterberg, H.; Hanrieder, J. Molecular imaging mass spectrometry for probing protein dynamics in neurodegenerative disease pathology. *J. Neurochem.* **2018**. [CrossRef] [PubMed]
22. Chughtai, K.; Heeren, R.M.A. Mass Spectrometric Imaging for biomedical tissue analysis. *Chem. Rev.* **2010**, *110*, 3237–3277. [CrossRef] [PubMed]
23. Jurchen, J.C.; Rubakhin, S.S.; Sweedler, J.V. MALDI-MS imaging of features smaller than the size of the laser beam. *J. Am. Soc. Mass Spectrom.* **2005**, *16*, 1654–1659. [CrossRef] [PubMed]
24. Passarelli, M.K.; Wang, J.; Mohammadi, A.S.; Trouillon, R.; Gilmore, I.; Ewing, A.G. Development of an Organic Lateral Resolution Test Device for Imaging Mass Spectrometry. *Anal. Chem.* **2014**, *86*, 9473–9480. [CrossRef] [PubMed]
25. Norris, J.L.; Caprioli, R.M. Analysis of Tissue Specimens by Matrix-Assisted Laser Desorption/Ionization Imaging Mass Spectrometry in Biological and Clinical Research. *Chem. Rev.* **2013**, *113*, 2309–2342. [CrossRef] [PubMed]
26. Laiko, V.V.; Baldwin, M.A.; Burlingame, A.L. Atmospheric Pressure Matrix-Assisted Laser Desorption/Ionization Mass Spectrometry. *Anal. Chem.* **2000**, *72*, 652–657. [CrossRef]
27. Schober, Y.; Guenther, S.; Spengler, B.; Römpp, A. Single Cell Matrix-Assisted Laser Desorption/Ionization Mass Spectrometry Imaging. *Anal. Chem.* **2012**, *84*, 6293–6297. [CrossRef]
28. Nguyen, J.; Russell, S.C. Targeted proteomics approach to species-level identification of Bacillus thuringiensis spores by AP-MALDI-MS. *J. Am. Soc. Mass Spectrom.* **2010**, *21*, 993–1001. [CrossRef]
29. Baker, T.C.; Han, J.; Borchers, C.H. Recent advancements in matrix-assisted laser desorption/ionization mass spectrometry imaging. *Curr. Opin. Biotechnol.* **2017**, *43*, 62–69. [CrossRef]
30. Fitzgerald, M.C.; Parr, G.R.; Smith, L.M. Basic matrices for the matrix-assisted laser desorption/ionization mass spectrometry of proteins and oligonucleotides. *Anal. Chem.* **1993**, *65*, 3204–3211. [CrossRef]
31. Cheng, S.-W.; Chan, T.-W.D. Use of Ammonium Halides as Co-matrices for Matrix-assisted Laser Desorption/Ionization Studies of Oligonucleotides. *Rapid Commun. Mass Spectrom.* **1996**, *10*, 907–910. [CrossRef]
32. Nordhoff, E.; Ingendoh, A.; Cramer, R.; Overberg, A.; Stahl, B.; Karas, M.; Hillenkamp, F.; Crain, P.F.; Chait, B. Matrix-assisted laser desorption/ionization mass spectrometry of nucleic acids with wavelengths in the ultraviolet and infrared. *Rapid Commun. Mass Spectrom.* **1992**, *6*, 771–776. [CrossRef] [PubMed]
33. Zhang, L.-K.; Gross, M.L. Matrix-assisted laser desorption/ionization mass spectrometry methods for oligodeoxynucleotides: Improvements in matrix, detection limits, quantification, and sequencing. *J. Am. Soc. Mass Spectrom.* **2000**, *11*, 854–865. [CrossRef]
34. Lecchi, P.; Pannell, L.K. The detection of intact double-stranded DNA by MALDI. *J. Am. Soc. Mass Spectrom.* **1995**, *6*, 972–975. [CrossRef]
35. Papac, D.I.; Wong, A.; Jones, A.J.S. Analysis of Acidic Oligosaccharides and Glycopeptides by Matrix-Assisted Laser Desorption/Ionization Time-of-Flight Mass Spectrometry. *Anal. Chem.* **1996**, *68*, 3215–3223. [CrossRef] [PubMed]
36. Wu, K.J.; Shaler, T.A.; Becker, C.H. Time-of-Flight Mass Spectrometry Of Underivatized Single-Stranded DNA Oligomers by Matrix-Assisted Laser Desorption. *Anal. Chem.* **1994**, *66*, 1637–1645. [CrossRef]
37. Shahgholi, M.; Garcia, B.A.; Chiu, N.H.L.; Heaney, P.J.; Tang, K. Sugar additives for MALDI matrices improve signal allowing the smallest nucleotide change (A:T) in a DNA sequence to be resolved. *Nucleic Acids Res.* **2001**, *29*, e91. [CrossRef]
38. Beavis, R.C.; Chaudhary, T.; Chait, B.T. α-Cyano-4-hydroxycinnamic acid as a matrix for matrixassisted laser desorption mass spectromtry. *J. Mass Spectrom.* **1992**, *27*, 156–158. [CrossRef]
39. Asara, J.M.; Allison, J. Enhanced detection of phosphopeptides in matrix-assisted laser desorption/ionization mass spectrometry using ammonium salts. *J. Am. Soc. Mass Spectrom.* **1999**, *10*, 35–44. [CrossRef]
40. Gobom, J.; Schuerenberg, M.; Mueller, M.; Theiss, D.; Lehrach, H.; Nordhoff, E. Alpha-cyano-4-hydroxycinnamic acid affinity sample preparation. A protocol for MALDI-MS peptide analysis in proteomics. *Anal. Chem.* **2001**, *73*, 434–438. [CrossRef]
41. Strupat, K.; Karas, M.; Hillenkamp, F. 2,5-Dihydroxybenzoic acid: A new matrix for laser desorption—Ionization mass spectrometry. *Int. J. Mass Spectrom. Ion Process.* **1991**, *111*, 89–102. [CrossRef]

42. Zhu, L.; Parr, G.R.; Fitzgerald, M.C.; Nelson, C.M.; Smith, L.M. Oligodeoxynucleotide Fragmentation in MALDI/TOF Mass Spectrometry Using 355-nm Radiation. *J. Am. Chem. Soc.* **1995**, *117*, 6048–6056. [CrossRef]
43. Macha, S.F.; Limbach, P.A.; Hanton, S.D.; Owens, K.G. Silver cluster interferences in matrix-assisted laser desorption/ionization (MALDI) mass spectrometry of nonpolar polymers. *J. Am. Soc. Mass Spectrom.* **2001**, *12*, 732–743. [CrossRef]
44. Vermillion-Salsbury, R.L.; Hercules, D.M.; Vermillion-Salsbury, R.L. 9-Aminoacridine as a matrix for negative mode matrix-assisted laser desorption/ionization. *Rapid Commun. Mass Spectrom.* **2002**, *16*, 1575–1581. [CrossRef]
45. Eberlin, L.S.; Ferreira, C.R.; Dill, A.L.; Ifa, D.R.; Cooks, R.G. Desorption electrospray ionization mass spectrometry for lipid characterization and biological tissue imaging. *Biochim. Biophys. Acta (BBA)* **2011**, *1811*, 946–960. [CrossRef] [PubMed]
46. Takats, Z.; Wiseman, J.M.; Gologan, B.; Cooks, R.G. Mass Spectrometry Sampling Under Ambient Conditions with Desorption Electrospray Ionization. *Science* **2004**, *306*, 471–473. [CrossRef] [PubMed]
47. Eberlin, L.S.; Ferreira, C.R.; Dill, A.L.; Ifa, D.R.; Cheng, L.; Cooks, R.G. Non-Destructive, Histologically Compatible Tissue Imaging by Desorption Electrospray Ionization Mass Spectrometry. *ChemBioChem* **2011**, *12*, 2129–2132. [CrossRef] [PubMed]
48. Dill, A.L.; Eberlin, L.S.; Costa, A.B.; Ifa, D.R.; Cooks, R.G. Data quality in tissue analysis using desorption electrospray ionization. *Anal. Bioanal. Chem.* **2011**, *401*, 1949–1961. [CrossRef] [PubMed]
49. Wang, X.; Hou, Y.; Hou, Z.; Xiong, W.; Huang, G. Mass Spectrometry Imaging of Brain Cholesterol and Metabolites with Trifluoroacetic Acid-Enhanced Desorption Electrospray Ionization. *Anal. Chem.* **2019**, *91*, 2719–2726. [CrossRef] [PubMed]
50. Wiseman, J.M.; Ifa, D.R.; Song, Q.; Cooks, R.G. Tissue Imaging at Atmospheric Pressure Using Desorption Electrospray Ionization (DESI) Mass Spectrometry. *Angew. Chem. Int. Ed.* **2006**, *45*, 7188–7192. [CrossRef] [PubMed]
51. Todd, P.J.; Schaaff, T.G.; Chaurand, P.; Caprioli, R.M. Organic ion imaging of biological tissue with secondary ion mass spectrometry and matrix-assisted laser desorption/ionization. *J. Mass Spectrom.* **2001**, *36*, 355–369. [CrossRef] [PubMed]
52. Falick, A.M.; Maltby, D.A. Derivatization of hydrophilic peptides for liquid secondary ion mass spectrometry at the picomole level. *Anal. Biochem.* **1989**, *182*, 165–169. [CrossRef]
53. Weibel, D.; Wong, S.; Lockyer, N.; Blenkinsopp, P.; Hill, R.; Vickerman, J.C. A C60Primary Ion Beam System for Time of Flight Secondary Ion Mass Spectrometry: Its Development and Secondary Ion Yield Characteristics. *Anal. Chem.* **2003**, *75*, 1754–1764. [CrossRef] [PubMed]
54. Hebert, L.E.; Weuve, J.; Scherr, P.A.; Evans, D.A. Alzheimer disease in the United States (2010–2050) estimated using the 2010 census. *Neurology* **2013**, *80*, 1778–1783. [CrossRef]
55. Wimo, A.; Jönsson, L.; Bond, J.; Prince, M.; Winblad, B. The worldwide economic impact of dementia 2010. *Alzheimer's Dement.* **2013**, *9*, 1–11. [CrossRef]
56. Marras, C.; Beck, J.C.; Bower, J.H.; Roberts, E.; Ritz, B.; Ross, G.W.; Abbott, R.D.; Savica, R.; Van Den Eeden, S.K.; Willis, A.W.; et al. Prevalence of Parkinson's disease across North America. *NPJ Parkinsons Dis.* **2018**, *4*, 21. [CrossRef]
57. Wang, X.; Xu, Z.; Xu, L. SP100B Expression Indexed Hemorrhage in Mouse Models of Cerebral Hemorrhage. *Indian J. Clin. Biochem.* **2018**, *33*, 361–364. [CrossRef]
58. Alzheimer's Association. 2018 Alzheimer's disease facts and figures. *Alzheimer's Dement.* **2018**, *14*, 367–429. [CrossRef]
59. Serrano-Pozo, A.; Frosch, M.P.; Masliah, E.; Hyman, B.T. Neuropathological Alterations in Alzheimer Disease. *Cold Spring Harb. Perspect. Med.* **2011**, *1*, a006189. [CrossRef]
60. Rodriguez, A.; Ehlenberger, D.B.; Dickstein, D.L.; Hof, P.R.; Wearne, S.L. Automated Three-Dimensional Detection and Shape Classification of Dendritic Spines from Fluorescence Microscopy Images. *PLoS ONE* **2008**, *3*, e1997. [CrossRef]
61. Román, E.G.D.S.; Manuel, I.; Giralt, M.; Ferrer, I.; Rodríguez-Puertas, R. Imaging mass spectrometry (IMS) of cortical lipids from preclinical to severe stages of Alzheimer's disease. *Biochim. Biophys. Acta (BBA)* **2017**, *1859*, 1604–1614. [CrossRef] [PubMed]

62. Chan, R.B.; Oliveira, T.G.; Cortes, E.P.; Honig, L.S.; Duff, K.E.; Small, S.A.; Wenk, M.R.; Shui, G.; Di Paolo, G. Comparative lipidomic analysis of mouse and human brain with Alzheimer disease. *J. Biol. Chem.* **2012**, *287*, 2678–2688. [CrossRef] [PubMed]
63. Donovan, M.H.; Yazdani, U.; Norris, R.D.; Games, D.; German, D.C.; Eisch, A.J. Decreased adult hippocampal neurogenesis in the PDAPP mouse model of Alzheimer's disease. *J. Comp. Neurol.* **2006**, *495*, 70–83. [CrossRef] [PubMed]
64. Semerci, F.; Maletic-Savatic, M. Transgenic mouse models for studying adult neurogenesis. *Front. Boil.* **2016**, *11*, 151–167. [CrossRef] [PubMed]
65. Beccari, S.; Valero, J.; Maletic-Savatic, M.; Sierra, A. A simulation model of neuroprogenitor proliferation dynamics predicts age-related loss of hippocampal neurogenesis but not astrogenesis. *Sci. Rep.* **2017**, *7*, 16528. [CrossRef] [PubMed]
66. Moreno-Jiménez, E.P.; Flor-García, M.; Terreros-Roncal, J.; Rábano, A.; Cafini, F.; Pallas-Bazarra, N.; Ávila, J.; Llorens-Martín, M. Adult hippocampal neurogenesis is abundant in neurologically healthy subjects and drops sharply in patients with Alzheimer's disease. *Nat. Med.* **2019**, *25*, 554–560. [CrossRef] [PubMed]
67. Crews, L.; Masliah, E. Molecular mechanisms of neurodegeneration in Alzheimer's disease. *Hum. Mol. Genet.* **2010**, *19*, R12–R20. [CrossRef] [PubMed]
68. Li, B.; Sierra, A.; Deudero, J.J.; Semerci, F.; Laitman, A.; Kimmel, M.; Maletic-Savatic, M. Multitype Bellman-Harris branching model provides biological predictors of early stages of adult hippocampal neurogenesis. *BMC Syst. Boil.* **2017**, *11*, 90. [CrossRef]
69. González-Domínguez, R.; García-Barrera, T.; Vitorica, J.; Gómez-Ariza, J.L. Region-specific metabolic alterations in the brain of the APP/PS1 transgenic mice of Alzheimer's disease. *Biochim. Biophys. Acta (BBA)* **2014**, *1842*, 2395–2402. [CrossRef]
70. González-Domínguez, R.; García-Barrera, T.; Vitorica, J.; Gómez-Ariza, J.L. Metabolomic screening of regional brain alterations in the APP/PS1 transgenic model of Alzheimer's disease by direct infusion mass spectrometry. *J. Pharm. Biomed. Anal.* **2015**, *102*, 425–435. [CrossRef]
71. Xu, J.; Begley, P.; Church, S.J.; Patassini, S.; Hollywood, K.A.; Jüllig, M.; Curtis, M.A.; Waldvogel, H.J.; Faull, R.L.; Unwin, R.D.; et al. Graded perturbations of metabolism in multiple regions of human brain in Alzheimer's disease: Snapshot of a pervasive metabolic disorder. *Biochim. Biophys. Acta (BBA)* **2016**, *1862*, 1084–1092. [CrossRef] [PubMed]
72. Kaya, I.; Zetterberg, H.; Blennow, K.; Hanrieder, J. Shedding Light on the Molecular Pathology of Amyloid Plaques in Transgenic Alzheimer's Disease Mice Using Multimodal MALDI Imaging Mass Spectrometry. *ACS Chem. Neurosci.* **2018**, *9*, 1802–1817. [CrossRef] [PubMed]
73. Mendis, L.H.S.; Grey, A.C.; Faull, R.L.M.; Curtis, M.A. Hippocampal lipid differences in Alzheimer's disease: A human brain study using matrix-assisted laser desorption/ionization-imaging mass spectrometry. *Brain Behav.* **2016**, *6*, e00517. [CrossRef] [PubMed]
74. Caughlin, S.; Maheshwari, S.; Agca, Y.; Agca, C.; Harris, A.J.; Jurcic, K.; Yeung, K.K.-C.; Cechetto, D.F.; Whitehead, S.N. Membrane-lipid homeostasis in a prodromal rat model of Alzheimer's disease: Characteristic profiles in ganglioside distributions during aging detected using MALDI imaging mass spectrometry. *Biochim. Biophys. Acta (BBA)* **2018**, *1862*, 1327–1338. [CrossRef] [PubMed]
75. Cruz-Alonso, M.; Fernandez, B.; Navarro, A.; Junceda, S.; Astudillo, A.; Pereiro, R. Laser ablation ICP-MS for simultaneous quantitative imaging of iron and ferroportin in hippocampus of human brain tissues with Alzheimer's disease. *Talanta* **2019**, *197*, 413–421. [CrossRef] [PubMed]
76. Hare, D.J.; Raven, E.P.; Roberts, B.R.; Bogeski, M.; Portbury, S.D.; McLean, C.A.; Masters, C.L.; Connor, J.R.; Bush, A.I.; Crouch, P.J.; et al. Laser ablation-inductively coupled plasma-mass spectrometry imaging of white and gray matter iron distribution in Alzheimer's disease frontal cortex. *NeuroImage* **2016**, *137*, 124–131. [CrossRef] [PubMed]
77. Markesbery, W.R. Oxidative Stress Hypothesis in Alzheimer's Disease. *Free. Radic. Boil. Med.* **1997**, *23*, 134–147. [CrossRef]
78. Dauer, W.; Przedborski, S. Parkinson's disease: Mechanisms and models. *Neuron* **2003**, *39*, 889–909. [CrossRef]
79. Stauber, J.; Lemaire, R.; Franck, J.; Bonnel, D.; Croix, D.; Day, R.; Wisztorski, M.; Fournier, I.; Salzet, M. MALDI Imaging of Formalin-Fixed Paraffin-Embedded Tissues: Application to Model Animals of Parkinson Disease for Biomarker Hunting. *J. Proteome Res.* **2008**, *7*, 969–978. [CrossRef]

80. Rose, S.; Chen, F.; Chalk, J.; Zelaya, F.; Strugnell, W.; Benson, M.; Semple, J.; Doddrell, D.; Rose, S.E.; Stuerenburg, H.J.; et al. Loss of connectivity in Alzheimer's disease: An evaluation of white matter tract integrity with colour coded MR diffusion tensor imaging. *J. Neurol. Neurosurg. Psychiatry* **2000**, *69*, 528–530. [CrossRef]
81. Gu, Y.; Hamajima, N.; Ihara, Y. Neurofibrillary Tangle-Associated Collapsin Response Mediator Protein-2 (CRMP-2) Is Highly Phosphorylated on Thr-509, Ser-518, and Ser-522. *Biochemistry* **2000**, *39*, 4267–4275. [CrossRef] [PubMed]
82. Sköld, K.; Svensson, M.; Nilsson, A.; Zhang, X.; Nydahl, K.; Caprioli, R.M.; Svenningsson, P.; Andrén, P.E. Decreased Striatal Levels of PEP-19 Following MPTP Lesion in the Mouse. *J. Proteome Res.* **2006**, *5*, 262–269. [CrossRef] [PubMed]
83. Ljungdahl, A.; Hanrieder, J.; Fälth, M.; Bergquist, J.; Andersson, M. Imaging Mass Spectrometry Reveals Elevated Nigral Levels of Dynorphin Neuropeptides in L-DOPA-Induced Dyskinesia in Rat Model of Parkinson's Disease. *PLoS ONE* **2011**, *6*, e25653. [CrossRef] [PubMed]
84. Schapira, A.H.; Agid, Y.; Barone, P.; Jenner, P.; Lemke, M.R.; Poewe, W.; Rascol, O.; Reichmann, H.; Tolosa, E. Perspectives on recent advances in the understanding and treatment of Parkinson's disease. *Eur. J. Neurol.* **2009**, *16*, 1090–1099. [CrossRef] [PubMed]
85. Shariatgorji, M.; Nilsson, A.; Goodwin, R.J.; Källback, P.; Schintu, N.; Zhang, X.; Crossman, A.R.; Bezard, E.; Svenningsson, P.; Andren, P.E. Direct Targeted Quantitative Molecular Imaging of Neurotransmitters in Brain Tissue Sections. *Neuron* **2014**, *84*, 697–707. [CrossRef] [PubMed]
86. Hunter, M.; DeMarais, N.J.; Faull, R.L.M.; Grey, A.C.; Curtis, M.A. Subventricular zone lipidomic architecture loss in Huntington's disease. *J. Neurochem.* **2018**, *146*, 613–630. [CrossRef] [PubMed]
87. Arribat, Y.; Talmat-Amar, Y.; Paucard, A.; Lesport, P.; Bonneaud, N.; Bauer, C.; Bec, N.; Parmentier, M.-L.; Benigno, L.; Larroque, C.; et al. Systemic delivery of P42 peptide: A new weapon to fight Huntington's disease. *Acta Neuropathol. Commun.* **2014**, *2*, 905. [CrossRef] [PubMed]
88. Maccarrone, G.; Nischwitz, S.; Deininger, S.-O.; Hornung, J.; König, F.B.; Stadelmann, C.; Turck, C.W.; Weber, F. MALDI imaging mass spectrometry analysis—A new approach for protein mapping in multiple sclerosis brain lesions. *J. Chromatogr. B* **2017**, *1047*, 131–140. [CrossRef]
89. Ekegren, T.; Hanrieder, J.; Andersson, M.; Bergquist, J. MALDI imaging of post-mortem human spinal cord in amyotrophic lateral sclerosis. *J. Neurochem.* **2013**, *124*, 695–707.
90. Cepeda, C.; Wu, N.; Andre, V.M.; Cummings, D.M.; Levine, M.S. The corticostriatal pathway in Huntington's disease. *Prog. Neurobiol.* **2007**, *81*, 253–271. [CrossRef]
91. Zuccato, C.; Valenza, M.; Cattaneo, E. Molecular Mechanisms and Potential Therapeutical Targets in Huntington's Disease. *Physiol. Rev.* **2010**, *90*, 905–981. [CrossRef] [PubMed]
92. Walker, T.M.; Lalor, M.K.; Broda, A.; Ortega, L.S.; Morgan, M.; Parker, L.; Churchill, S.; Bennett, K.; Golubchik, T.; Giess, A.P.; et al. Assessment of Mycobacterium tuberculosis transmission in Oxfordshire, UK, 2007–12, with whole pathogen genome sequences: An observational study. *Lancet Respir. Med.* **2014**, *2*, 285–292. [CrossRef]
93. Patassini, S.; Begley, P.; Reid, S.J.; Xu, J.; Church, S.J.; Curtis, M.; Dragunow, M.; Waldvogel, H.J.; Unwin, R.D.; Snell, R.G.; et al. Identification of elevated urea as a severe, ubiquitous metabolic defect in the brain of patients with Huntington's disease. *Biochem. Biophys. Res. Commun.* **2015**, *468*, 161–166. [CrossRef] [PubMed]
94. Patassini, S.; Begley, P.; Xu, J.; Church, S.J.; Reid, S.J.; Kim, E.H.; Curtis, M.A.; Dragunow, M.; Waldvogel, H.J.; Snell, R.G.; et al. Metabolite mapping reveals severe widespread perturbation of multiple metabolic processes in Huntington's disease human brain. *Biochim. Biophys. Acta (BBA)* **2016**, *1862*, 1650–1662. [CrossRef] [PubMed]
95. Graham, S.F.; Pan, X.; Yilmaz, A.; Macias, S.; Robinson, A.; Mann, D.; Green, B.D. Targeted biochemical profiling of brain from Huntington's disease patients reveals novel metabolic pathways of interest. *Biochim. Biophys. Acta (BBA)* **2018**, *1864*, 2430–2437. [CrossRef] [PubMed]
96. International Multiple Sclerosis Genetics Consortium. Refining genetic associations in multiple sclerosis. *Lancet Neurol.* **2008**, *7*, 567–569. [CrossRef]
97. Goldenberg, M.M. Multiple sclerosis review. *Pharm. Ther.* **2012**, *37*, 175–184.
98. Wheeler, D.; Bandaru, V.V.R.; Calabresi, P.A.; Nath, A.; Haughey, N.J. A defect of sphingolipid metabolism modifies the properties of normal appearing white matter in multiple sclerosis. *Brain* **2008**, *131*, 3092–3102. [CrossRef]

99. Mangalam, A.K.; Poisson, L.M.; Nemutlu, E.; Datta, I.; Denic, A.; Dzeja, P.; Rodriguez, M.; Rattan, R.; Giri, S. Profile of Circulatory Metabolites in a Relapsing-remitting Animal Model of Multiple Sclerosis using Global Metabolomics. *J. Clin. Cell. Immunol.* **2013**, *4*, 1–12.
100. Rowland, L.P.; Shneider, N.A. Amyotrophic Lateral Sclerosis. *N. Engl. J. Med.* **2001**, *344*, 1688–1700. [CrossRef]
101. Clement, A.M. Wild-Type Nonneuronal Cells Extend Survival of SOD1 Mutant Motor Neurons in ALS Mice. *Science* **2003**, *302*, 113–117. [CrossRef] [PubMed]
102. Cudkowicz, M.E.; van den Berg, L.H.; Shefner, J.M.; Mitsumoto, H.; Mora, J.S.; Ludolph, A.; Hardiman, O.; Bozik, M.E.; Ingersoll, E.W.; Archibald, D.; et al. Dexpramipexole versus placebo for patients with amyotrophic lateral sclerosis (EMPOWER): A randomised, double-blind, phase 3 trial. *Lancet Neurol.* **2013**, *12*, 1059–1067. [CrossRef]

metabolites

MDPI

Article

In Vivo Microdialysis of Endogenous and ^{13}C-labeled TCA Metabolites in Rat Brain: Reversible and Persistent Effects of Mitochondrial Inhibition and Transient Cerebral Ischemia

Jesper F. Havelund [1,*,†], Kevin H. Nygaard [2,3,†], Troels H. Nielsen [2,3], Carl-Henrik Nordström [2,3], Frantz R. Poulsen [2,3], Nils. J. Færgeman [1], Axel Forsse [2,3,†] and Jan Bert Gramsbergen [4,†]

1 VILLUM Center for Bioanalytical Sciences, Department of Biochemistry and Molecular Biology, University of Southern Denmark, Campusvej 55, 5230 Odense M, Denmark; nils.f@bmb.sdu.dk
2 Department of Neurosurgery, Odense University Hospital, University of Southern Denmark, Sdr. Boulevard 29, 5000 Odense C, Denmark; kevin.nygaard@hotmail.com (K.H.N.); troels.nielsen@rsyd.dk (T.H.N.); carl-henrik.nordstrom@med.lu.se (C.-H.N.); frantz.r.poulsen@rsyd.dk (F.R.P.); axel.forsse@gmail.com (A.F.)
3 BRIDGE—Brain ResearchE—Inter-Disciplinary Guided Excellence, Institute of Clinical Research, University of Southern Denmark, Winsløwparken 19, 5000 Odense C, Denmark
4 Institute of Molecular Medicine, University of Southern Denmark, 55, 5230 Odense C, Denmark; jbgramsbergen@health.sdu.dk
* Correspondence: jhav@bmb.sdu.dk; Tel.: +45-65509029
† These authors contributed equally to this work.

Received: 31 July 2019; Accepted: 25 September 2019; Published: 27 September 2019

Abstract: Cerebral micro-dialysis allows continuous sampling of extracellular metabolites, including glucose, lactate and pyruvate. Transient ischemic events cause a rapid drop in glucose and a rise in lactate levels. Following such events, the lactate/pyruvate (L/P) ratio may remain elevated for a prolonged period of time. In neurointensive care clinics, this ratio is considered a metabolic marker of ischemia and/or mitochondrial dysfunction. Here we propose a novel, sensitive microdialysis liquid chromatography-mass spectrometry (LC-MS) approach to monitor mitochondrial dysfunction in living brain using perfusion with ^{13}C-labeled succinate and analysis of ^{13}C-labeled tricarboxylic acid cycle (TCA) intermediates. This approach was evaluated in rat brain using malonate-perfusion (10–50 mM) and endothelin-1 (ET-1)-induced transient cerebral ischemia. In the malonate model, the expected changes upon inhibition of succinate dehydrogenase (SDH) were observed, i.e., an increase in endogenous succinate and decreases in fumaric acid and malic acid. The inhibition was further elaborated by incorporation of ^{13}C into specific TCA intermediates from ^{13}C-labeled succinate. In the ET-1 model, increases in non-labeled TCA metabolites (reflecting release of intracellular compounds) and decreases in ^{13}C-labeled TCA metabolites (reflecting inhibition of de novo synthesis) were observed. The analysis of ^{13}C incorporation provides further layers of information to identify metabolic disturbances in experimental models and neuro-intensive care patients.

Keywords: ^{13}C-labeled succinate; cerebral ischemia; energy metabolism; endothelin-1; LC-MS; malonate; micro-dialysis; mitochondrial dysfunction; reperfusion; tricarboxylic acid cycle

1. Introduction

In neuro-intensive care units, cerebral microdialysis is routinely used to monitor interstitial levels of glucose, lactate and pyruvate, the main energy substrates of the brain. In patients with traumatic brain injury (TBI) or aneurysmal subarachnoid hemorrhage (aSAH), periods of compromised cerebral blood flow are characterized by a decrease in glucose, a rise in lactate and elevated L/P ratios [1,2].

Elevated L/P ratios with normal or increased levels of pyruvate after an ischemic insult are considered a metabolic marker of mitochondrial dysfunction, which has been associated with delayed neurological deterioration (DND) [3,4]. Early detection and monitoring of mitochondrial dysfunction, as well as understanding the underlying mechanisms of disturbed energy metabolism post-injury are essential to improve treatments and outcome in neuro-intensive care patients.

Brain energy metabolism has been studied in vitro and in vivo by perfusion with ^{14}C-labeled (radioactive) energy substrates, including glucose, lactate, pyruvate, glutamate and glutamine, and monitoring $^{14}CO_2$ production under various experimental conditions. Studies using fluorocitrate, which at lower concentrations inhibits glial, but not neuronal TCA cycle activity, combined with ^{14}C-labeled microdialysis allowed assessment of oxidation rates of different energy substrates in glial cells or neurons [5].

More recently cerebral microdialysis and perfusion with ^{13}C-labeled substrates and analysis of ^{13}C-labeled TCA intermediates has been used to study brain energy metabolism in head injury patients [6–8]. ^{13}C-labeled energy substrates included 1,2-$^{13}C_2$ glucose to study glycolysis and the pentose phosphate pathway [6], 3-^{13}C-lactate to study lactate metabolism via the TCA cycle [8] and 2,3-$^{13}C_2$-succinate to study enhancement of TCA cycle metabolism [7]. In the clinical microdialysis studies mentioned above, the recovered ^{13}C-labeled metabolites were analyzed by nuclear magnetic resonance (NMR). Although NMR is a standard technique for identifying ^{13}C-labeled molecules, the sensitivity is not very high, meaning that large volumes of dialysate are needed and only abundant metabolites in the millimolar range can be detected. In the study by Jalloh et al. 2016, perfusing with 12 mM 2,3-$^{13}C_2$-succinate (a pharmacologically active dose enhancing local brain metabolism), micro-dialysate samples were pooled over a 24 h period (180 μL pooled dialysates) to allow detection of ^{13}C-labeled fumarate, malate, lactate and glutamine. As that study showed, exogenous succinate was taken up by brain cells (astrocytes and neurons) and metabolized via the TCA cycle within mitochondria.

In the present experimental study in rats, we used a similar approach of perfusing brain tissue with ^{13}C-labeled succinate through the dialysis probe (Figure 1) and measuring ^{13}C-labeled TCA-centered metabolites in the dialysate. However, in contrast to the study by Jalloh et al., we used highly sensitive LC-MS to identify and quantify ^{13}C-labeled TCA-centered metabolites upon continuous perfusion with a tracer dose (1 mM) of uniformly ^{13}C-labeled succinate. Because of the higher sensitivity of LC-MS, we were able to measure both endogenous (^{12}C) and ^{13}C-labeled TCA-centered metabolites in striatal dialysates with a temporal resolution as low as 30 min (30 μL samples). We only show TCA-centered metabolites, however, many other endogenous compounds, e.g., amino acids, purines and pyrimidines were detected. The major improvement in time resolution and extra layers of information (acute release of endogenous metabolites and acute changes of de novo synthesis) that this approach offers, allows detailed biochemical monitoring and better understanding of mechanisms causing mitochondrial dysfunction and DND in TBI and aSAH patients. Finally, we want to emphasize that the method described here can be adapted to monitor metabolic disturbances following physical or medical interventions in other tissues, for instance subcutaneously or intramuscularly, and under various pathological conditions, such as diabetes or cancer.

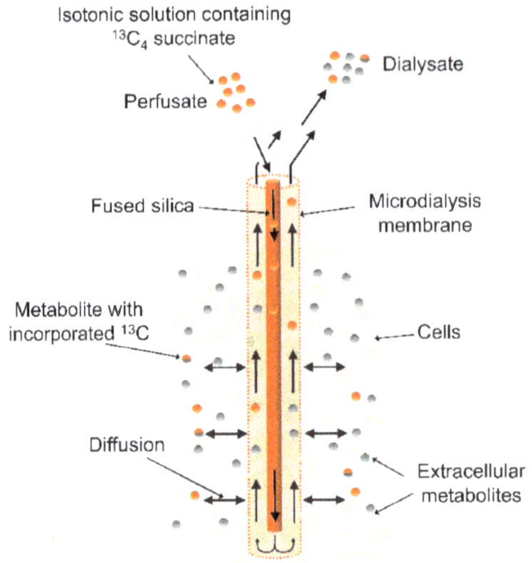

Figure 1. Microdialysis set-up. Ringer's solution with labeled ^{13}C-succinate is delivered to the microdialysis probe in the rat brain. Diffusion through the membrane allows ^{13}C-succinate to be taken up and metabolized by the surrounding cells.

In this study in living brain the ^{13}C-labeled microdialysis LC-MS approach was validated using two different rat models:

(a) Mitochondrial dysfunction induced by local perfusion with malonate, a reversible inhibitor of SDH.

(b) Transient cerebral ischemia induced by intracerebral application of the potent vasoconstrictor ET-1.

2. Results

2.1. The malonate Model

Perfusion with the SDH inhibitor malonate (10 and 50 mM at 0 h and 15 h, respectively) caused a very clear dose-dependent increase in endogenous succinate (Figure 2). Other TCA metabolites showed the opposite effect: A dose-dependent decrease in abundance. Changes in glutamine were related to those in alpha-ketoglutarate. Changes in ^{13}C-incorporated metabolites showed tendencies similar to endogenous metabolites, but the effects were generally much more pronounced (Figure 2).

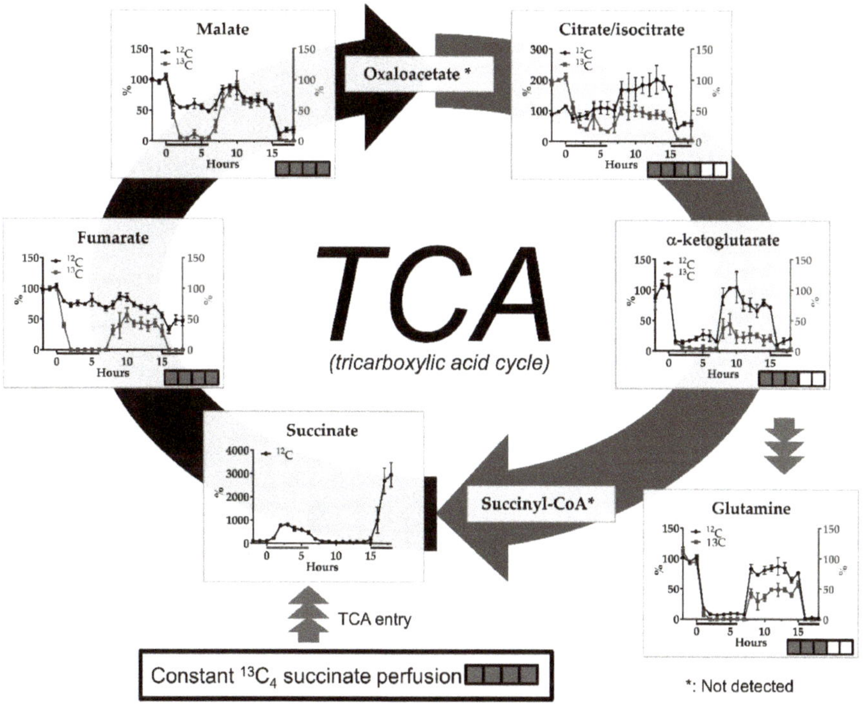

Figure 2. Effects of perfusion with 10 mM (0–6 h) or 50 mM malonate (14–18 h) on interstitial levels of endogenous (^{12}C) and ^{13}C-labeled tricarboxylic acid cycle (TCA) metabolites succinate, fumarate, malate, isocitrate, alpha-ketoglutarate and related glutamine upon constant perfusion with 1 mM $^{13}C_4$ succinate. Data are mean +/− SEM of 5 rats and expressed as % change of baseline abundance. The number of ^{13}C or ^{12}C atoms in the different metabolites after uptake of $^{13}C_4$ succinate and one turn of the TCA cycle is indicated with the number of filled or open squares. Black bars under the x-axis illustrate time periods of malonate administration. * not detected.

2.2. *Transient Cerebral Ischemia Model*

In the ET-1-induced cerebral ischemia-reperfusion model we observed a glucose drop below 50% of baseline levels in the first 30 min fraction. Large increases of glucose-6-phosphate (up to 800%) and lactate (up to 300%) were observed within in the first 2 h after induction of ischemia and did not normalize completely in the subsequent hours of monitoring (Figure 3).

We observed large differences in the magnitude and timing of change in endogenous and ^{13}C-labeled metabolites following ischemia-reperfusion (Figure 3). Changes in the abundance of ^{13}C-fumarate and ^{13}C-malate show the opposite tendency compared to their ^{12}C analogues, which illustrates that the subcellular source and de novo synthesis of ^{12}C or ^{13}C-labeled compounds differ, i.e., ^{12}C can be derived from both cytosol or mitochondrial compartments whereas ^{13}C is only derived from mitochondria. Further down-stream citrate/isocitrate showed increases in both endogenous and ^{13}C-labeled compounds with the latter showing the most dramatic changes. There were no apparent differences in alterations between endogenous and ^{13}C-labeled alpha-ketoglutarate (increases up to 200%) and the related compound glutamine (no changes after the insult).

Alterations in pyruvate, which can be derived from glycolysis or formed by decarboxylation of malate, differed according to labeling pattern, i.e., an increase in ^{13}C-pyruvate was observed after ischemia-reperfusion whereas the level of the ^{12}C form was not altered (Figure 3).

2.3. ^{13}C-labeling %

In Table 1, ^{13}C-labeling %, defined as ^{13}C-labeled/total compound (^{13}C + ^{12}C) × 100% for several TCA intermediates and the monocarboxylates pyruvate and lactate are shown during baseline conditions, the ischemic period and 30 min reperfusion, as well as after longer reperfusion time following ET-1-induced vasospasm. Fumarate and malate show the highest labeling % whereas isocitrate, alpha-ketoglutarate and glutamine show a considerably lower labeling %. The labeling pattern of TCA metabolites is in agreement with the biochemical distance to the labeling source, i.e., ^{13}C-succinate.

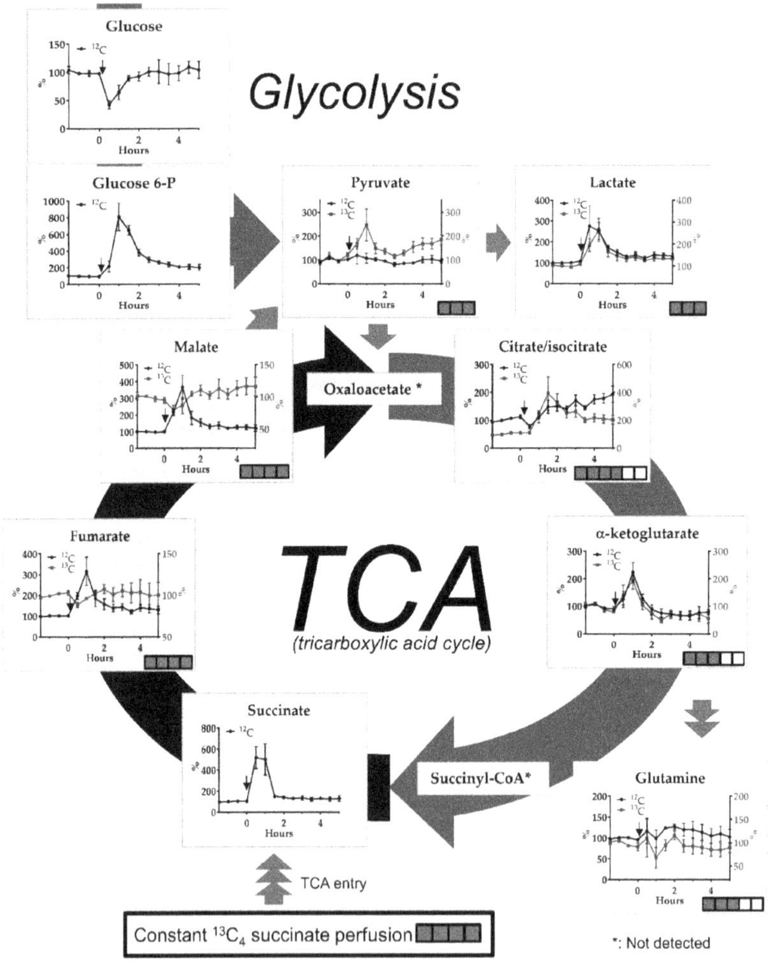

Figure 3. Effect of ET-1-induced cerebral ischemia-reperfusion on interstitial levels of endogenous (^{12}C) glucose and glucose-6-phosphate and endogenous (^{12}C) and ^{13}C-labeled pyruvate, lactate and TCA metabolites succinate, fumarate, malate, isocitrate, alpha-ketoglutarate and related glutamine upon constant perfusion with 1 mM $^{13}C_4$ succinate. Data are mean +/− SEM of 4 rats and expressed as % change of baseline abundance. Number of ^{13}C or ^{12}C atoms in the different metabolites is indicated with number of filled or open squares. Black arrows indicate time point of ET-1 infusion. * means not detected.

After ischemia and 30 min reperfusion there is a significant drop in labeling % for fumarate and malate, indicating mitochondrial dysfunction, which has largely recovered after > 4 h reperfusion. Interestingly, labeling % for pyruvate increased after ischemia, but did not reach statistical significance.

Table 1. ^{13}C-labeling % for selected metabolites in the ET-1 model. Data is shown as mean +/− SEM of 4 rats. Significant differences using nonparametric Kruskal Wallis test with Dunn's multiple comparison's test (vs. baseline) are shown: ** $p < 0.005$ and **** $p < 0.0001$.

% Labeling	Fumaric Acid	Malic Acid	(Iso) Citrate	α-Ketogluarate	Glutamine	Pyruvate	Lactic Acid
Baseline	88.1 (+/− 0.2)	91.3 (+/− 0.4)	8.4 (+/− 0.9)	14.0 (+/− 2.0)	6.1 (+/− 0.6)	6.3 (+/− 1.3)	0.3 (+/− 0.04)
Ischemia + 30 min of Reperfusion	79.2 (+/− 2.1) **	75.3 (+/− 2.6) ****	11.9 (+/− 2.8)	15.9 (+/− 2.7)	4.7 (+/− 0.7)	8.7 (+/− 1.4)	0.2 (+/− 0.02)
After >4 h Reperfusion	87.5 (+/− 0.5)	91.1 (0.4)	9.6 (+/− 1.0)	14.1 (+/− 3.0)	5.3 (+/− 0.8)	8.4 (+/− 1.3)	0.3 (+/− 0.03)

2.4. Histological Brain Damage

In Figure 4 placement of the microdialysis probes in the malonate (A) and ET-1 model (B) is shown, as well as the guide cannula for ET-1 infusion (B). Malonate perfusion did not cause histological brain damage whereas ET-1 infusion caused ischemic damage in the ipsilateral striatum.

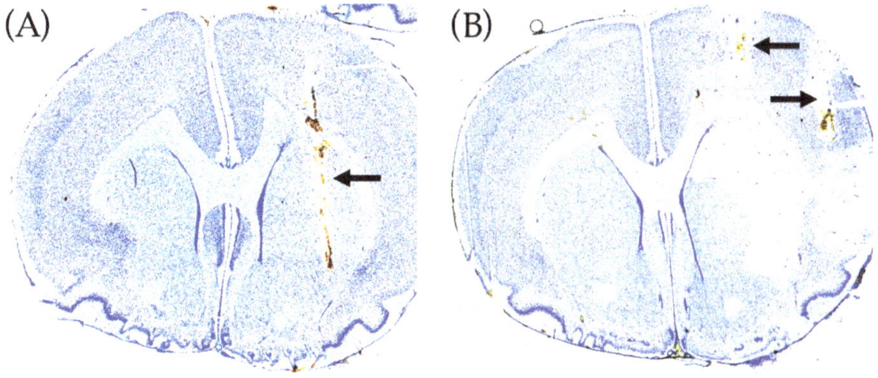

Figure 4. Histology of rat brains using toluidine blue staining. (**A**) Malonate perfusion model. The position of the microdialysis probe in striatum is shown by the arrow. (**B**) ET-1 rat model of transient cerebral ischemia. The position of the guide cannula for the microdialysis probe in the ipsilateral striatum is shown by the upper arrow (the microdialysis probe track in striatum is not visible). The position of the guide cannula for ET-1 infusion in the pirifom cortex is shown by the lower arrow. ET-1 infusion caused histological damage in the ipsilateral striatum.

3. Discussion

3.1. General

In this study we showed the potential of studying acute mitochondrial dysfunction in living brain by perfusion with a tracer dose of ^{13}C-labeled succinate through a microdialysis probe and subsequent LC-MS analysis of TCA-centered metabolites in the dialysates. Since LC-MS allowed the detection and relative quantification of both ^{13}C-labeled and endogenous (^{12}C) TCA metabolites, we can distinguish the efflux of endogenous metabolites as a consequence of cellular damage from changes in *de novo* synthesis (^{13}C-labeled metabolites) as a result of mitochondrial inhibition.

We assume that the efflux of ^{13}C-succinate from the probe and delivery to the cells is similar under baseline and experimental conditions, because perfusion with malonate or induction of cerebral ischemia did not cause any changes in ^{13}C-succinate levels in the dialysates. In contrast, ^{13}C-labeling of other TCA metabolites was strongly influenced by the experimental conditions. Succinate uptake into glial cells and neurons is mediated by the SLC13 family of Na^+-coupled dicarboxylate and tricarboxylate transporters [9]. Such transporters are found both on cell membranes and the mitochondrial inner membrane. Succinate is metabolized to fumarate by SDH localized on the inner mitochondrial membrane and is also known as electron transport chain complex II. Thus, all ^{13}C-labeled metabolites found in the dialysate are the result of succinate metabolism in the TCA cycle.

Perfusion with $^{13}C_4$-labeled succinate resulted in a labeling % of ^{13}C-labeled versus endogenous metabolites in accordance with the direction of the TCA cycle and "biochemical distance" (relationship) to succinate (see Figure 3 for full labeling patterns after one round of the TCA cycle and Table 1 for labeling efficacy, % full labeling). Thus, the highest labeling % was found for the TCA metabolites fumarate and malate, followed by citrate/isocitrate and alpha-ketogluarate. ^{13}C-alpha-ketogluarate can be converted to ^{13}C-glutamate (not detected), which again can be converted to ^{13}C-glutamine by glutamine synthase in astrocytes. ^{13}C-pyruvate can be formed by decarboxylation of ^{13}C-labeled malate by malic enzyme (see arrow in Figure 3) or conversion of ^{13}C-labeled oxaloacetate by other enzymes [10,11]. The labeling % for lactate was almost negligible.

The large drop in labeling efficacy between fumarate/malate and citrate/isocitrate may be explained by (a) the distance in the biochemical pathway to the labeling source ^{13}C-succinate, (b) dilution of labeled malate (precursor for subsequent TCA intermediates) in the interstitial space and diffusion away from the probe, and (c) that trafficking of energy substrates through the intercellular space is very limited—estimated to be less than 12% for glucose and lactate (see [12]) for a discussion of this topic).

The rapid decline in labeling efficacy using retrograde dialysis in vivo is unlike in vitro experiments where ^{13}C-labeled precursors are added to the culture medium, yielding much higher labeling efficacy in subsequent TCA metabolites [13].

3.2. *The Malonate Model*

Malonic acid (malonate) is a reversible inhibitor of SDH, the enzyme converting succinate to fumarate. Another microdialysis study in rat brain using flow injection analysis with biosensors and perfusion with malonate (5–50 mM for 1 h) through the probe, reported rapidly increasing lactate and decreasing glucose levels in the dialysates [14], which is in line with mitochondrial inhibition and increased glycolysis. In the present study using similar doses of malonate (perfusion with 10 and 50 mM), however, we observed dramatic reductions in ^{13}C-labeled and endogenous lactate and pyruvate, suggesting a strong inhibition of glycolysis. This discrepancy may be due to the different microdialysis membranes (15 kD cut-off PES membrane in the previous study versus 50 kD cut-off polyacrylonitrile (PAN) membrane in this study, which resulted in different recoveries and thus different interstitial malonate concentrations), and the different rat strains (Wistar versus Sprague Dawley) used in these studies. It has been reported that lower doses of malonate (i.e., 30 mM) inhibit the TCA cycle with only a partial effect on glycolysis whereas higher doses of malonate (i.e., 60 mM) inhibit both glycolysis and TCA cycle in rat skeletal muscle [15].

In our study, the effect of malonate perfusion is clearly illustrated by the dramatic rise in endogenous succinate levels and return to baseline levels when perfusion is switched to normal Ringer's solution. The dose-dependent rise in endogenous succinate is perfectly in line with inhibition of SDH. The inhibition of *de novo* synthesis of fumarate and malate is most clearly illustrated by the complete inhibition of ^{13}C incorporation in fumarate and malate during malonate perfusion, whereas reductions in endogenous levels are more modest with 10 mM malonate. The finding that endogenous malate levels are more affected by malonate perfusion than fumarate, although fumarate is the next intermediate in the TCA cycle after succinate, suggests that some back-cycling occurs between malate and fumarate [13], or that fumarate is produced from other sources, e.g., via the urea cycle [16]. Differences in the percentage change of ^{13}C-labeled and endogenous TCA cycle intermediate were also apparent for citrate/isocitrate and illustrate that monitoring of ^{13}C-labeled metabolites during perfusion with ^{13}C-succinate is a much more sensitive tool to detect mitochondrial dysfunction than monitoring endogenous metabolite levels. Levels of ^{13}C- and endogenous citrate during recovery after the first period of SDH inhibition also showed differences: A rebound effect (above baseline) for endogenous citrate, but still reduced levels for ^{13}C-labeled citrate, suggesting increased activity of pyruvate dehydrogenase and pyruvate carboxylation to enhance levels of non-labeled oxaloacetate. Enhanced pyruvate carboxylation in neural tissue has been reported following irreversible inhibition of SDH using 3-nitropropionic acid [17]. Under normal conditions, pyruvate carboxylation only occurs in astrocytes, which has been studied previously using ^{13}C-labeled bicarbonate [18].

Glutamine, which is related to the TCA cycle via glutamate (not detected) and alpha-ketoglutarate is released by astrocytes and taken up by neurons for glutamate synthesis and energy metabolism [19]. In the present study, ^{13}C-incorporation in glutamine was completely blocked by malonate perfusion, reflecting strong inhibition of the TCA cycle in astrocytes (no uptake and no production of the precursor ^{13}C -glutamate), as well as in neurons (no release of ^{13}C -glutamate).

3.3. *The ET-1-induced Transient Ischemia Model*

(ET-1) is a potent vasoconstrictor, which has been associated with cerebral vasospasms and subsequent transient ischemic events in subarachnoid hemorrhage patients [20]. In rodents, intracerebral application of ET-1 in the vicinity of the medial cerebral artery has been used as an animal model for transient focal cerebral ischemia [21–23]. In this model, transient occlusion of the medial cerebral artery can be induced in awake, freely moving animals, causing ischemia-reperfusion injury in the ipsilateral striatum.

Recently, we described mitochondrial dysfunction in the ET-1 rat model, which was characterized by a prolonged elevation of the L/P ratio and concomitant normal or elevated levels of pyruvate following ischemia-reperfusion using an enzymatic assay with sampling time intervals of 15 min [23]. Here, using sampling intervals of 30 min and LC-MS, we observed a more than 50% drop of glucose in the first dialysate fraction following ET-1 application (at 30 min). In addition, we observed increases in glucose-6 phosphate (maximum increase about 7.5-fold at 30 min reperfusion, i.e., 60 min after ET-1 infusion) and lactate (maximum increase about 2.5-fold at 30 min after ET-1 infusion) lasting for up to 5 h after the insult, indicating degradation of brain glycogen and downstream glycolysis during and after the ischemic insult.

The ischemic insult caused a dramatic rise in endogenous succinate—up to a 5-fold increase at 30–60 min after onset of the insult. Ischemic succinate accumulation arises from reversal of SDH activity, which is driven by fumarate overflow from purine nucleotide breakdown and partial reversal of the malate/aspartate shuttle [24]. After reperfusion, re-oxidation of succinate by SDH may drive extensive reactive oxygen species production because of reverse electron transport at complex I. It has been reported that decreasing succinate accumulation by an SDH inhibitor, such as malonate (see above), can reduce ischemia-reperfusion injury in mouse models of heart attack and stroke [24,25]. In this context, it is interesting that cerebral perfusion with high doses of succinate has been proposed as a treatment to improve outcome in head injury patients [7].

We saw a large increase in endogenous malate levels peaking at 60 min after ET-1, concomitant with a significant drop in ^{13}C-labeled malate at 30 min after ET-1, followed by elevated ^{13}C-malate levels following reperfusion. Changes in ^{13}C- and ^{12}C-fumarate showed a similar pattern. The clinical significance of these changes may be as follows: If extracellular levels of ^{13}C labeled metabolites are decreased and endogenous metabolites are increased, this may indicate compromised TCA cycle function (because of reduced labeling %) and damage to mitochondrial and cellular membranes (because of the increase in endogenous metabolite levels).

Most pyruvate is formed by glycolysis, but a minor part can be formed by conversion of malate by malic enzyme (ME) activity. In this study, ^{13}C-labeled pyruvate is thus derived from ^{13}C-labeled malate. In contrast to endogenous pyruvate levels, which were not significantly changed by the ET-1 insult, ^{13}C-labeled pyruvate levels were increased up to a maximum of 2.5-fold of baseline at 30 min of reperfusion (i.e., 60 min after ET-1 infusion) and were still elevated above baseline levels for the next 4 h (Figure 3). However, these changes, expressed as % labeling did not reach significance (Table 1). Increased ME activity shortly after the insult may play a role in combating oxidative stress [26].

In contrast to fumarate and malate, endogeneous citrate/isocitrate levels decreased in the first 30 min after onset of ischemia, which is in line with reduced influx of acetyl-CoA into the TCA cycle during the period of compromised cerebral blood flow. However, immediately after the period of ischemia, endogeneous citrate/isocitrate levels started to rise to about 2-fold of baseline, whereas ^{13}C-labeled citrate increased by up to 3–4 times, indicating a faster running TCA cycle after ischemia.

Opposite to the malonate model, where dramatic changes in alpha-ketoglutarate are paralleled by changes in glutamine (Figure 2), in the ET-1-induced ischemia-reperfusion model glutamine levels did not follow the elevations in alpha-ketoglutarate levels (Figure 3). Glutamate-glutamine cycling, i.e., glial uptake of glutamate, conversion to glutamine and subsequent release of glutamine, is an energy demanding process, which is known to be impaired following ischemic events. A microdialysis study in neurointensive care patients with subarachnoid hemorrhage reported an inverse correlation between low glutamine/glutamate ratios and elevated L/P ratios [27]. Thus, the lack of significant changes in endogenous or ^{13}C-labeled glutamine while alpha-ketoglutarate is transiently increased after the insult, may be explained by impaired energy metabolism in the reperfusion phase.

4. Conclusions

Following perfusion with ^{13}C-succinate, changes in ^{13}C-labeled TCA metabolites provide a more sensitive index of TCA cycle dysfunction, than changes in endogenous TCA metabolites. Discrepancies between extracellular changes in ^{13}C-labeled and endogenous metabolites under pathological conditions may be explained by the loss of cell membrane integrity (cell death). The differential response of endogenous versus ^{13}C-labeled malate can be used to monitor metabolic perturbations following cerebral ischemia-reperfusion.

Microdialysis-perfusion with a tracer dose of ^{13}C-succinate and subsequent LC-MS analysis of dialysate fractions is a promising research tool to monitor neurointensive care patients and get in-depth information on TCA cycle dysfunction following vascular or traumatic brain insults.

5. Materials and Methods

5.1. Animals

The animal experiments were approved by the local ethics committee and in accordance with the Danish Animal Experiment Inspectorate and EU legislation (lic. Nr. 2017-15-0201-01256). A total of 18 adult Sprague Dawley rats were used, weighing on average 273 g (range 216–379 g) with a mean age of 7.5 weeks (range 6–9 weeks) on the day of stereotaxic surgery (see 5.2 below). After surgery the rats were individually housed in a 12 h light/dark cycle with free access to food and water.

Eight rats (purchased from Janvier labs, Saint-Berthevin, France) were used for the malonate perfusion experiments (see 5.4 below). Two malonate-treated rats were used for optimization of LC-MS analysis of dialysates (see 5.7 below) and the data for one other rat were useless because of technical problems with the LC-MS, leaving 5 malonate-treated rats for statistical analysis.

Ten rats (purchased from Taconic Biosciences A/S, Ejby, Denmark) were used for ET-1 experiments (see 5.2 and 5.3 below). Four ET-1 treated rats were discarded, because glucose levels were not reduced following ET-1 administration, indicating that induction of cerebral ischemia was unsuccessful. Two other ET-1 rats were discarded from the statistical analysis because of flow problems during microdialysis. Thus, four ET-1 rats were included in the statistical analysis of the data.

5.2. Stereotaxic Surgery

For the ET-1 experiments, two microdialysis guide cannulas (shaft: 4 mm, Brainlink®, Groningen, Netherlands) were implanted in the left cerebral hemisphere using a stereotaxic frame (Kopf Instruments, Tujunga, CA, USA). One guide was placed for microinjection of ET-1 into the piriform cortex close to the proximal part of the medial cerebral artery (MCA) and one guide was placed for microdialysis in the ipsilateral striatum (see Figure 4B). For experiments with the SDH inhibitor malonate only one microdialysis guide was placed in striatum. In these experiments, malonate was administered by perfusion through the microdialysis probe (retrograde dialysis). The stereotaxic coordinates relative to bregma, with the nose bar at −3.9 mm (according to the atlas of Paxinos and Watson, 1986), were as follows:

Guide cannula for ET-1 injection: A + 0.9 mm; L 5.2 mm; V 4.6 mm (ET-1 experiment only)

Guide cannula for microdialysis probe: A + 0.5 mm; L 2.5 mm; V 3.2 mm

Stereotaxic surgery was done under Hypnorm/Dormicum anesthesia (Hypnorm: 0.315 mg/mL fentanyl and 10 mg/mL fluanisone, Janssen Pharmaceutica, Beerse, Belgium; 0.3 mL/kg s.c. Dormicum: 5 mg/mL midazolam, Hoffmann-La Roche, Basel, Switzerland; 5 mg/kg s.c.). Lidocaine (20 mg/mL Farmaplus AS, Oslo, Norway) was used as a local anesthetic. Body temperature was kept at 37.5 °C with a thermostatically regulated heating pad (Bosch CTKI3, München, Germany).

The guide cannulas and a slotted screw for head block tethering (Instech labs Inc., Plymouth Meeting, PA, USA) were fixed to the skull using glass ionomer luting cement (GC Fuji plus capsule, GC corporation, Tokyo, Japan). A slow release oral formulation of 0.4 mg/kg buprenorphin (Temgesic 0,2 mg sublingual tab., RB Pharmaceuticals, Slough, UK) was used as postoperative analgesia and rehydration was administered as a subcutaneous injection of 5 mL 0.9% NaCl immediately after surgery.

5.3. Microdialysis Setup for ET-1-Experiments

One day after stereotaxic surgery, the microdialysis probe (50kDa cut-off, 3mm polyacrylonitrile (PAN) membrane, BrainLink®, Groningen, The Netherlands) and probe for ET-1 injection (see above) were inserted through the guide cannulas under brief anesthesia (ca. 5 min) using inhalation of isoflurane (Baxter A/S, Allerød, Denmark). The inlet and outlet tubing (FEP tubing, 1.2 mL/10 cm, AgnTho's AB, Stockholm, Sweden) was connected to a swivel (AgnTho's AB, Stockholm, Sweden), syringe pump (22 Harvard Apparatus, Inc., Holliston, MA, USA) and fraction collector (CMA 142, Stockholm, Sweden) using 0.38 mm IDEX silicon connectors.

Within approximately 15 min after insertion through the guide, the microdialysis probe was perfused with 1 mM ^{13}C-labeled succinate ($^{13}C_4$ 99%, Sigma-Aldrich, Denmark A/S, Copenhagen) in sterile Ringer's solution at a flow rate of 1.0 μL/min. Microdialysis fractions were collected at 30 min intervals and stored at −20 °C within 2 h after collection. Microdialysis experiments were performed in awake, freely moving animals.

The ET-1 injection cannula (an old microdialysis probe of which the membrane was removed) was connected to a 100 μL Hamilton syringe using FEP tubing. The Hamilton syringe and FEP-tubing were filled with an ET-1 solution (Endothelin-1 ≥ 97%, Sigma-Aldrich Denmark A/S, Copenhagen, 10 pmol/μL dissolved in sterile Ringer's solution (147 mM NaCl, 4 mM KCl, 1.1 mM $CaCl_2$, 1.0 mM $MgCl_2$) for manual infusion of 15 μL ET-1 solution. After insertion of the ET-1 injection cannula, the tip of the fused silica tubing ended 3.0 mm below the guide cannula, i.e., 7.6 mm ventral to bregma. ET-1 was infused after collection of the first six 30 min fractions (baseline monitoring).

5.4. Mitochondrial Inhibition by Malonate Perfusion

Malonate perfusion experiments were done one day after stereotaxic placement of a guide cannula in striatum (unilaterally) and after inserting the microdialysis probe through the guide cannula as described above. In the malonate experiments, 60 min fractions (flow rate 1.0 μL/min) were collected and baseline levels were monitored for six hours using Ringer's solution with ^{13}C-labeled succinate (6 samples of 60 µL) before starting perfusion with 10 mM malonate (malonic acid, disodium salt monohydrate, Sigma-Aldrich, dissolved in Ringer's containing ^{13}C-succinate) for another six hours, followed by regular Ringer's with 1 mM ^{13}C-succinate for eight hours and finally six hours perfusion with 50 mM malonate in Ringer's containing 1 mM ^{13}C-succinate.

5.5. ET-1 Induced Transient Cerebral Ischemia

After three hours of baseline monitoring collecting 30 min fractions (6 samples of 30 µL), focal transient ischemia was induced by infusing 150 pmol ET-1 (Sigma-Aldrich) in 15 min (10 pmol/µL, 1 µL/min; ET-1 dissolved in sterile Ringer's solution; 60 µL aliquots of 10 pmol/µL were stored at −20 °C). Microdialysis was continued for at least six hours after the insult.

Histology

One day after microdialysis, the rats were killed by a lethal dose of pentobarbital (pentobarbital 200 mg/mL with lidocainehydrochloride 20 mg/mL, Glostrup Apotek, Denmark, 0.2–0.3 mL pr. rat) and decapitation before cardiac arrest. The brains were rapidly removed from the skull, and frozen using high pressure CO_2 and stored at −80 °C until histological processing (cryostat sectioning and toluidine blue staining) for analysis of the placement of guides and infarct size.

5.6. Statistical Analysis

Metabolite data are expressed as mean +/− SEM of percentage change of each rat's own baseline and visualized as a time-line. Statistical differences between groups were analyzed using nonparametric Kruskal Wallis test with Dunn's multiple comparison's test (vs. baseline). The XY graphs were generated in GraphPad Prism (GraphPad Software, inc. San Diego, CA, USA).

5.7. LC-MS Sample Preparation and Analysis

Each microdialysis fraction of 30 µL was lyophilized prior to resuspension in 12 µL 1% formic acid (FA) and transfer to vial including 100 µL insert. A pool (quality control) of the samples was constructed by transferring 2 µL of each sample to a new vial, which was injected in every sixth sample to monitor signal drift and system reproducibility. Ten µL from each sample was injected in random order using a 1290 Infinity high pressure liquid chromatography system (Agilent Technologies) equipped with a Supelco Discovery® HS F5-3 (2.1 × 150 mm and 3 µm particle size) column kept at 40 °C. Compounds were eluted using a flow rate of 300 µL/min and the following gradient composition of A (0.1% FA) and B (0.1% FA, acetonitrile) solvents: 100% A from 0–3 min, 100–60% A from 3–10 min, 60–0% A from 10–11 and 100% B isocratic from 11–12 min before equilibration for 5 min with the initial conditions. Eluting metabolites were detected by a 6530B quadrupole time of flight mass spectrometer (Agilent Technologies) operated in negative ion mode scanning from 40–1050 m/z with the following settings: 3 scans/sec., gas temp at 325 °C, drying gas at 8 L/min, nebulizer at 35 psi, sheath gas temp at 350 °C, sheath gas flow at 11 L/min, VCap at 3500 V, fragmentor at 125 V and skimmer at 65 V. Each spectrum was internally calibrated during analysis using the signals of purine (119.03632) and Hexakis 1H,1H,3H-tetrafluoropropoxy phosphazine with formate adduct (966.000725), which was delivered to a second needle in the ion source by an isocratic pump running with a flow of 20 µL/min. A library containing molecular formula and retention time of the metabolites of interest was constructed using MassHunter PCDL Manager v. B.08.00 (Agilent Technologies). All reported annotations, except lactate and glutamine, were based on accurate mass and co-elution with synthetic standards and their fragments (Metabolomics Standards Initiative (MSI) [28] level 1 annotation). Lactate and glutamine were annotated based on the existence of co-eluting fragments from a pooled sample analyzed in "all-ion" mode using 0, 10 and 40 V in collision energy (MSI level 3 annotation). The ion fragments from the known compounds were obtained from METLIN [29]. Chromatograms for all compounds were extracted and the areas were quantified using Profinder v. B.08.00 (Agilent Technologies) in

"Batch isotopologue extraction" mode, which extracts the signal from the isotopes and corrects for their natural abundance, with a mass tolerance of 10 ppm and retention time tolerance of 0.1 min. Quality control samples were used to evaluate system reproducibility, and potentially, to exclude compounds with a relative standard deviation (RSD) above 30%, however, all shown compounds had a RSD < 15%.

Author Contributions: Conceptualization, J.F.H., A.F., J.B.G.; Methodology, J.F.H., K.H.N., A.F., J.B.G.; Software, J.F.H.; Formal Analysis, J.F.H.; Investigation, K.H.N., J.F.H.; Resources, N.J.F., J.B.G.; Writing-Original Draft Preparation, J.F.H., J.B.G.; Writing-Review & Editing, J.F.H., K.H.N., T.H.N., C.-H.N., F.R.P., N.J.F., A.F., J.B.G.; Visualization, J.F.H.; Supervision, N.J.F., A.F., J.B.G.; Project Administration, J.F.H., N.J.F., T.H.N., C.-H.N., F.R.P., A.F., J.B.G.; Funding Acquisition, F.R.P., T.H.N., A.F., N.J.F., J.B.G.

Funding: This research was funded by The Lundbeck Foundation (scholar stipend to K.H.N. supported by A.F., T.H.N. and F.R.P.) and The Danish Parkinson Society Foundation (research grant to J.F.H., N.J.F. and J.B.G.).

Acknowledgments: We thank The Lundbeck Foundation and The Danish Parkinson Society Foundation for funding.

Conflicts of Interest: The authors declare no conflict of interest. The funders had no role in the design of the study; in the collection, analyses, or interpretation of data; in the writing of the manuscript, and in the decision to publish the results.

References

1. Nordstrom, C.H. Cerebral microdialysis in TBI-limitations and possibilities. *Acta Neurochir.* **2017**, *159*, 2275–2277. [CrossRef] [PubMed]
2. Nordstrom, C.H.; Nielsen, T.H.; Schalen, W.; Reinstrup, P.; Ungerstedt, U. Biochemical indications of cerebral ischaemia and mitochondrial dysfunction in severe brain trauma analysed with regard to type of lesion. *Acta Neurochir.* **2016**, *158*, 1231–1240. [CrossRef] [PubMed]
3. Jacobsen, A.; Nielsen, T.H.; Nilsson, O.; Schalen, W.; Nordstrom, C.H. Bedside diagnosis of mitochondrial dysfunction in aneurysmal subarachnoid hemorrhage. *Acta Neurol. Scand.* **2014**, *130*, 156–163. [CrossRef] [PubMed]
4. Sarrafzadeh, A.S.; Sakowitz, O.W.; Kiening, K.L.; Benndorf, G.; Lanksch, W.R.; Unterberg, A.W. Bedside microdialysis: A tool to monitor cerebral metabolism in subarachnoid hemorrhage patients? *Crit. Care Med.* **2002**, *30*, 1062–1070. [CrossRef] [PubMed]
5. Zielke, H.R.; Zielke, C.L.; Baab, P.J. Direct measurement of oxidative metabolism in the living brain by microdialysis: A review. *J. Neurochem.* **2009**, *109* (Suppl. 1), 24–29. [CrossRef]
6. Carpenter, K.L.; Jalloh, I.; Gallagher, C.N.; Grice, P.; Howe, D.J.; Mason, A.; Timofeev, I.; Helmy, A.; Murphy, M.P.; Menon, D.K.; et al. (13)C-labelled microdialysis studies of cerebral metabolism in TBI patients. *Eur. J. Pharm. Sci.* **2014**, *57*, 87–97. [CrossRef] [PubMed]
7. Jalloh, I.; Helmy, A.; Howe, D.J.; Shannon, R.J.; Grice, P.; Mason, A.; Gallagher, C.N.; Stovell, M.G.; van der Heide, S.; Murphy, M.P.; et al. Focally perfused succinate potentiates brain metabolism in head injury patients. *J. Cereb. Blood Flow Metab.* **2016**. [CrossRef] [PubMed]
8. Jalloh, I.; Helmy, A.; Howe, D.J.; Shannon, R.J.; Grice, P.; Mason, A.; Gallagher, C.N.; Murphy, M.P.; Pickard, J.D.; Menon, D.K.; et al. A Comparison of Oxidative Lactate Metabolism in Traumatically Injured Brain and Control Brain. *J. Neurotrauma* **2018**, *35*, 2025–2035. [CrossRef] [PubMed]
9. Pajor, A.M. Sodium-coupled dicarboxylate and citrate transporters from the SLC13 family. *Pflugers Arch.* **2014**, *466*, 119–130. [CrossRef] [PubMed]
10. Waagepetersen, H.S.; Qu, H.; Hertz, L.; Sonnewald, U.; Schousboe, A. Demonstration of pyruvate recycling in primary cultures of neocortical astrocytes but not in neurons. *Neurochem. Res.* **2002**, *27*, 1431–1437. [CrossRef] [PubMed]
11. Olstad, E.; Olsen, G.M.; Qu, H.; Sonnewald, U. Pyruvate recycling in cultured neurons from cerebellum. *J. Neurosci. Res.* **2007**, *85*, 3318–3325. [CrossRef] [PubMed]
12. Huinink, K.; Korf, J.; Gramsbergen, J.B. Microdialysis and microfiltration: Technology and cerebral applications for energy substrates. In *Neural Metabolism In Vivo*; Gruetter, R., Choi, I.-Y., Eds.; Springer: New York, NY, USA, 2012; Volume 4. [CrossRef]

13. Walls, A.B.; Bak, L.K.; Sonnewald, U.; Schousboe, A.; Waagepetersen, H.S. Metabolic Mapping of Astrocytes and Neurons in Culture Using Stable Isotopes and Gas Chromatography-Mass Spectrometry (GC-MS). In *Brain Energy Metabolism*; Hirrlinger, J., Waagepetersen, H.S., Eds.; Humana Press: New York, NY, USA, 2014.
14. Skjoeth-Rasmussen, J.; Lambertsen, K.; Gramsbergen, J.B. On-line glucose and lactate monitoring in rat striatum: Effect of malonate and correlation with histological damage. *J. Neurochem.* **2003**, *85*, 35. [CrossRef]
15. Fawaz, E.N.; Fawaz, G. Inhibition of glycolysis in rat skeletal muscle by malonate. *Biochem. J.* **1962**, *83*, 438–445. [CrossRef] [PubMed]
16. Siegel, G.J.; Agranoff, B.W. *Basic Neurochemistry: Molecular, Cellular, and Medical Aspects*; Lippincott Williams Wilkins: Philadelphia, PA, USA, 1999.
17. Gramsbergen, J.B.; Sandberg, M.; Kornblit, B.; Zimmer, J. Pyruvate protects against 3-nitropropionic acid neurotoxicity in corticostriatal slice cultures. *Neuroreport* **2000**, *11*, 2743–2747. [CrossRef] [PubMed]
18. Sonnewald, U.; Rae, C. Pyruvate carboxylation in different model systems studied by (13)C MRS. *Neurochem. Res.* **2010**, *35*, 1916–1921. [CrossRef] [PubMed]
19. Bak, L.K.; Schousboe, A.; Waagepetersen, H.S. The glutamate/GABA-glutamine cycle: Aspects of transport, neurotransmitter homeostasis and ammonia transfer. *J. Neurochem.* **2006**, *98*, 641–653. [CrossRef] [PubMed]
20. Petzold, G.C.; Einhaupl, K.M.; Dirnagl, U.; Dreier, J.P. Ischemia triggered by spreading neuronal activation is induced by endothelin-1 and hemoglobin in the subarachnoid space. *Ann. Neurol.* **2003**, *54*, 591–598. [CrossRef] [PubMed]
21. Van Hemelrijck, A.; Vermijlen, D.; Hachimi-Idrissi, S.; Sarre, S.; Ebinger, G.; Michotte, Y. Effect of resuscitative mild hypothermia on glutamate and dopamine release, apoptosis and ischaemic brain damage in the endothelin-1 rat model for focal cerebral ischaemia. *J. Neurochem.* **2003**, *87*, 66–75. [CrossRef]
22. Gramsbergen, J.B.; Skjoth-Rasmussen, J.; Rasmussen, C.; Lambertsen, K.L. On-line monitoring of striatum glucose and lactate in the endothelin-1 rat model of transient focal cerebral ischemia using microdialysis and flow-injection analysis with biosensors. *J. Neurosci. Methods* **2004**, *140*, 93–101. [CrossRef]
23. Forsse, A.; Nielsen, T.H.; Nygaard, K.H.; Nordstrom, C.H.; Gramsbergen, J.B.; Poulsen, F.R. Cyclosporin A ameliorates cerebral oxidative metabolism and infarct size in the endothelin-1 rat model of transient cerebral ischaemia. *Sci. Rep.* **2019**, *9*, 3702. [CrossRef]
24. Chouchani, E.T.; Pell, V.R.; Gaude, E.; Aksentijevic, D.; Sundier, S.Y.; Robb, E.L.; Logan, A.; Nadtochiy, S.M.; Ord, E.N.J.; Smith, A.C.; et al. Ischaemic accumulation of succinate controls reperfusion injury through mitochondrial ROS. *Nature* **2014**, *515*, 431–435. [CrossRef] [PubMed]
25. Tretter, L.; Patocs, A.; Chinopoulos, C. Succinate, an intermediate in metabolism, signal transduction, ROS, hypoxia, and tumorigenesis. *Biochim. Biophys. Acta* **2016**, *1857*, 1086–1101. [CrossRef] [PubMed]
26. Singh, R.; Lemire, J.; Mailloux, R.J.; Appanna, V.D. A novel strategy involved in [corrected] anti-oxidative defense: The conversion of NADH into NADPH by a metabolic network. *PLoS ONE* **2008**, *3*, e2682. [CrossRef]
27. Samuelsson, C.; Hillered, L.; Zetterling, M.; Enblad, P.; Hesselager, G.; Ryttlefors, M.; Kumlien, E.; Lewen, A.; Marklund, N.; Nilsson, P.; et al. Cerebral glutamine and glutamate levels in relation to compromised energy metabolism: A microdialysis study in subarachnoid hemorrhage patients. *J. Cereb. Blood Flow Metab.* **2007**, *27*, 1309–1317. [CrossRef] [PubMed]
28. Salek, R.M.; Steinbeck, C.; Viant, M.R.; Goodacre, R.; Dunn, W.B. The role of reporting standards for metabolite annotation and identification in metabolomic studies. *Gigascience* **2013**, *2*, 13. [CrossRef] [PubMed]
29. Guijas, C.; Montenegro-Burke, J.R.; Domingo-Almenara, X.; Palermo, A.; Warth, B.; Hermann, G.; Koellensperger, G.; Huan, T.; Uritboonthai, W. METLIN: A Technology Platform for Identifying Knowns and Unknowns. *Anal. Chem.* **2018**, *90*, 3156–3164. [CrossRef] [PubMed]

MDPI
St. Alban-Anlage 66
4052 Basel
Switzerland
Tel. +41 61 683 77 34
Fax +41 61 302 89 18
www.mdpi.com

Metabolites Editorial Office
E-mail: metabolites@mdpi.com
www.mdpi.com/journal/metabolites

www.ingramcontent.com/pod-product-compliance
Lightning Source LLC
LaVergne TN
LVHW071950080526
838202LV00064B/6715